Arnold F. Nikiforov
Vasilii B. Uvarov

Special Functions of Mathematical Physics

A Unified Introduction with Applications

Translated from the Russian by Ralph P. Boas

1988

Springer Basel AG

Authors' address:
Arnold F. Nikiforov
Vasilii B. Uvarov
M.V. Keldish Institute of Applied Mathematics
of the Academy of Sciences of the USSR
Miusskaja Square
Moscow 125 047
USSR

Originally published as
Specjal'nye funkcii matematičeskoj fiziki
by Science, Moscow, 1978.

Library of Congress Cataloging in Publication Data

Nikiforov, A. F.
Special functions of mathematical physics.

Translation of: Spetsial´nye funktsii matematicheskoi fiziki.
Bibliography: p.
Includes index.
1. Functions, Special. 2. Mathematical physics.
3. Quantum theory. I. Uvarov, V. B. (Vasilii Borisovich)
II. Title.
QC20.7.F87N5513 1988 530.1'5 84-14959
ISBN 978-1-4757-1597-2 ISBN 978-1-4757-1595-8 (eBook)
DOI 10.1007/978-1-4757-1595-8

CIP-Kurztitelaufnahme der Deutschen Bibliothek

Nikiforov, A. F.:
Special functions of mathematical physics : a unified introd. with applications / A. F. Nikiforov ; V. B. Uvarov. Transl. by R. P.

Einheitssacht.: Specjial'nye funkcii matematičeskoj fiziki <dt.>
ISBN 978-1-4757-1597-2

NE: Uvarov, Vasilij B.:

Originally published by Birkhäuser Verlag Basel in 1988

Typesetting and Layout: mathScreen *online*, Basel

ISBN 978-1-4757-1597-2

Table of Contents

Preface to the American edition

With students of Physics chiefly in mind, we have collected the material on special functions that is most important in mathematical physics and quantum mechanics. We have not attempted to provide the most extensive collection possible of information about special functions, but have set ourselves the task of finding an exposition which, based on a unified approach, ensures the possibility of applying the theory in other natural sciences, since it provides a simple and effective method for the independent solution of problems that arise in practice in physics, engineering and mathematics.

For the American edition we have been able to improve a number of proofs; in particular, we have given a new proof of the basic theorem (§3). This is the fundamental theorem of the book; it has now been extended to cover difference equations of hypergeometric type (§§12, 13). Several sections have been simplified and contain new material.

We believe that this is the first time that the theory of classical orthogonal polynomials of a discrete variable on both uniform and nonuniform lattices has been given such a coherent presentation, together with its various applications in physics.

Acknowledgements

The authors are grateful to Professor Boas for his skillful and lucid translation. As a result of this work, some portions of the book seem to be more clear-cut and precise in English than they appear in Russian. We thank all those who have contributed to the production of this edition.

Foreword to the Russian edition

Interest in special functions has greatly increased as a result of the extensive development of numerical methods and the growing role of computer simulation.

There are two reasons for this trend. In the first place, for many physical processes a mathematical description based on "first principles" leads to differential, integral, or integro-differential equations of rather complex form. Consequently the original problem usually has to be considerably simplified in order to clarify its most important qualitative features and to understand the relative roles played by various factors. If a solution of the simplified problem can be obtained in an explicit mathematical form that can easily be analyzed, one may be able to obtain a qualitative picture, without much expenditure of time and effort (but possibly with aid of a computer). Then one can analyze how the behavior of the solution depends on the parameters of the problem.

In the second place, in the solution of complicated problems on a computer it is convenient to make use of simplified problems in order to select reliable and economical numerical algorithms. Here it is seldom possible to restrict one's self to problems that lead to elementary functions. Moreover, a knowledge of special functions is essential for understanding many important problems of theoretical and mathematical physics.

The most commonly encountered special functions are those that are known as the "special functions of mathematical physics": the classical orthogonal polynomials (Jacobi, Laguerre, Hermite), spherical harmonics, and the Bessel and hypergeometric functions. Much basic research has been devoted to the theory of these functions and their applications. Unfortunately this research involves rather cumbersome mathematical techniques and many special devices. Consequently there long been a need for a theory of special functions based on general but simple ideas.

The authors of the present book have discovered an easily comprehended way of presenting the theory of special functions, based on a generalization of the Rodrigues formula for the classical orthogonal polynomials. Their approach makes it possible to obtain explicit integral representations of all the special functions of mathematical physics and to derive their basic properties. In particular, this method can be used to solve the second-order linear differential equations that are usually solved by Laplace's method. The construction of the theory of special functions uses a minimal amount of mathematical apparatus: it requires only the elements of the theory of ordinary differential equations and of complex analysis. This is a significant advantage, since it is well known that the large amount of essential mathematical knowledge, in particular that involving special functions, is a fundamental obstacle to the study of theoretical and mathematical physics.

In the process of working through the book the reader will gain experience in the development of asymptotic formulas, expansions in series, recursion formulas, estimates of various kinds, and computational formulas, and will come to see the intrinsic logical connections among special functions that at first sight seem completely different.

The book discusses connections with other branches of mathematics and physics. Considerable attention is paid to applications in quantum mechanics. The main interest here is in the study of problems on the determination of discrete energy spectra and the corresponding wave functions in problems that can be solved by means of the classical orthogonal polynomials. The authors have succeeded in presenting these problems without the traditional use of generalized power series. Hence they have been able, in Chapter V, to give elegant and easy solutions of such fundamental problems of quantum mechanics as the problems of the harmonic oscillator and of the motion of particles in a central field, and solutions of the Schrödinger, Dirac and Klein-Gordon equations for the Coulomb potential. We also call attention to the presentation, based on the method of V.A. Steklov, of the Wentzel-Kramers-Brillouin method.

The authors discuss the addition theorems for spherical harmonics and Bessel functions, which are widely applied in the theory of atomic spectra, in scattering theory, and in the design of nuclear reactors. In the study of generalized spherical harmonics the authors really come to grips with the theory of representations of the rotation group and the general theory of angular momentum. Later on, readers will be able to deepen their knowledge of special functions by consulting books in which special functions are studied by group-theoretical methods. The classical orthogonal polynomials of a discrete variable are of interest in the theory of difference methods. From the point of view of numerical calculation, it is instructive to apply quadrature formulas of Gaussian type for calculating sums and constructing approximate formulas for special functions. We note that this book presents a number of problems that are needed in applications but are touched on only lightly, or not at all, in textbooks.

The authors are specialists in mathematical physics and quantum mechanics. The book originated in the course of their work on a current problem of plasma physics in the M.V. Keldysh Institute of Applied Mathematics of the Academy of Sciences of the USSR.

The book contains a large amount of material, presented concisely in a lucid and well-organized way. It is certain to be useful to a wide circle of readers — to both undergraduate and graduate students, and to workers in mathematical and theoretical physics.

A.A. Samarskiĭ

Member of the Academy of Sciences of the USSR

Preface to the Russian edition

In solving many problems of theoretical and mathematical physics one is led to use various special functions. Such problems arise, for example, in connection with heat conduction, the interaction between radiation and matter, the propagation of electromagnetic or acoustic waves, the theory of nuclear reactors, and the internal structure of stars.

In practice, special functions usually arise as solutions of differential equations. Consequently the natural approach for mathematical physics is to deduce the properties of the functions directly from the differential equations that arise in natural mathematical formulations of physical problems. For this reason the authors have developed a method which makes it possible to present the theory of special functions by starting from a differential equation of the form

$$u'' + \frac{\tilde{\tau}(z)}{\sigma(z)}u' + \frac{\tilde{\sigma}(z)}{\sigma^2(z)}u = 0\,, \tag{1}$$

where $\sigma(z)$ and $\tilde{\sigma}(z)$ are polynomials, at most of second degree, and $\tilde{\tau}(z)$ is a polynomial, at most of first degree. The differential equations of most of the special functions that occur in mathematical physics and quantum mechanics are particular cases of (1).

The book is organized as follows. The first chapter discusses a class of transformations $u = \phi(z)y$ by means of which, for special choices of $\phi(z)$, equation (1) is transformed into an equation of the same type. We can select transformations that carry (1) into an equation of the simpler form

$$\sigma(z)y'' + \tau(z)y' + \lambda y = 0\,, \tag{2}$$

where $\tau(z)$ is a polynomial of at most the first degree and λ is a constant.

We shall call (2) an *equation of hypergeometric type*;* and its solutions, *functions of hypergeometric type*. The theory of these functions is developed in the following stages. First we show that the derivatives of functions of hypergeometric type are again functions of hypergeometric type. This property lets us construct a family of particular solutions of (2) corresponding to particular values of λ, starting from the obvious solution of (2), namely $y(z)$ =const. for $\lambda = 0$. Such solutions are polynomials in z; they may be written in explicit form by means of the Rodrigues formula. A natural generalization of the integral representation of these polynomials follows from the Rodrigues formula and makes it possible to obtain an integral representation of all functions of hypergeometric type corresponding to arbitrary values of λ. By using this integral representation and the transformation of (2) into other equations of the same type, we can obtain all the basic properties of the functions in question: power series expansions, asymptotic representations, recursion relations and functional equations. This approach lets us obtain the complete family of solutions of (1).

The second, third and fourth chapters are devoted to carrying out this program for particular functions of hypergeometric type: the classical orthogonal polynomials, spherical harmonics, Bessel functions, and hypergeometric functions. Consequently chapters II–IV can be read in any order after chapter I.

Chapter II, devoted to classical orthogonal polynomials, has been extended to include the theory of classical orthogonal polynomials of a discrete variable. Here our methodology of the theory of the classical orthogonal polynomials is applied to a difference equation instead of to a differential equation. There then arise various families of classical orthogonal polynomials of a discrete variable on both uniform and nonuniform lattices. It is interesting to observe that the study of classical orthogonal polynomials of a discrete variable was initiated by Chebyshev as early as the middle of the nineteenth century; it was then continued by many eminent investigators. However, no books have developed the theory of these polynomials as solutions of a difference equation. It has not even been clear until recently which polynomials, among those introduced by various authors and arising from various considerations, belong to the class described above.

Chapter V is devoted to applications. It should be noticed that we have discussed practically all the basic problems of quantum mechanics that can be solved in explicit form, and have constructed their solutions by a unified method. Physicists will be interested in our presentation of the remarkably simple connection between the Clebsch-Gordan coefficients, so extensively

* This name is used because the particular solutions of equation (2) are hypergeometric functions when $\sigma(z)=z(1-z)$, and confluent hypergeometric functions when $\sigma(z)=z$ (see Chapter IV).

used in quantum mechanics, and orthogonal polynomials of a discrete variable — the Hahn polynomials. We also discuss the connection between the Racah coefficients and the classical orthogonal polynomials of a discrete variable, which are orthogonal on a discrete lattice.

Since familiarity with the properties of Euler's gamma function is a necessary prerequisite for the study of special functions, we present the theory of the gamma function in an appendix. There we also discuss the properties of Laplace integrals, which are used to obtain analytic continuations and asymptotic representations of special functions. At the end of the book we have provided a list of the basic formulas. If a reader requires more detailed information, we recommend the three volumes of the Bateman Project ([E2]), which contain all the formulas from the theory of special functions up to the middle 1940's, and also [A1] and [O1]. A more detailed idea of the contents of the book can be obtained from the table of contents and the following diagram of the connections among the chapters:

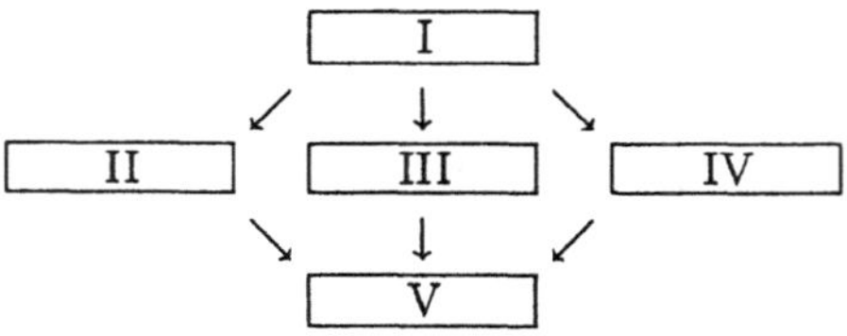

The method of studying special functions presented here is a further development of the method followed in the authors' book [N2]. In particular, it enables the reader to form a rather good idea of the theory of special functions after having studied only the first three sections of the book.

The basic material of the book was presented in a course of lectures on methods of mathematical physics given for several years in the Faculty of Theoretical and Experimental Physics of the Moscow Institute of Engineering Physics, and also in special courses in the Physical and Chemical Faculties and the Faculty of Numerical Mathematics and Cybernetics of the Moscow State University.

The authors thank T.T.Tsirulis, V.Ya. Arsenin, B.L. Rozhdestvenskiĭ and S.K. Suslov, as well as the staff of the Department of Theoretical and Nuclear Physics of the Moscow Institute of Engineering Physics, for helpful comments on the content of the book.

A.F. Nikiforov, V.B. Uvarov

Translator's preface

The book is divided, in the conventional Russian manner, into glavy (chapters), paragrafy (sections, abbreviated §), and punkty (abbreviated p. in Russian; I have used "part" for these). Sections (§§) are numbered consecutively through the book; the numbering of formulas begins again with each section. I have retained the authors' references to Russian sources, but I have added references to the corresponding English versions whenever I could find them, and otherwise to similar English books. References are cited in the form [Ln], where L is a letter and n is a numeral. When Russian terminology differs markedly from that customarily used in American English, I have silently adopted the latter (for example, "Bessel functions" instead of "cylinder functions").

For this translation, the authors have substantially rearranged and amplified the material, particularly in §13, where they discuss recent work on the connection between the Hahn polynomials and the Clebsch-Gordan coefficients, and between the Wigner $6j$-symbols and the Racah polynomials. They have also added (§27, parts 2 and 3) applications of orthogonal polynomials of a discrete variable to the compression of information, and of Bessel functions to laser sounding of the atmosphere.

I am indebted to Professor Mary L. Boas of the DePaul University Physics Department for help with the physics vocabulary.

RPB

Chapter I

Foundations of the Theory of Special Functions

§ 1 A differential equation for special functions

Many important problems of theoretical and mathematical physics lead to the differential equation

$$u'' + \frac{\tilde{\tau}(z)}{\sigma(z)}u' + \frac{\tilde{\sigma}(z)}{\sigma^2(z)}u = 0, \tag{1}$$

where $\sigma(z)$ and $\tilde{\sigma}(z)$ are polynomials of degree at most 2, and $\tilde{\tau}(z)$ is a polynomial of degree at most 1. Equations of this form arise, for example, in solving the Laplace and Helmholtz equations in curvilinear coordinate systems by the method of separation of variables, and in the discussion of such fundamental problems of quantum mechanics as the motion of a particle in a spherically symmetric field, the harmonic oscillator, the solution of the Schrödinger, Dirac and Klein-Gordon equations for a Coulomb potential, and the motion of a particle in a homogeneous electric or magnetic field. Moreover, equation (1) also arises in typical problems of atomic, molecular and nuclear physics.

Among the solutions of equations of the form (1) are several classes of special functions: the classical orthogonal polynomials (Jacobi, Laguerre, Hermite), spherical harmonics, Bessel and hypergeometric functions. These are often referred to as *the special functions of mathematical physics.*

We shall always suppose that z and the coefficients of $\sigma(z), \tilde{\sigma}(z)$ and $\tilde{\tau}(z)$ can have any real or complex values.

We now try to reduce (1) to a simpler form by taking $u = \phi(z)y$ and choosing an appropriate $\phi(z)$. We have

$$y'' + \left(2\frac{\phi'}{\phi} + \frac{\tilde{\tau}}{\sigma}\right)y' + \left(\frac{\phi''}{\phi} + \frac{\phi'}{\phi}\frac{\tilde{\tau}}{\sigma} + \frac{\tilde{\sigma}}{\sigma^2}\right)y = 0. \tag{2}$$

To keep (2) from being more complicated than the original equation (1), it is natural to require that the coefficient of y' has the form $\tau(z)/\sigma(z)$, where $\tau(z)$ is a polynomial of degree at most 1. This leads to

$$\phi'(z)/\phi(z) = \pi(z)/\sigma(z), \tag{3}$$

where

$$\pi(z) = \frac{1}{2}[\tau(z) - \tilde{\tau}(z)] \tag{4}$$

is a polynomial of degree at most 1. Since

$$\frac{\phi''}{\phi} = \left(\frac{\phi'}{\phi}\right)' + \left(\frac{\phi'}{\phi}\right)^2 = \left(\frac{\pi}{\sigma}\right)' + \left(\frac{\pi}{\sigma}\right)^2,$$

equation (2) takes the form

$$y'' + \frac{\tau(z)}{\sigma(z)}y' + \frac{\bar{\sigma}(z)}{\sigma^2(z)}y = 0, \tag{5}$$

where

$$\tau(z) = \tilde{\tau}(z) + 2\pi(z), \tag{6}$$

$$\bar{\sigma}(z) = \tilde{\sigma}(z) + \pi^2(z) + \pi(z)[\tilde{\tau}(z) - \sigma'(z)] + \pi'(z)\sigma(z). \tag{7}$$

The functions $\tau(z)$ and $\bar{\sigma}(z)$ are polynomials of degrees at most 1 and 2, respectively. Consequently (5) is an equation of the same type as (1). Hence we have found a class of transformations that do not change the type of the equation, namely the transformations induced by the substitution $u = \phi(z)y$, where $\phi(z)$ satisfies (3) with an arbitrary linear polynomial $\pi(z)$.

We now try to choose $\pi(z)$ so that (5) will be both as simple as possible and convenient for studying the properties of the solutions. We shall choose the coefficients of $\pi(z)$ so that the polynomial $\bar{\sigma}(z)$ in (5) will be divisible by $\sigma(z)$, i.e.

$$\bar{\sigma}(z) = \lambda\sigma(z), \tag{8}$$

where λ is a constant. This is possible because if we equate coefficients of powers of z on both sides of (8), we obtain three equations in three unknowns, the constant λ and the two coefficients of $\pi(z)$. Consequently (5) can be reduced to the form

$$\sigma(z)y'' + \tau(z)y' + \lambda y = 0. \tag{9}$$

We shall refer to (9) as an *equation of hypergeometric type*, and its solutions as *functions of hypergeometric type*. Correspondingly, it is natural to call (1) a *generalized equation of hypergeometric type.**

To determine $\pi(z)$ and λ we rewrite (8) in the form

$$\pi^2 + (\tilde{\tau} - \sigma')\pi + \tilde{\sigma} - k\sigma = 0,$$

where

$$k = \lambda - \pi'(z). \tag{10}$$

If we assume that k is known, solving the quadratic equation for $\pi(z)$ yields

$$\pi(z) = \frac{\sigma' - \tilde{\tau}}{2} \pm \sqrt{\left(\frac{\sigma' - \tilde{\tau}}{2}\right)^2 - \tilde{\sigma} + k\sigma}\,. \tag{11}$$

Since $\pi(z)$ is a polynomial, the expression under the square root sign must be the square of a polynomial. This is possible only if its discriminant is zero. Hence we obtain an equation, in general quadratic, for k.

After determining k, we obtain $\pi(z)$ from (11), and then $\phi(z), \tau(z)$ and λ by using (3), (6) and (10). It is clear that the reduction of (1) to an equation of hypergeometric type (9) can be made in several ways corresponding to different choices of k and of the ambiguous sign in formula (11) for $\pi(z)$.

Our transformation allows us to replace the study of the original equation (1) by the study of the simpler equation (9).

Example. Let us transform the Bessel equation

$$z^2u'' + zu' + (z^2 - \nu^2)u = 0$$

into the form (9) by the substitution $u = \phi(z)y$. The Bessel equation is a special case of (1) with $\sigma(z) = z$, $\tilde{\tau}(z) = 1$, $\tilde{\sigma}(z) = z^2 - \nu^2$. In this case the expression under the square root in (11) has the form $-z^2 + \nu^2 + kz$. Setting the discriminant of this quadratic equal to zero, we obtain the equation

$$k^2 + 4\nu^2 = 0$$

for the constant k. Hence $k = \pm 2i\nu$, and consequently by (11)

$$\pi(z) = \pm\sqrt{-z^2 + \nu^2 \pm 2i\nu z} = \pm(iz \pm \nu).$$

* If $\sigma(z)$ is a quadratic polynomial, (1) is a special case of the Riemann equation with three different singular points, one of them at infinity. The Riemann equation is studied in courses on the analytic theory of differential equations (see [M4], [T1] or [W3].)

In this case there are four possibilities for $\pi(z)$. Consider, for example, the case $k = 2i\nu$, $\pi(z) = iz + \nu$. Using (3), (6) and (10), we find

$$\phi(z) = z^{\nu} e^{iz},$$
$$\tau(z) = 2iz + 2\nu + 1,$$
$$\lambda = k + \pi'(z) = i(2\nu + 1).$$

Then (9) has the form

$$zy'' + (2iz + 2\nu + 1)y' + i(2\nu + 1)y = 0.$$

Remarks. 1) Since (1) is not changed by replacing $\sigma(z), \tilde{\tau}(z)$ and $\tilde{\sigma}(z)$ by $c\sigma(z), c\tilde{\tau}(z)$, and $c^2\tilde{\sigma}(z)$, where c is any constant, the coefficient of the leading term of $\sigma(z)$ can be chosen arbitrarily. A similar remark applies to (9).

2) From now on we shall consider only cases when $\sigma(z)$ in (1) and (9) does not have a double root. Actually, if $\sigma(z)$ has a double root, i.e. $\sigma(z) = (z-a)^2$, equation (1) can be transformed into

$$\frac{d^2u}{ds^2} + \frac{2 - s\tilde{\tau}(a + 1/s)}{s}\frac{du}{ds} + \frac{s^2\tilde{\sigma}(a + 1/s)}{s^2}u = 0 \tag{12}$$

by the substitution $z - a = 1/s$.

Since $s\tilde{\tau}(a + 1/s)$ and $s^2\tilde{\sigma}(a + 1/s)$ are polynomials in s of degree at most 1 and 2, respectively, (12) is an equation of the form (1) with $\sigma(s) = s$, which does not have a double root.

3) It is not possible to transform (1) into the form (9) if $\sigma(z) = 1$ and $(\tilde{\tau}/2)^2 - \tilde{\sigma}$ is linear. In this case we can transform (1) into a simpler form by taking $\pi(z)$ in (3) so that $\tau(z)$ is zero. Then $\bar{\sigma}(z)$ will be linear and (5) takes the form

$$y'' + (az + b)y = 0. \tag{13}$$

A linear transformation $s = az + b$ takes (13) into a special case of

$$\frac{d^2y}{ds^2} + \frac{1 - 2\alpha}{s}\frac{dy}{ds} + \left[(\beta\gamma s^{\gamma-1})^2 + \frac{\alpha^2 - \nu^2\gamma^2}{s^2}\right] y = 0, \tag{14}$$

where $\alpha, \beta, \gamma, \nu$ are constants. This equation is studied in §14 (see Lommel's equation). The solutions of (14) can be expressed in terms of Bessel functions.

4) A problem that can be reduced to the solution of an equation of hypergeometric type is the solution of a system of first-order differential equations

$$\begin{aligned} u_1' &= a_{11}(z)u_1 + a_{12}(z)u_2, \\ u_2' &= a_{21}(z)u_1 + a_{22}(z)u_2 \end{aligned} \tag{15}$$

in the case when the $a_{ik}(z)$ have the form

$$a_{ik}(z) = \frac{\tau_{ik}(z)}{\sigma(z)}, \tag{16}$$

where $\tau_{ik}(z)$ are polynomials of degree at most 1, and $\sigma(z)$ is a polynomial of degree at most 2*. If we eliminate $u_2(z)$ from (15), we obtain the equation

$$u_1'' - (a_{11} + a_{22} + \frac{a_{12}'}{a_{12}})u_1' + (a_{11}a_{22} - a_{12}a_{21} + a_{11}\frac{a_{12}'}{a_{12}} - a_{11}')u_1 = 0 \tag{17}$$

for $u_1(z)$. Since

$$\frac{a_{12}'}{a_{12}} = -\frac{\sigma'}{\sigma} + \frac{\tau_{12}'}{\tau_{12}},$$

equation (17) is of hypergeometric type when $\tau_{12}' = 0$. If $\tau_{12}' \neq 0$, we can first apply a linear transformation

$$\begin{aligned} v_1 &= \alpha u_1 + \beta u_2, \\ v_2 &= \gamma u_1 + \delta u_2, \end{aligned}$$

where $\alpha, \beta, \gamma, \delta$ are constants. We then obtain a system of the form

$$\left.\begin{aligned} v_1' &= \tilde{a}_{11}(z)v_1 + \tilde{a}_{12}(z)v_2, \\ v_2' &= \tilde{a}_{21}(z)v_1 + \tilde{a}_{22}(z)v_2, \end{aligned}\right\} \tag{18}$$

where the $\tilde{a}_{ik}(z)$ are linear combinations of the $a_{ik}(z)$ with constant coefficients, depending on $\alpha, \beta, \gamma, \delta$, and consequently have the form

$$\tilde{a}_{ik}(z) = \frac{\tilde{\tau}_{ik}(z)}{\sigma(z)}$$

($\tilde{\tau}_{ik}(z)$ are polynomials of degree at most 1). If the coefficients $\alpha, \beta, \gamma, \delta$ are chosen so that $\tilde{\tau}_{12}' = 0$, as is always possible, then after eliminating

---

* System (15) arises for example, when the energy spectrum of an electron moving in a Coulomb field is determined by solving the Dirac equation (see §26, part 3).

$v_2(z)$ from (18) we obtain a generalized equation of hypergeometric type for $v_1(z)$.

If $\sigma(z)$ is a polynomial of degree 1, we can get from (15) to a generalized equation of hypergeometric type in a different way, by choosing $\alpha, \beta, \gamma, \delta$ so that $\tilde{a}_{12}$ is independent of z, i.e. so that $\tilde{\tau}_{12} = \nu\sigma(z)$ (ν constant).

§ 2 Polynomials of hypergeometric type. The Rodrigues formula.

We now investigate the properties of the solutions of the equation of hypergeometric type,

$$\sigma(z)y'' + \tau(z)y' + \lambda y = 0. \tag{1}$$

Let us show that *all the derivatives of functions of hypergeometric type are also of hypergeometric type.*

To prove this, we differentiate (1). We then find that $v_1(z) = y'(z)$ satisfies the equation

$$\sigma(z)v_1'' + \tau_1(z)v_1' + \mu_1 v_1 = 0, \tag{2}$$

where

$$\tau_1(z) = \tau(z) + \sigma'(z),$$
$$\mu_1 = \lambda + \tau'(z).$$

Since $\tau_1(z)$ is a polynomial of degree 1 at most, and μ_1 is independent of z, equation (2) is an equation of hypergeometric type.

The converse is also true: *Every solution of* (2) *with* $\lambda \neq 0$ *is the derivative of a solution of* (1).

Let $v_1(z)$ be a solution of (2). If $v_1(z)$ is to be the derivative of a solution $y(z)$ of (1), these functions must be related in the following way (see (1)):

$$y(z) = -\frac{1}{\lambda}[\sigma(z)v_1' + \tau(z)v_1].$$

We can show that the function $y(z)$ defined by this formula satisfies (1), and that its derivative is $v_1(z)$. We have

$$\lambda y' = -[\sigma(z)v_1'' + \tau_1(z)v_1' + \tau'(z)v_1] = \lambda v_1,$$

i.e. $y' = v_1(z)$. Substituting $v_1 = y'$ in the original expression for $y(z)$, we obtain (1) for $y(z)$.

In a similar way, by induction, we can obtain an equation of hypergeometric type for $v_n(z) = y^{(n)}(z)$:

$$\sigma(z)v_n'' + \tau_n(z)v_n' + \mu_n v_n = 0, \tag{3}$$

where

$$\tau_n(z) = \tau(z) + n\sigma'(z),$$
$$\mu_n = \lambda + n\tau' + \frac{n(n-1)}{2}\sigma''.$$

Moreover, every solution of (3) for $\mu_k \neq 0$ ($k = 0, 1, \ldots, n-1$) can be represented in the form $v_n(z) = y^{(n)}(z)$, where $y(z)$ is a solution of (1).

This property lets us construct a family of particular solutions of (1) corresponding to a given λ. In fact, when $\mu_n = 0$ equation (3) has the particular solution $v_n(z) = \text{const}$. Since $v_n(z) = y^{(n)}(z)$, this means that when

$$\lambda = \lambda_n = -n\tau' - \frac{n(n-1)}{2}\sigma''$$

the equation of hypergeometric type has a particular solution of the form $y(z) = y_n(z)$ which is a polynomial of degree n. We shall call such solutions *polynomials of hypergeometric type*. The polynomials $y_n(z)$ are, in a sense, the simplest solutions of (1).*

To find the polynomials $y_n(z)$ explicitly, we multiply (1) and (3) by appropriate functions $\rho(z)$ and $\rho_n(z)$ so that they can be written in self-adjoint form:

$$(\sigma\rho y')' + \lambda\rho y = 0, \tag{4}$$
$$(\sigma\rho_n v_n')' + \mu_n\rho_n v_n = 0. \tag{5}$$

Here $\rho(z)$ and $\rho_n(z)$ satisfy the differential equations

$$(\sigma\rho)' = \tau\rho, \tag{6}$$
$$(\sigma\rho_n)' = \tau_n\rho_n. \tag{7}$$

Now using the explicit form of $\tau_n(z)$ we can easily establish the connection between $\rho_n(z)$ and $\rho_0(z) \equiv \rho(z)$.

* In fact, the existence of polynomial solutions of (1) follows from the fact that the operator $\sigma(z)d^2/dz^2 + \tau(z)d/dz$ carries polynomials of degree n into polynomials of the same degree.

We have

$$(\sigma\rho_n)'/\rho_n = \tau + n\sigma' = (\sigma\rho)'/\rho + n\sigma'$$

whence

$$\rho_n'/\rho_n = \rho'/\rho + n\sigma'/\sigma,$$

and consequently

$$\rho_n(z) = \sigma^n(z)\rho(z) \qquad (n = 0, 1. \ldots). \tag{8}$$

Since $\sigma\rho_n = \rho_{n+1}$ and $v_n'(z) = v_{n+1}(z)$, we can rewrite (5) in the form

$$\rho_n v_n = -\frac{1}{\mu_n}(\rho_{n+1}v_{n+1})'.$$

Hence when $m < n$ we obtain successively

$$\begin{aligned}\rho_m v_m &= -\frac{1}{\mu_m}(\rho_{m+1}v_{m+1})' \\ &= \left(-\frac{1}{\mu_m}\right)\left(-\frac{1}{\mu_{m+1}}\right)(\rho_{m+2}v_{m+2})'' = \ldots = \frac{A_m}{A_n}(\rho_n v_n)^{(n-m)},\end{aligned}$$

where

$$A_n = (-1)^n \prod_{k=0}^{n-1} \mu_k, \quad A_0 = 1. \tag{9}$$

We now proceed to obtain an explicit form for the polynomials of hypergeometric type. If $y(z)$ is a polynomial of degree n, i.e. $y = y_n(z)$, then

$$v_m(z) = y_n^{(m)}(z), \qquad v_n(z) = y^{(n)}(z) = \text{const.},$$

and we obtain the following expression for $y_n^{(m)}(z)$:

$$y_n^{(m)}(z) = \frac{A_{mn}B_n}{\rho_m(z)}[\rho_n(z)]^{(n-m)}, \tag{10}$$

where

$$A_{mn} = A_m(\lambda)|_{\lambda=\lambda_n}, \qquad B_n = \frac{1}{A_{nn}}y_n^{(n)}(z). \tag{11}$$

Hence, in particular, when $m = 0$ we have an explicit representation for the polynomials $y_n(z)$ of hypergeometric type:

$$y_n(z) = \frac{B_n}{\rho(z)}[\sigma^n(z)\rho(z)]^{(n)} \quad (n = 0, 1, \ldots). \tag{12}$$

Consequently the polynomial solutions of (1) are defined by (12) up to a normalizing factor. These solutions correspond to the values $\mu_n = 0$, i.e.

$$\lambda = \lambda_n = -n\tau' - \frac{n(n-1)}{2}\sigma'' \quad (n = 0, 1, \ldots). \tag{13}$$

We call (12) the *Rodrigues formula*, since it was established in 1814 by B.O.Rodrigues for special polynomials of hypergeometric type, namely the Legendre polynomials, for which $\sigma(z) = 1 - z^2, \rho(z) = 1$.

§ 3 Integral representation for functions of hypergeometric type

We now generalize the Rodrigues formula to find particular solutions of the equation (2.1)* of hypergeometric type for arbitrary values of λ. For this purpose we first write equation (2.12) for the polynomial solutions in a different form by using Cauchy's integral formula for analytic functions:

$$y_n(z) = \frac{C_n}{\rho(z)} \int\limits_C \frac{\sigma^n(s)\rho(s)}{(s-z)^{(n+1)}} ds. \tag{1}$$

Here $c_n = B_n n!/(2\pi i)$, where C is a closed contour surrounding the point $s = z$, and $\rho(z)$ is a solution of $(\sigma\rho)' = \tau\rho$.

This representation of a particular solution of (2.1) with $\lambda = \lambda_n$ lets us guess that when λ is arbitrary we should look for a particular solution of the form

$$y(z) = y_\nu(z) = \frac{C_\nu}{\rho(z)} \int\limits_C \frac{\sigma^\nu(s)\rho(s)}{(s-z)^{\nu+1}} ds, \tag{2}$$

where C_ν is a normalizing constant and ν is connected with λ by an equation analogous to (2.13):

$$\lambda = -\nu\tau' - \frac{\nu(\nu-1)}{2}\sigma''. \tag{3}$$

Let us show that for an appropriate choice of the contour C, in general not closed, our guess is correct.

* When a formula from a different section is cited, the section number is given; thus, (2.1) is formula (1) of §2 of this chapter.

Theorem 1. *Let $\rho(z)$ satisfy the equation*

$$[\sigma(z)\rho(z)]' = \tau(z)\rho(z),$$

where ν is a root of the equation $\lambda + \nu\tau' + \frac{1}{2}\nu(\nu-1)\sigma'' = 0$, and let

$$u(z) = \int_C \frac{\rho_\nu(s)}{(s-z)^{\nu+1}}ds, \quad \rho_\nu(s) = \sigma^\nu \rho(s).$$

Then the equation

$$\sigma(z)y'' + \tau(z)y' + \lambda y = 0$$

of hypergeometric type has a particular solution of the form

$$y(z) \equiv y_\nu(z) = \frac{C_\nu}{\rho(z)}u(z)$$

(where C_ν is a normalizing constant) provided that
1) in calculating $u'(z)$ and $u''(z)$ we can interchange differentiation with respect to z and integration with respect to s, i.e.

$$u'(z) = (\nu+1)\int_C \frac{\rho_\nu(s)}{(s-z)^{\nu+2}}ds, \quad u''(z) = (\nu+1)(\nu+2)\int_C \frac{\rho_\nu(s)}{(s-z)^{\nu+3}}ds;$$

2) the contour C is chosen so that

$$\left.\frac{\sigma^{\nu+1}(s)\rho(s)}{(s-z)^{\nu+2}}\right|_{s_1}^{s_2} = 0, \tag{4}$$

where s_1 and s_2 are the endpoints of C.

Proof. Let us obtain a differential equation for $u(z)$. For this purpose we use the equation for $\rho_\nu(s)$,

$$[\sigma(s)\rho_\nu(s)]' = \tau_\nu(s)\rho_\nu(s),$$

where $\tau_\nu(s) = \tau(s) + \nu\sigma'(s)$ (compare (2.7)). We multiply this equation by $(s-z)^{-\nu-2}$, integrate both sides over C, and then integrate by parts:

$$\left.\frac{\sigma(s)\rho_\nu(s)}{(s-z)^{\nu+2}}\right|_{s_1}^{s_2} + (\nu+2)\int_C \frac{\sigma(s)\rho_\nu(s)}{(s-z)^{\nu+3}}ds = \int_C \frac{\tau_\nu(s)\rho_\nu(s)}{(s-z)^{\nu+2}}ds\,.$$

By hypothesis, the integrated terms reduce to 0. Let us expand $\sigma(s)$ and $\tau_\nu(s)$ in powers of $s-z$:

$$\sigma(s)=\sigma(z)+\sigma'(z)(s-z)+\frac{1}{2}\sigma''(z)(s-z)^2,$$
$$\tau_\nu(s)=\tau_\nu(z)+\tau_\nu'(z)(s-z).$$

If we then use our formulas for $u(z), u'(z)$ and $u''(z)$, we obtain the equation

$$\frac{1}{\nu+1}\sigma(z)u''+\frac{\nu+2}{\nu+1}\sigma'(z)u'+\frac{\nu+2}{2}\sigma''u=\frac{1}{\nu+1}\tau_\nu(z)u'+\tau_\nu' u.$$

If we insert the explicit form of $\tau_\nu(z)$, we can write the preceding equation in the form

$$\sigma(z)u''+\left[2\sigma'(z)-\tau(z)\right]u'-(\nu+1)\left(\tau'+\frac{\nu-2}{2}\sigma''\right)u=0. \tag{5}$$

We now use (5) to obtain an equation for $y(z)$. We have

$$(\sigma\rho y)'=(\sigma\rho)'y+\sigma\rho y'$$

whence it follows that

$$\sigma\rho y'=(\sigma\rho y)'-\tau\rho y=C_\nu\left[(\sigma u)'-\tau u\right].$$

After differentiating and using (5), we obtain

$$(\sigma\rho y')'=C_\nu\left[(\sigma u)''-(\tau u)'\right]=C_\nu\left[\sigma u''+(2\sigma'-\tau)u'+(\sigma''-\tau')u\right]$$
$$=C_\nu\left[(\nu+1)\left(\tau'+\frac{\nu-2}{2}\sigma''\right)+(\sigma''-\tau')\right]u.$$

From this and (3), we obtain

$$(\sigma\rho y')'=-\lambda\rho y.$$

This equation is the same as (2.4), which is equivalent to (2.1).

This theorem is of fundamental importance in the study of particular special functions.

Observe that hypothesis (4) in the theorem will be satisfied, in particular, if the ends of C are chosen so that $\sigma^{\nu+1}(s)\rho(s)/(s-z)^{\nu+2}$ is zero at both of them, i.e.

$$\left.\frac{\sigma^{\nu+1}(s)\rho(s)}{(s-z)^{\nu+2}}\right|_{s=s_1,s_2}=0. \tag{6}$$

Let us consider some possible forms of C for which (6) is satisfied.

a) Let s_0 be a root of the equation $\sigma(s) = 0$. If $\sigma^{\nu+1}(s)\rho(s)|_{s=s_0} = 0$, then one end of the contour can be taken at $s = s_0$.

b) If $\mathrm{Re}(\nu + 2) < 0$, one end of the contour can be taken at $s = z$.

c) We can also take one end of the contour at $s = \infty$ if

$$\lim_{s\to\infty} \frac{\sigma^{\nu+1}(s)\rho(s)}{(s-z)^{\nu+2}} = 0.$$

In this way we can construct many particular solutions of an equation of hypergeometric type, corresponding to different contours C and different values of ν. In addition, the number of particular solutions can be increased by using the transformation discussed in §1. In fact, equation (2.1) can be considered as a generalized equation of hypergeometric type (1.1), for which $\tilde{\sigma}(z) = \lambda\sigma(z), \tilde{\tau}(z) = \tau(z)$. After the transformation the original equation becomes another equation of hypergeometric type. After constructing the particular solutions of the latter, the inverse transformation yields new particular solutions for the original equation. Since an equation of hypergeometric type has only two linearly independent solutions, every solution must be a linear combination of two linearly independent solutions. In this way we can, in particular, obtain functional equations for functions of hypergeometric type.

In constructing solutions of an equation of hypergeometric type, we shall restrict ourselves to simple contours: straight lines or segments of straight lines, connecting points s_1 and s_2 for which (6) holds. Contours of this kind can be found, in general, only under certain restrictions on the coefficients of the differential equation. We extend the results so obtained to more general cases by using analytic continuation.

Let us review the notion of analytic continuation, which plays an important role in later work (see [D2], [E3], [S2], or [S8]). Let $f(z)$ be given on a set E belonging to a region D. If $F(z)$ is analytic in D and coincides with $f(z)$ on E, then $F(z)$ is an *analytic continuation* of $f(z)$ to D. We have the following proposition.

Principle of analytic continuation. *If E contains at least one limit point of D, then $f(z)$ has at most one analytic continuation to D.* In particular, the continuation is unique if E is a line segment in D.

Here and later, analytic functions are understood to be single-valued; such functions are sometimes called *regular*. If a function that we have to consider is not single-valued, we introduce cuts along suitable lines in the complex plane so as to restrict attention to a single-valued branch of the function.

In evaluating expressions of the form $(z - a)^\alpha$ the expression that is being raised to the power is taken to have the angle of smallest absolute value compatible with the given cut. For example, in choosing a branch of the function $(1-z)^\alpha(1+z)^\beta$, which has branch points at $z=-1$ and $z=+1$, it is sufficient to make a cut along the real axis for $z \geq -1$. Correspondingly, $(1-z)^\alpha$ is evaluated on the cut with $|\arg(1-z)| < \pi$, and $(1+z)^\beta$ with $0 < \arg z < 2\pi$.

Since we are going to use the integral representation (2) for solutions of an equation of hypergeometric type, we shall need, for analytic continuation of the solutions of the equation, to rely on the following theorem on the analyticity of an integral that depends on a parameter (see [B5], [E3], [S2]).

Theorem 2. *Let C be a piecewise smooth curve, of finite length, in the complex s plane, and D a region of the complex z plane. If $f(s,z)$ is continuous as a function of two variables for $s \in C$ and $z \in D$, and in addition is an analytic function of z in D for every $s \in C$, then the function*

$$F(z) = \int_C f(z,s)ds$$

is analytic in D, and

$$F'(z) = \int_C f'_z(z,s)ds.$$

The conclusion of the theorem also remains valid for uniformly convergent improper integrals $F(z)$. In studying integral representations for various special functions, it is convenient to use the following simple test for the uniform convergence of integrals: *if the continuous function $f(z,s)$ satisfies $|f(z,s)| \leq \phi(s)$ for all $s \in C$ and $z \in D$, and the integral $\int_C \phi(s)|ds|$ converges, then $\int_C f(z,s)ds$ converges uniformly for z in D.*

Since the derivative of a function $y = y(z)$ of hypergeometric type is again a function of the same type, it follows that by continuing a function of hypergeometric type analytically we obtain analytic continuations of $y'(z)$ and $y''(z)$ with respect to z and with respect to their parameters. The integral representation of a function $y(z)$ of hypergeometric type was constructed on the hypothesis that the function satisfies an equation of hypergeometric type (2.1) under certain restrictions on z and on the parameters of $y(z)$. By the principle of analytic continuation, $y(z)$ will satisfy the same equation in the whole region in which the left-hand side of the equation is analytic (the right-hand side, zero, is analytic in every region).*

* However, if we use the analytic theory of differential equations (see, for example, [T1]), the region of analyticity of the solutions of an equation (2.1) can be determined directly from the form of the equation (the singular points of an equation of hypergeometric type are the roots of $\sigma(z)=0$ and the point at infinity).

In the following discussion we shall study the solutions of particular equations of hypergeometric type by using the integral representation (2), and the results will be extended to a wider domain by means of the principle of analytic continuation.

§ 4 Recursion relations and differentiation formulas

Let us consider a general method of obtaining various relationships for functions $y_\nu(z)$ defined by the integral representation (3.2). We begin by establishing relationships among functions of the form

$$\phi_{\nu\mu}(z) = \int\limits_C \frac{\sigma^\nu(s)\rho(s)}{(s-z)^{\mu+1}}ds,$$

which appear in the definitions of the $y_\nu(z)$ and their derivatives.

Lemma. *Any three functions $\phi_{\nu_i\mu_i}(z)$ are connected by a linear relation*

$$\sum_{i=1}^{3} A_i(z)\phi_{\nu_i\mu_i}(z) = 0$$

with polynomial coefficients $A_i(z)$, provided that the differences $\nu_i - \nu_j$ and $\mu_i - \mu_j$ are integers and that

$$\left.\frac{\sigma^{\nu_0+1}(s)\rho(s)}{(s-z)^{\mu_0}}s^m\right|_{s_1}^{s_2} = 0 \qquad (m = 0, 1, 2, \ldots),$$

where s_1 and s_2 are the endpoints of the contour C; ν_0 is the ν_i with the smallest real part; μ_0, the μ_i with the largest real part.

Proof. Consider the sum $\sum_{i=1}^{3} A_i\phi_{\nu_i\mu_i}(z)$. We show that the coefficients $A_i = A_i(z)$ can be chosen so that this linear combination is zero. For any fixed z we have

$$\sum_i A_i\phi_{\nu_i\mu_i}(z) = \int\limits_C \frac{\sigma^{\nu_0}(s)\rho(s)}{(s-z)^{\mu_0+1}}P(s)ds,$$

where μ_0 and ν_0 are defined in the statement of the lemma, and

$$P(s) = \sum_i A_i\sigma^{\nu_i-\nu_0}(s)(s-z)^{\mu_0-\mu_i}.$$

Since the differences $\nu_i - \nu_0$ and $\mu_0 - \mu_i$ are nonnegative integers, $P(s)$ is a polynomial in s. We choose the A_i so that

$$\frac{\sigma^{\nu_0}(s)\rho(s)}{(s-z)^{\mu_0+1}}P(s) = \frac{d}{ds}\left[\frac{\sigma^{\nu_0+1}(s)\rho(s)}{(s-z)^{\mu_0}}Q(s)\right], \tag{1}$$

where $Q(s)$ is a polynomial (we show below that such a choice of the coefficients is possible). We obtain

$$\sum_i A_i \phi_{\nu_i\mu_i}(z) = \left.\frac{\sigma^{\nu_0+1}(s)\rho(s)}{(s-z)^{\mu_0}}Q(s)\right|_{s_1}^{s_2}.$$

If we require that the condition

$$\left.\frac{\sigma^{\nu_0+1}(s)\rho(s)}{(s-z)^{\mu_0}}s^m\right|_{s_1}^{s_2} = 0 \qquad (m = 0, 1, 2, \ldots),$$

which is similar to (3.4), holds at the endpoints of C, then the endpoint terms become zero, and with the A_i determined in this way we have the linear relation

$$\sum_i A_i \phi_{\nu_i\mu_i}(z) = 0. \tag{2}$$

Let us show that it is always possible to choose the coefficients of $Q(s)$ and the coefficients A_i so that (1) in fact holds. To do this, we rewrite (1) in a more convenient form, using the differential equation $(\sigma\rho_\nu)' = \tau_\nu\rho_\nu$ for $\rho_\nu(s) = \sigma^\nu(s)\rho(s)$, where $\tau_\nu(s) = \tau(s) + \nu\sigma'(s)$. We obtain

$$P(s) = Q(s)\left[(s-z)\tau_{\nu_0}(s) - \mu_0\sigma(s)\right] + \sigma(s)(s-z)Q'(s). \tag{3}$$

If we compare the left-hand and the right-hand sides of this equation, it is easy to see that the degree of $Q(s)$ is two less than the degree of $P(s)$.

If we equate coefficients of powers of s on the two sides of (3), we obtain a system of homogeneous linear equations in the coefficients of $Q(s)$ and the coefficients $A_i (i = 1, 2, 3)$ that appear in the expression for $P(s)$. The number of equations is two more than the number of unknown coefficients of $Q(s)$. Hence the number of unknowns is at least one more than the number of equations, and consequently one of the unknown coefficients can be assigned arbitrarily. In the case when $P(s)$ is at most of degree 1, the relation we are considering remains valid if we take $Q(s) = 0$. In the resulting system of equations, the coefficients of the unknowns are polynomials in z, so that in this case after one coefficient is selected the remaining coefficients are rational functions of z. After multiplying (2) by the common denominator of the $A_i(z)$ we obtain a linear relation with polynomial coefficients. This completes the proof of the lemma.

In practical applications of the method the degree of $P(s)$ can sometimes be reduced by integrating by parts in some of the functions $\phi_{\nu_i\mu_i}(z)$. We have

$$\begin{aligned}\phi_{\nu\mu}(z) &= \int_C \frac{\sigma^\nu(s)\rho(s)}{(s-z)^{\mu+1}}ds \\ &= -\frac{1}{\mu}\frac{\sigma^\nu(s)\rho(s)}{(s-z)^\mu}\bigg|_{s_1}^{s_2} + \frac{1}{\mu}\int_C \frac{\tau_{\nu-1}(s)\sigma^{\nu-1}(s)\rho(s)}{(s-z)^\mu}ds,\end{aligned}$$

where $\tau_{\nu-1}(s) = \tau(s) + (\nu-1)\sigma'(s)$. Supposing, as usual, that the integrated terms yield zero, we obtain

$$\phi_{\nu\mu}(z) = \frac{1}{\mu}\int_C \frac{\tau_{\nu-1}(s)\sigma^{\nu-1}(s)\rho(s)}{(s-z)^\mu}ds. \tag{4}$$

Example 1. Find a relation among $\phi_{\nu,\nu-1}(z)$, $\phi_{\nu\nu}(z)$ and $\phi_{\nu,\nu+1}(z)$.

In this case $\nu_0 = \nu$, $\mu_0 = \nu+1$, $P(s) = A_1(s-z)^2 + A_2(s-z) + A_3$, $Q(s) = q_0$ (a constant); the endpoint condition

$$\frac{\sigma^{\nu_0+1}(s)\rho(s)}{(s-z)^{\mu_0}}Q(s)\bigg|_{s_1}^{s_2} = 0,$$

which arises in the proof of the lemma, is equivalent to

$$\frac{\sigma^{\nu+1}(s)\rho(s)}{(s-z)^{\nu+1}}\bigg|_{s_1}^{s_2} = 0.$$

Equation (3) has the form

$$A_1(s-z)^2 + A_2(s-z) + A_3 = q_0\left[(s-z)\tau_\nu(s) - (\nu+1)\sigma(s)\right].$$

Taking $q_0 = 1$, expanding the right-hand side in powers of $s-z$, and comparing coefficients, we obtain

$$\begin{aligned} A_1 &= \tau_\nu' - \frac{\nu+1}{2}\sigma'' = \tau' + \frac{\nu-1}{2}\sigma'', \\ A_2 &= \tau_\nu(z) - (\nu+1)\sigma'(z) = \tau(z) - \sigma'(z), \\ A_3 &= -(\nu+1)\sigma(z). \end{aligned} \tag{5}$$

Therefore

$$A_1(z)\phi_{\nu,\nu-1}(z) + A_2(z)\phi_{\nu\nu}(z) + A_3(z)\phi_{\nu,\nu+1}(z) = 0, \tag{6}$$

where the coefficients $A_i(z)$ are defined by (5). It is convenient to rewrite the last equation in a different form. Since

$$y_\nu(z) = \frac{C_\nu}{\rho(z)}\phi_{\nu\nu}(z), \quad \phi_{\nu,\nu+1}(z) = \frac{1}{\nu+1}\phi'_{\nu\nu}(z),$$
$$[\sigma(z)\rho(z)]' = \tau(z)\rho(z),$$

relation (6) yields a convenient integral representation for the derivatives of functions of hypergeometric type:

$$y'_\nu(z) = \frac{C_\nu^{(1)}}{\sigma(z)\rho(z)}\int_C \frac{\sigma^\nu(s)\rho(s)}{(s-z)^\nu}ds, \tag{7}$$

where

$$C_\nu^{(1)} = \left(\tau' + \frac{\nu-1}{2}\sigma''\right)C_\nu.$$

Generalization of the relation (7) deduced in Example 1 enables us to obtain a convenient integral representation for the derivatives of any order of functions of hypergeometric type.

In fact, (7) can be interpreted in the following way: an integral representation for the first derivative of a function of hypergeometric type

$$y_\nu(z) = \frac{C_\nu}{\rho(z)}\int_C \frac{\sigma^\nu(s)\rho(s)}{(s-z)^{\nu+1}}ds = \frac{C_\nu}{\rho(z)}\phi_{\nu\nu}(z) \tag{8}$$

can be obtained from the original representation by replacing ν by $\nu-1$, $\rho(z)$ by $\rho_1(z) = \sigma(z)\rho(z)$, and multiplying by the additional factor $\tau' + \frac{1}{2}(\nu-1)\sigma''$. It is then clear that

$$y_\nu^{(k)}(z) = \frac{C_\nu^{(k)}}{\sigma^k(z)\rho(z)}\phi_{\nu,\nu-k}(z), \tag{9}$$

where

$$C_\nu^{(k)} = \left(\tau'_{k-1} + \frac{\nu-k}{2}\sigma''\right)C_\nu^{(k-1)} = \left(\tau' + \frac{\nu+k-2}{2}\sigma''\right)C_\nu^{(k-1)}$$
$$= \prod_{s=0}^{k-1}\left(\tau' + \frac{\nu+s-1}{2}\sigma''\right)C_\nu.$$

If we use (9) and the lemma proved above, we obtain the following theorem.

Theorem. *Any three functions* $y_{\nu_i}^{(k_i)}(z)$ *are connected by a relation of the form*

$$\sum_{i=1}^{3} A_i(z) y_{\nu_i}^{(k_i)}(z) = 0$$

with polynomial coefficients $A_i(z)$, *provided that the differences* $\nu_i - \nu_j$ *are integers and that*

$$\left. \frac{\sigma^{\nu_0+1}(s)\rho(s)}{(s-z)^{\mu_0+1}} s^m \right|_{s_1}^{s_2} = 0 \quad (m = 0, 1, 2, \ldots).$$

Here s_1 *and* s_2 *are the endpoints of* C; ν_0 *is the* ν_i *of smallest real part; and* μ_0 *is the* ν_i *of largest real part.*

We note that the equations that determine the coefficients $A_i(z)$ are linear and homogeneous in the unknowns and independent of the contour C used in defining $y_\nu(z)$. Consequently *two functions* $y_\nu(z)$ *of hypergeometric type which differ only by factors independent of* ν *and by the choice of* C *will satisfy relations of the kind under consideration with the same coefficients.*

Example 2. Let us obtain a formula

$$A_1 y_\nu'(z) + A_2 y_{\nu+1}(z) + A_3 y_\nu(z) = 0, \tag{10}$$

connecting $y_\nu'(z), y_\nu(z)$ and $y_{\nu+1}(z)$. We shall refer to formulas that express derivatives of functions of hypergeometric type in terms of the functions themselves as *differentiation formulas.*

To obtain (10) we use the integral representations (7) and (8) for $y_\nu'(z)$ and $y_\nu(z)$, and a preliminary transformation of $y_{\nu+1}(z)$, using (4). Then we can write the left-hand side of (10) in the form

$$A_1 y_\nu'(z) + A_2 y_{\nu+1}(z) + A_3 y_\nu(z) = \frac{1}{\rho(z)} \int_C \frac{\sigma^\nu(s)\rho(s)}{(s-z)^{\nu+1}} P(s) ds,$$

where

$$P(s) = \left[A_1 \frac{C_\nu \kappa_\nu}{\sigma(z)}(s-z) + A_2 \frac{C_{\nu+1}\tau_\nu(s)}{\nu+1} + A_3 C_\nu \right],$$

$$\kappa_\nu = \tau' + \frac{\nu-1}{2}\sigma''.$$

Since $P(s)$ is a linear polynomial, $Q(s) = 0$ and consequently

$$A_1 \frac{C_\nu \kappa_\nu}{\sigma(z)}(s-z) + A_2 \frac{C_{\nu+1}\tau_\nu(s)}{\nu+1} + A_3 C_\nu = 0.$$

In determining A_1, A_2, and A_3 from this equation it is convenient to take $A_1 = \sigma(z)$, expand the left-hand side in powers of $s-z$, and equate coefficients of powers of $s-z$. We find

$$A_2 = -(\nu+1)\frac{\kappa_\nu}{\tau'_\nu}\frac{C_\nu}{C_{\nu+1}}, \qquad A_3 = \kappa_\nu \frac{\tau_\nu(z)}{\tau'_\nu}.$$

As a result we obtain the differentiation formula

$$\sigma(z)y'_\nu(z) = \frac{\kappa_\nu}{\tau'_\nu}\left[(\nu+1)\frac{C_\nu}{C_{\nu+1}}y_{\nu+1}(z) - \tau_\nu(z)y_\nu(z)\right]. \tag{11}$$

In particular, for the polynomials

$$y_n(z) = \frac{B_n}{\rho(z)}\frac{d^n}{dz^n}[\sigma^n(z)\rho(z)]$$

of hypergeometric type, we have $\nu = n (n = 0, 1, 2, \ldots)$ and $C_n = n!B_n/(2\pi i)$ (see formula (3.1)). Hence in this case the differentiation formula (11) can be written in the form

$$\sigma(z)y'_n(z) = \frac{\kappa_n}{\tau'_n}\left[\frac{B_n}{B_{n+1}}y_{n+1}(z) - \tau_n(z)y_n(z)\right], \tag{12}$$

where

$$\tau_n(z) = \tau(z) + n\sigma'(z), \qquad \kappa_n = \tau' + \frac{n-1}{2}\sigma''.$$

Thus in Chapter 1 we have considered a method of constructing integral representations for particular solutions of the generalized equation of hypergeometric type and indicated ways of studying the properties of these solutions. With this we complete our discussion of the general theory of special functions; in the next chapter we turn to the study of the classical orthogonal polynomials, which form an important subclass of the special functions of mathematical physics.

Chapter II

The Classical Orthogonal Polynomials

§ 5 Basic properties of polynomials of hypergeometric type.

1. Jacobi, Laguerre and Hermite polynomials. In §2 we introduced the polynomials $y_n(z)$ of hypergeometric type, which are solutions of

$$\sigma(z)y'' + \tau(z)y' + \lambda y = 0 \tag{1}$$

with $\lambda = \lambda_n = -n\tau' - \frac{1}{2}n(n-1)\sigma''$. They are given explicitly by the *Rodrigues formula*

$$y_n(z) = \frac{B_n}{\rho(z)}[\sigma^n(z)\rho(z)]^{(n)}, \tag{2}$$

where B_n is a normalizing constant and $\rho(z)$ satisfies the differential equation

$$[\sigma(z)\rho(z)]' = \tau(z)\rho(z). \tag{3}$$

Solving (3), we obtain, up to constant factors, the possible forms for $\rho(z)$ corresponding to the possible degrees of $\sigma(z)$:

$$\rho(z) = \begin{cases} (b-z)^\alpha (z-a)^\beta & \text{for } \sigma(z) = (b-z)(z-a), \\ (z-a)^\alpha e^{\beta z} & \text{for } \sigma(z) = z - a, \\ e^{\alpha z^2 + \beta z} & \text{for } \sigma(z) = 1. \end{cases}$$

Here a, b, α and β are constants (in general, complex). By linear changes of variable, the expressions for $\sigma(z)$ and $\rho(z)$ can be reduced (up to constant multipliers) to the following canonical forms:

$$\rho(z) = \begin{cases} (1-z)^\alpha (1+z)^\beta & \text{for } \sigma(z) = 1 - z^2, \\ z^\alpha e^{-z} & \text{for } \sigma(z) = z, \\ e^{-z^2} & \text{for } \sigma(z) = 1. \end{cases}$$

Under these transformations equations (1) and (3) become equations of the same form, and the corresponding polynomials $y_n(z)$ of hypergeometric type remain polynomials in the new variable and are, as before, defined by the Rodrigues formula (2).

According to the form of $\sigma(z)$ we obtain the following systems of polynomials:

1) Let $\sigma(z) = 1 - z^2$, $\rho(z) = (1-z)^\alpha(1+z)^\beta$. Then

$$\tau(z) = -(\alpha+\beta+2)z + \beta - \alpha.$$

The corresponding polynomials $y_n(z)$ with* $B_n = (-1)^n/(2^n n!)$ are called the *Jacobi polynomials* and denoted by $P_n^{(\alpha,\beta)}(z)$:

$$P_n^{(\alpha,\beta)}(z) = \frac{(-1)^n}{2^n n!}(1-z)^{-\alpha}(1+z)^{-\beta}\frac{d^n}{dz^n}\left[(1-z)^{n+\alpha}(1+z)^{n+\beta}\right].$$

Important special cases of the Jacobi polynomials are:

a) the *Legendre polynomials* $P_n(z) = P_n^{(0,0)}(z)$;

b) the *Chebyshev polynomials of the first and second kinds*:

$$T_n(z) = \cos n\phi,$$
$$U_n(z) = \frac{1}{n+1}T'_{n+1}(z) = \frac{\sin(n+1)\phi}{\sin\phi},$$

where $\phi = \cos^{-1}(z)$. It will be shown later (§6, part 2) that

$$T_n(z) = \frac{n!}{(1/2)_n}P_n^{(-1/2,-1/2)}(z),$$
$$U_n(z) = \frac{(n+1)!}{(3/2)_n}P_n^{(1/2,1/2)}(z);$$

c) the *Gegenbauer polynomials*, also known as *ultraspherical polynomials*,

$$C_n^\lambda(z) = \frac{(2\lambda)_n}{(\lambda+1/2)_n}P_n^{(\lambda-1/2,\lambda-1/2)}(z).$$

We have used the notation

$$(\alpha)_n = \alpha(\alpha+1)\dots(\alpha+n-1) = \frac{\Gamma(\alpha+n)}{\Gamma(\alpha)},$$

where $\Gamma(z)$ is the gamma function (see Appendix A).

---

* The values B_n are chosen for historical reasons but could be arbitrary. They agree with the normalization in [E2].

2) Let $\sigma(z) = z$, $\rho(z) = z^\alpha e^{-z}$. Then

$$\tau(z) = -z + \alpha + 1.$$

The polynomials $y_n(z)$ with $B_n = 1/n!$ are the *Laguerre polynomials* $L_n^\alpha(z)$:

$$L_n^\alpha(z) = \frac{1}{n!} e^z z^{-\alpha} \frac{d^n}{dz^n}(z^{\alpha+n} e^{-z}).$$

3) Let $\sigma(z) = 1$, $\rho(z) = e^{-z^2}$. Then $\tau(z) = -2z$. The polynomials $y_n(z)$ with $B_n = (-1)^n$ are the Hermite polynomials $H_n(z)$:

$$H_n(z) = (-1)^n e^{z^2} \frac{d^n}{dz^n}(e^{-z^2}).$$

We have not considered the case when $\sigma(z)$ has a double zero, i.e. $\sigma(z) = (z-a)^2$. The polynomials of hypergeometric type corresponding to $\sigma(z) = (z-a)^2$ can be expressed in terms of the Laguerre polynomials. It was shown in §1 that when $\sigma(z) = (z-a)^2$ the substitution $s = 1/(z-a)$ carries the generalized equation of hypergeometric type

$$u'' + \frac{\tilde{\tau}(z)}{\sigma(z)} u' + \frac{\tilde{\sigma}(z)}{\sigma^2(z)} u = 0$$

into an equation of the same type with $\sigma(s) = s$. In particular, the equation

$$\sigma(z)y'' + \tau(z)y' + \lambda_n y = 0 \qquad \left(\lambda_n = -n\tau' - \frac{1}{2}n(n-1)\sigma''\right)$$

for polynomials of hypergeometric type with $\sigma(z) = (z-a)^2$ becomes, under the substitution $s = 1/(z-a)$,

$$\frac{d^2y}{ds^2} + \frac{2 - s\tau(a + 1/s)}{s}\frac{dy}{ds} + \frac{\lambda_n}{s^2} y = 0.$$

As we showed in §1, the substitution $y = \phi(s)u$ carries this equation into an equation of hypergeometric type if $\phi(s)$ is chosen appropriately. A possible form for $\phi(s)$ is $1/s^n$. Since

$$\tau(z) = \tau(a) + \tau'(z)(z-a),$$

we have $\tau(a + 1/s) = \tau(a) + \tau'/s$ and we obtain the following equation for $u(s)$:

$$su'' - [s\tau(a) + \tau' + 2(n-1)]u' + n\tau(a)u = 0.$$

Since $u(s) = s^n y$ and y is a polynomial of degree n in $z = a + 1/s$, the function $u(s)$ is a polynomial of degree n in s. Consequently $u(s)$ is a polynomial of hypergeometric type. Since, in the present case, the function $\rho(s)$ that determines the polynomials of hypergeometric type according to the Rodrigues formula has the form

$$\rho(s) = s^{-\tau' - 2n + 1} e^{-\tau(a)s},$$

the polynomials $u(s) = u_n(s)$ are, if $\tau(a) \neq 0$, the same up to a normalizing factor as the Laguerre polynomials $L_n^\alpha(t)$, with $\alpha = -\tau' - 2n + 1$, $t = \tau(a)s$. Therefore the polynomials $y_n(z)$ are connected with the Laguerre polynomials as follows:

$$y_n(z) = C_n (z-a)^n L_n^{-\tau' - 2n + 1}\left(\frac{\tau(a)}{z-a}\right).$$

This formula is still valid when $\tau(a) = 0$.

The best known polynomials of hypergeometric type in the case $\sigma(z) = (z-a)^2$ are the *Bessel polynomials*, for which

$$\sigma(z) = z^2, \quad \tau(z) = 2(z+1), \quad \rho(z) = e^{-2/z}.$$

Their Rodrigues formula is

$$y_n(z) = 2^{-n} e^{2/z} \frac{d^n}{dz^n}\left(z^{2n} e^{-2/z}\right).$$

The Bessel polynomials are normalized by $y_n(0) = 1$. Their explicit form is

$$y_n(z) = \frac{(-1)^n n!}{2^n} z^n L_n^{-(2n+1)}\left(\frac{2}{z}\right).$$

2. Consequences of the Rodrigues formula. We have shown that the derivatives of all orders of polynomials $y_n(z)$ of hypergeometric type are also polynomials of hypergeometric type (see section 2). The Rodrigues formula for $y_n^{(m)}(z)$ has the form

$$y_n^{(m)}(z) = \frac{A_{mn} B_n}{\sigma^m(z)\rho(z)} \frac{d^{n-m}}{dz^{n-m}} \left[\sigma^n(z)\rho(z)\right], \tag{4}$$

where

$$A_{mn} = (-1)^m \prod_{k=0}^{m-1} \mu_{kn}, \quad A_{0n} = 1, \quad \mu_{kn} = \mu_k(\lambda)\Big|_{\lambda=\lambda_n} = \lambda_n - \lambda_k.$$

Since $\mu_{kn} = -(n-k)\left(\tau' + \frac{1}{2}(n+k-1)\sigma''\right)$, we have

$$A_{mn} = \frac{n!}{(n-m)!}\prod_{k=0}^{m-1}\left(\tau' + \frac{1}{2}(n+k-1)\sigma''\right). \tag{5}$$

Notice that the Rodrigues formula for $y_n^{(m)}(z)$ can be obtained up to a normalizing factor from the Rodrigues formula for $y_n(z)$ by replacing n by $n-m$ and $\rho(z)$ by $\rho_m(z) = \sigma^m(z)\rho(z)$. Let us consider some corollaries of equation (4).

1) From the Rodrigues formulas for $y_n(z)$ and $y_n'(z)$ we obtain the following *differentiation formulas* for the Jacobi, Laguerre and Hermite polynomials:

$$\begin{aligned}\frac{dP_n^{(\alpha,\beta)}(z)}{dz} &= \frac{1}{2}(n+\alpha+\beta+1)P_{n-1}^{(\alpha+1,\beta+1)}(z),\\ \frac{dL_n^{(\alpha)}(z)}{dz} &= -L_{n-1}^{(\alpha+1)}(z),\\ \frac{dH_n(z)}{dz} &= 2nH_{n-1}(z).\end{aligned} \tag{6}$$

2) By using the Rodrigues formula we can express the derivatives $y_n'(z)$ in terms of the $y_n(z)$ themselves. In fact, since

$$\begin{aligned}y_{n+1}(z) &= \frac{B_{n+1}}{\rho(z)}\frac{d^{n+1}}{dz^{n+1}}\left[\sigma(z)\rho_n(z)\right]\\ &= \frac{B_{n+1}}{\rho(z)}\frac{d^n}{dz^n}\left[\tau_n(z)\rho_n(z)\right],\end{aligned}$$

we have, by Leibniz's formula for differentiating a product,

$$\begin{aligned}y_{n+1}(z) &= \frac{B_{n+1}}{\rho(z)}\left\{\tau_n(z)\frac{d^n}{dz^n}\left[\sigma^n(z)\rho(z)\right] + n\tau_n'\frac{d^{n-1}}{dz^{n-1}}\left[\sigma^n(z)\rho(z)\right]\right\}\\ &= \frac{B_{n+1}}{B_n}\left[\tau_n(z)y_n(z) - \frac{n}{\lambda_n}\tau_n'\sigma(z)y_n'(z)\right].\end{aligned}$$

Hence

$$\sigma(z)y_n'(z) = \frac{\lambda_n}{n\tau_n'}\left[\tau_n(z)y_n(z) - \frac{B_n}{B_{n+1}}y_{n+1}(z)\right], \tag{7}$$

in agreement with (4.12).

3) From (4) for $m = n - 1$ it is easy to find the coefficients a_n and b_n of the highest powers of z in the expansion

$$y_n(z) = a_n z^n + b_n z^{n-1} + \dots .$$

Since

$$y_n^{(n-1)}(z) = n! a_n z + (n-1)! b_n,$$
$$\frac{d}{dz}[\sigma^n(z)\rho(z)] = \frac{d}{dz}[\sigma(z)\rho_{n-1}(z)] = \tau_{n-1}(z)\rho_{n-1}(z),$$

we have

$$A_{n-1,n} B_n \tau_{n-1}(z) = n! a_n z + (n-1)! b_n .$$

Hence

$$a_n = \frac{A_{n-1,n} B_n}{n!}\tau'_{n-1} = B_n \prod_{k=0}^{n-1}\left(\tau' + \frac{1}{2}(n+k-1)\sigma''\right), \quad a_0 = B_0;$$
$$\frac{b_n}{a_n} = \frac{n\tau_{n-1}(0)}{\tau'_{n-1}}. \tag{8}$$

4) By using the Rodrigues formula we can easily find the values of the Laguerre and Jacobi polynomials for special values of z by applying Leibniz's rule for the derivative of a product. For example,

$$P_n^{(\alpha,\beta)}(1) = L_n^{\alpha}(0) = \frac{\Gamma(n+\alpha+1)}{n!\Gamma(\alpha+1)},$$
$$P_n^{(\alpha,\beta)}(-1) = (-1)^n \frac{\Gamma(n+\beta+1)}{n!\Gamma(\beta+1)}. \tag{9}$$

Hence we have, for the Legendre polynomials,

$$P_n(1) = 1, \quad P_n(-1) = (-1)^n. \tag{9a}$$

3. Generating functions. A *generating function* for a system of polynomials $y_n(z)$ of hypergeometric type is a function $\Phi(z,t)$ whose expansion in powers of t has, for sufficiently small $|t|$, the form

$$\Phi(z,t) = \sum_{n=0}^{\infty} \frac{\bar{y}_n(z)}{n!} t^n, \tag{10}$$

where $\bar{y}_n(z)$ is a polynomial of hypergeometric type for which the constant B_n in the Rodrigues formula (2) is 1, i.e.

$$\bar{y}_n(z) = \frac{1}{\rho(z)} \frac{d^n}{dz^n} [\sigma^n(z)\rho(z)]$$

(evidently $y_n(z) = B_n \bar{y}(z)$). By (3.1),

$$\bar{y}_n(z) = \frac{1}{\rho(z)} \frac{n!}{2\pi i} \int_C \frac{\sigma^n(s)\rho(s)}{(s-z)^{n+1}} ds, \tag{11}$$

where C is a closed contour surrounding the point $s = z$ (it assumed that $\rho(s)$ is analytic in the region inside C). We substitute the expression (11) for $\bar{y}_n(z)$ into (10) and interchange summation and integration:

$$\Phi(z,t) = \frac{1}{2\pi i \rho(z)} \int_C \frac{\rho(s)}{s-z} \left\{ \sum_{n=0}^{\infty} \left[\frac{\sigma(s)t}{s-z} \right]^n \right\} ds.$$

The interchange of summation and integration can easily be justified for sufficiently small $|t|$ and fixed z. The geometric series in the integrand can be summed, and we obtain

$$\Phi(z,t) = \frac{1}{2\pi i \rho(z)} \int_C \frac{\rho(s)ds}{s-z-\sigma(s)t}.$$

The denominator of the integrand has, in general, two zeros. If $t \to 0$, one of the zeros tends to $s = z$. At the same time, the second zero, if it exists, tends to infinity. Therefore when $|t|$ is sufficiently small we may suppose that there is just one zero $s = \xi(z,t)$ of the denominator inside C, and the integrand has a single pole inside C with residue

$$C_{-1} = \left. \frac{\rho(s)}{1-\sigma'(s)t} \right|_{s=\xi(z,t)}.$$

Consequently we have the representation

$$\Phi(z,t) = \left. \frac{\rho(s)}{\rho(z)} \frac{1}{1-\sigma'(s)t} \right|_{s=\xi(z,t)}. \tag{12}$$

Here $\xi(z,t)$ is the zero of the quadratic equation (in s) $s - z - \sigma(s)t = 0$ which is near $s = z$ for small $|t|$.

Formula (12) for the function $\Phi(z,t)$ in (10) was established for sufficiently small $|t|$. By the principle of analytic continuation, (10) remains valid in the region of analyticity (with respect to t) of the two sides of (10), i.e. in the disk $|t| < R$, where R is the distance from the origin to the nearest singular point of $\Phi(z,t)$ (for fixed z).

As an example we obtain the generating function for the Legendre polynomials. In this case

$$\xi(z,t) = \frac{-1+\sqrt{1+4t(t+z)}}{2t}$$

and consequently, by (12),

$$\Phi(z,t) = \left.\frac{1}{1+2st}\right|_{s=\xi(z,t)} = \frac{1}{\sqrt{1+4tz+4t^2}}.$$

Since $B_n = (-1)^n/(2^n n!)$ for the Legendre polynomials, we have

$$\frac{1}{\sqrt{1+4tz+4t^2}} = \sum_{n=0}^{\infty} P_n(z)(-2t)^n.$$

If we replace t by $-t/2$, we obtain the usual form of the generating function:

$$\frac{1}{\sqrt{1-2tz+t^2}} = \sum_{n=0}^{\infty} P_n(z)t^n. \tag{13}$$

The expansion (13) converges for $|t| < 1$ if $-1 \le z \le 1$, since the singular points of the generating function, which are at the roots of the equation $1-2tz+t^2 = 0$, are given by $t_{1,2} = e^{\pm i\phi}$ $(\cos\phi = z)$ and lie on $|t| = 1$.

Formula (13) is often used in theoretical physics in the form

$$\frac{1}{|\mathbf{r}_1-\mathbf{r}_2|} = \sum_{n=0}^{\infty} \frac{r_<^n}{r_>^{n+1}} P_n(\cos\theta), \tag{14}$$

where $r_< = \min(r_1,r_2)$, $r_> = \max(r_1,r_2)$, and θ is the angle between the vectors $\mathbf{r}_1$ and $\mathbf{r}_2$. In fact,

$$|\mathbf{r}_1-\mathbf{r}_2| = \sqrt{r_1^2+r_2^2-2r_1r_2\cos\theta}$$
$$= \begin{cases} r_1\sqrt{1+\left(\dfrac{r_2}{r_1}\right)^2 - 2\dfrac{r_2}{r_1}\cos\theta} & (r_2 < r_1), \\ r_2\sqrt{1+\left(\dfrac{r_1}{r_2}\right)^2 - 2\dfrac{r_1}{r_2}\cos\theta} & (r_1 < r_2). \end{cases}$$

Hence

$$|\mathbf{r}_1 - \mathbf{r}_2| = r_> \sqrt{1 + \left(\frac{r_<}{r_>}\right)^2 - 2\frac{r_<}{r_>}\cos\theta}$$

and consequently

$$\frac{1}{|\mathbf{r}_1 - \mathbf{r}_2|} = \frac{1}{r_> \sqrt{1 + \left(\frac{r_<}{r_>}\right)^2 - 2\frac{r_<}{r_>}\cos\theta}} = \sum_{n=0}^{\infty} \frac{r_<^n}{r_>^{n+1}} P_n(\cos\theta).$$

By using (10) and (12) we can obtain the following expansions for the Laguerre and Hermite polynomials:

$$(1-t)^{-\alpha-1} \exp\left(-\frac{zt}{1-t}\right) = \sum_{n=0}^{\infty} L_n^{\alpha}(z) t^n, \tag{15}$$

$$\exp(2zt - t^2) = \sum_{n=0}^{\infty} H_n(z) \frac{t^n}{n!}. \tag{16}$$

4. Orthogonality of polynomials of hypergeometric type. If $\rho(z)$ is considered on the real axis $z = x$, and satisfies some additional conditions, we can obtain a number of special properties of the polynomials $y_n(z)$ of hypergeometric type.

Theorem. *Let $\rho(x)$ satisfy*

$$\sigma(x)\rho(x)x^k\Big|_{x=a,b} = 0 \qquad (k = 0, 1, \ldots). \tag{17}$$

at the endpoints of an interval (a, b). Then the polynomials $y_n(x)$ of hypergeometric type corresponding to different λ_n's are orthogonal on (a, b) with weight $\rho(x)$, i.e.

$$\int_a^b y_n(x) y_m(x) \rho(x) dx = 0 \qquad (\lambda_m \neq \lambda_n).$$

Proof. Consider the differential equations for $y_n(x)$ and $y_m(x)$:

$$[\sigma(x)\rho(x)y_n']' + \lambda_n\rho(x)y_n = 0,$$
$$[\sigma(x)\rho(x)y_m']' + \lambda_m\rho(x)y_m = 0.$$

Multiply the first equation by $y_m(x)$ and the second by $y_n(x)$, subtract the second from the first, and integrate from a to b. Since

$$y_m(x)\left[\sigma(x)\rho(x)y_n'(x)\right]' - y_n(x)\left[\sigma(x)\rho(x)y_m'(x)\right]' = \frac{d}{dx}\left\{\sigma(x)\rho(x)W\left[y_m(x), y_n(x)\right]\right\},$$

where $W(u,v) = \begin{vmatrix} u & v \\ u' & v' \end{vmatrix}$ is the Wronskian, we obtain

$$(\lambda_m - \lambda_n) \leq \int_a^b y_m(x)y_n(x)\rho(x)dx = \sigma(x)\rho(x)W\left[y_m(x), y_n(x)\right]\Big|_a^b.$$

Since $W[y_m(x), y_n(x)]$ is a polynomial in x, the right-hand side is zero, by (17). Hence when $\lambda_m \neq \lambda_n$ we have

$$\int_a^b y_m(x)y_n(x)\rho(x)dx = 0, \tag{18}$$

as was to be proved.

The polynomials $y_n(x)$ for which $\rho(x)$ satisfies (17) are known as the *classical orthogonal polynomials*. They are usually considered under the auxiliary conditions that $\rho(x) > 0$ and $\sigma(x) > 0$ on (a,b). These conditions are satisfied by the Jacobi polynomials $P_n^{\alpha,\beta}(x)$ for $a = -1$, $b = 1$, $\alpha > -1$, $\beta > -1$; by the Laguerre polynomials $L_n^{\alpha}(x)$ for $a = 0, b = +\infty, \alpha > -1$; by the Hermite polynomials $H_n(x)$ for $a = -\infty$, $b = +\infty$. We observe that in these cases the condition $\lambda_m \neq \lambda_n$ can be replaced by $m \neq n$.

From the properties of the derivatives of polynomials of hypergeometric type (see part 2) it follows that *the derivatives $y_n^{(k)}(x)$ of the classical orthogonal polynomials are also classical polynomials, orthogonal with weight $\sigma^k(x)\rho(x)$ on (a,b)*:

$$\int_a^b y_n^{(k)}(x)y_m^{(k)}(x)\rho_k(x)dx = \delta_{mn}d_{kn}^2, \tag{18a}$$

where

$$\delta_{mn} = \begin{cases} 0 & \text{for } m \neq n, \\ 1 & \text{for } m = n. \end{cases}$$

It is easy to express the squared norm d_{kn}^2 of $y_n^{(k)}(x)$ in terms of the squared norm $d_n^2 = d_{0n}^2$ of $y_n(x)$ by using the differential equation for $y_n^{(k)}(x)$:

$$\frac{d}{dx}\left[\rho_{k+1}(x)y_n^{(k+1)}(x)\right] + \mu_{kn}\rho_k(x)y_n^{(k)}(x) = 0. \tag{19}$$

If we multiply (19) by $y_n^{(k)}(x)$ and integrate over (a, b), after integrating by parts we obtain

$$\rho_{k+1}(x)y_n^{(k+1)}(x)y_n^{(k)}(x)\Big|_a^b - \int_a^b \left[y_n^{(k+1)}(x)\right]^2 \rho_{k+1}(x)dx$$

$$+ \mu_{kn}\int_a^b \left[y_n^{(k)}(x)\right]^2 \rho_k(x)dx = 0.$$

The integrated terms are zero because of (17), and consequently

$$d_{k+1,n}^2 = \mu_{kn}d_{kn}^2.$$

Hence, by induction, we obtain

$$d_{mn}^2 = d_n^2 \prod_{k=0}^{m-1} \mu_{kn}. \tag{20}$$

We can calculate d_n^2 for the Jacobi, Laguerre, and Hermite polynomials from (20) with $m = n$, since

$$y_n^{(n)}(x) = n!a_n, \quad d_{nn}^2 = (n!a_n)^2 \int_a^b \sigma^n(x)\rho(x)dx. \tag{20a}$$

The integral $\int_a^b \sigma^n(x)\rho(x)dx$ can be evaluated in terms of gamma functions (see Appendix A, part 5):

$$d_n^2 = \begin{cases} \dfrac{2^{\alpha+\beta+1}\Gamma(n+\alpha+1)\Gamma(n+\beta+1)}{n!(2n+\alpha+\beta+1)\Gamma(n+\alpha+\beta+1)} & \text{for } P_n^{(\alpha,\beta)}(x), \\ \dfrac{1}{n!}\Gamma(n+\alpha+1) & \text{for } L_n^{\alpha}(x), \\ 2^n n!\sqrt{\pi} & \text{for } H_n(x). \end{cases}$$

The basic information about the Jacobi, Laguerre, and Hermite polynomials is given in Table 1.

Table 1. Data for the classical orthogonal polynomials

$y_n(x)$	$P_n^{(\alpha,\beta)}(x)$ $(\alpha > -1, \beta > -1)$	$L_n^{\alpha}(x)$ $(\alpha > -1)$	$H_n(x)$
(a, b)	$(-1, 1)$	$(0, \infty)$	$(-\infty, \infty)$
$\rho(x)$	$(1-x)^{\alpha}(1+x)^{\beta}$	$x^{\alpha}e^{-x}$	e^{-x^2}
$\sigma(x)$	$1 - x^2$	x	1
$\tau(x)$	$\beta - \alpha - (\alpha + \beta + 2)x$	$1 + \alpha - x$	$-2x$
λ_n	$n(n + \alpha + \beta + 1)$	n	$2n$
B_n	$\dfrac{(-1)^n}{2^n n!}$	$\dfrac{1}{n!}$	$(-1)^n$
a_n	$\dfrac{\Gamma(2n + \alpha + \beta + 1)}{2^n n! \Gamma(n + \alpha + \beta + 1)}$	$\dfrac{(-1)^n}{n!}$	2^n
b_n	$\dfrac{(\alpha - \beta)\Gamma(2n + \alpha + \beta)}{2^n (n-1)! \Gamma(n + \alpha + \beta + 1)}$	$(-1)^{n-1}\dfrac{n + \alpha}{(n-1)!}$	0
d_n^2	$\dfrac{2^{\alpha+\beta+1}\Gamma(n + \alpha + 1)\Gamma(n + \beta + 1)}{n!(2n + \alpha + \beta + 1)\Gamma(n + \alpha + \beta + 1)}$	$\dfrac{\Gamma(n + \alpha + 1)}{n!}$	$2^n n! \sqrt{\pi}$

§ 6 Some general properties of orthogonal polynomials

The classical orthogonal polynomials have many properties that follow directly from their orthogonality. Similar properties are possessed by any polynomials that are orthogonal on an interval (a, b) with a weight $\rho(x) > 0$.

Let us consider some general properties of polynomials $p_n(x)$ that are orthogonal on an interval (a, b) with weight function $\rho(x)$, that is, that satisfy

$$\int_a^b p_n(x)p_m(x)\rho(x)dx = 0 \quad (m \neq n).$$

We shall suppose that the leading coefficient of $p_n(x)$ is real and different from zero (n is the degree of the polynomial).

1. Expansion of an arbitrary polynomial in terms of the orthogonal polynomials. Let us show that *every polynomial $q_n(x)$ of degree n is a linear combination of the orthogonal polynomials $p_k(x) (k = 0, 1, \ldots, n)$*, i.e.

$$q_n(x) = \sum_{k=0}^{n} c_{kn}p_k(x). \tag{1}$$

This is evident for $n = 0$. For general $n > 0$ we proceed by induction. Suppose that every polynomial $q_{n-1}(x)$ has the representation

$$q_{n-1}(x) = \sum_{k=0}^{n-1} c_{k,n-1}p_k(x). \tag{2}$$

For $q_n(x)$ we determine the constant c_{nn} so that the highest coefficient of $q_n(x) - c_{nn}p_n(x)$ is zero, i.e. $q_n(x) - c_{nn}p_n(x) = q_{n-1}(x)$. If we use (2), we obtain (1) for $q_n(x)$.

The coefficients c_{kn} in (1) are easily determined by the orthogonality property

$$\int_a^b p_n(x)p_m(x)\rho(x)dx = 0 \quad (m \neq n) \tag{3}$$

which leads to the formula

$$c_{kn} = \frac{1}{d_k^2}\int_a^b q_n(x)p_k(x)\rho(x)dx, \tag{4}$$

where

$$d_k^2 = \int_a^b p_k^2(x)\rho(x)dx$$

is the square of the norm.

Let us show that the orthogonality condition (3) is equivalent to

$$\int_a^b p_n(x)x^m\rho(x)dx = 0 \quad (m < n). \tag{5}$$

In fact, if we insert the expansion of $p_m(x)$ in powers of x into (3), with $m < n$, then (3) follows from (5). Similarly, if we expand x^m in terms of the $p_k(x)$, we obtain (5) from (3).

It follows from (5) that $p_n(x)$ is *orthogonal to every polynomial of lower degree.*

2. Uniqueness of the system of orthogonal polynomials corresponding to a given weight. We can show that *the interval* (a, b) *and the weight* $\rho(x)$ *determine the polynomials* $p_n(x)$ *that satisfy the orthogonality condition* (5), *up to a normalizing factor.*

Suppose that there are two sets of polynomials $p_n(x)$ and $\tilde{p}_n(x)$ that satisfy (5). We have

$$\tilde{p}_n(x) = \sum_{k=0}^{n} c_{kn}p_k(x).$$

By (4) and (5), we have $c_{kn} = 0$ for $k < n$, so that $p_n(x)$ and $\tilde{p}_n(x)$ must be proportional.

There is an explicit representation for $p_n(x)$ as a determinant:

$$p_n(x) = A_n \begin{vmatrix} C_0 & C_1 & \dots & C_n \\ C_1 & C_2 & \dots & C_{n+1} \\ \dots & \dots & \dots & \dots \\ C_{n-1} & C_n & \dots & C_{2n-1} \\ 1 & x & \dots & x^n \end{vmatrix}, \tag{6}$$

where A_n is a normalizing constant, and $C_k = \int_a^b x^k\rho(x)dx$ are the moments of $\rho(x)$. In fact, it can be verified that the polynomial (6) satisfies the orthogonality condition (5).*

* The coefficient of x^n in (6) is different from zero when $A_n \neq 0$, since it is proportional to the Gram determinant of the functions $1, x, \dots, x^{n-1}$ (see [E2] and [G1]).

Example. It is known that the system of functions $\{\cos n\phi\}$ has the orthogonality property:

$$\int_0^\pi \cos n\phi \cos m\phi \, d\phi = 0, \qquad m \neq n.$$

By using the relation

$$\cos(n+1)\phi + \cos(n-1)\phi = 2\cos\phi\cos n\phi$$

and mathematical induction it is easy to show that $\cos n\phi$ is a polynomial of degree n in $x = \cos\phi$ with leading coefficient $a_n = 2^{n-1}(a_0 = 1)$. This polynomial is the Chebyshev polynomial of the first kind, $T_n(x)$. Making the substitution $x = \cos\phi$, we obtain the following orthogonality relation for $T_n(x)$:

$$\int_{-1}^1 T_n(x)T_m(x)\frac{dx}{\sqrt{1-x^2}} = 0, \qquad m \neq n.$$

From this we see that the polynomials $T_n(x) = \cos n\phi$ in $x = \cos\phi$ satisfy the same orthogonality relation as the Jacobi polynomials $P_n^{(-\frac{1}{2},-\frac{1}{2})}(x)$. Therefore, by virtue of the uniqueness property of the orthogonal polynomials for a given weight, we have

$$T_n(x) = c_n P_n^{(-\frac{1}{2},-\frac{1}{2})}(x).$$

Equating the coefficients of x^n, we obtain $c_n = n!/(1/2)_n$.

If we use the differentiation formula (5.6) for the Jacobi polynomials then the Chebyshev polynomials of the second kind

$$U_n(x) = \frac{\sin(n+1)\phi}{\sin\phi}$$

may be expressed in terms of the polynomials $P_n^{(\frac{1}{2},\frac{1}{2})}(x)$ as

$$U_n(x) = \frac{(n+1)!}{(3/2)_n} P_n^{(\frac{1}{2},\frac{1}{2})}(x).$$

The polynomials $\bar{T}(x) = 2^{1-n}T_n(x)$ are extensively used in problems of approximation theory and in the general theory of iterative methods. The reason for this is that they are the polynomials deviating least from zero on

$[-1,1]$ among the polynomials $q_n(x)$ of degree n with leading coefficient 1. That is, the minimum of

$$\max_{-1\le x\le 1} |q_n(x)|$$

is attained for

$$\bar{T}_n(x) = 2^{1-n}\cos n\phi, \quad \phi = \arccos x.$$

In fact, at the points where $x_k = \cos(k\pi/n)$ $(k = 0, 1, \ldots, n)$, the polynomial $\bar{T}_n(x)$ takes, with alternating signs, values of modulus

$$\max_{-1\le x\le 1} |\bar{T}_n(x)| = \frac{1}{2^{n-1}}.$$

Hence if we suppose that there is a polynomial $p_n(x)$ with leading coefficient 1, for which

$$-\max_{-1\le x\le 1} |\bar{T}_n(x)| < p_n(x) < \max_{-1\le x\le 1} |\bar{T}_n(x)|,$$

for $x \in [-1,1]$, then the polynomial $\bar{T}_n(x) - p_n(x)$, of degree $n-1$, would alternate in sign at the points $x_0, x_1, \ldots, x_n$ and would have to have n zeros, which is impossible.

One can show in the same way that the polynomial

$$\tilde{T}_n(x) = \frac{T_n(x)}{T_n(x_0)} \quad \text{with } x_0 \notin [-1,1]$$

is the polynomial deviating least from zero on $[-1,1]$ among polynomials $q_n(x)$ of degree n that have $q_n(x_0) = 1$.

3. Recursion relations. *All orthogonal polynomials satisfy recursion formulas connecting three consecutive polynomials* $p_{n-1}(x), p_n(x), p_{n+1}(x)$:

$$xp_n(x) = \alpha_n p_{n+1}(x) + \beta_n p_n(x) + \gamma_n p_{n-1}(x), \tag{7}$$

where α_n, β_n *and* γ_n *are constants.*

For the proof, we use the expansion

$$xp_n(x) = \sum_{k=0}^{n+1} c_{kn} p_k(x). \tag{8}$$

By (4),

$$c_{kn} = \frac{1}{d_k^2}\int_a^b p_k(x)xp_n(x)\rho(x)dx. \tag{9}$$

Since $xp_k(x)$ is a polynomial of degree $k+1$, by the orthogonality properties of $p_n(x)$ we have $c_{kn} = 0$ when $k+1 < n$. Hence (8) can be written in the form

$$xp_n = \alpha_n p_{n+1} + \beta_n p_n + \gamma_n p_{n-1},$$

where $\alpha_n = c_{n+1,n}$, $\beta_n = c_{nn}$, $\gamma_n = c_{n-1,n}$, as required.

The coefficients α_n, β_n, γ_n can be expressed in terms of d_n^2, i.e. the square of the norm, and the coefficients a_n, b_n of the highest terms in $p_n(x)$:

$$p_n(x) = a_n x^n + b_n x^{n-1} + \dots \qquad (a_n \neq 0).$$

It is clear from (9) that $d_k^2 c_{kn} = d_n^2 c_{nk}$. Since $\alpha_{n-1} = c_{n,n-1}$, $\gamma_n = c_{n-1,n}$, if we put $k = n-1$ we obtain

$$\gamma_n = \alpha_{n-1} d_n^2 / d_{n-1}^2. \tag{10}$$

On the other hand, comparing the coefficients of the highest terms on the left-hand and right-hand sides of (8), we have $a_n = \alpha_n a_{n+1}$, $b_n = \alpha_n b_{n+1} + \beta_n a_n$. Hence

$$\alpha_n = \frac{a_n}{a_{n+1}}, \quad \beta_n = \frac{b_n}{a_n} - \frac{b_{n+1}}{a_{n+1}}, \quad \gamma_n = \frac{a_{n-1}}{a_n} \frac{d_n^2}{d_{n-1}^2}. \tag{11}$$

Therefore if we know the coefficients a_n and b_n and the squared norms d_n^2 of any orthogonal polynomials $p_n(x)$, we can successively determine the polynomials.

The coefficients $\alpha_n, \beta_n, \gamma_n$ obtained from (11) for the Jacobi, Laguerre and Hermite polynomials are presented in Table 2. In addition, Table 2 (p. 38) shows the coefficients $\tilde{\alpha}_n, \tilde{\beta}_n, \tilde{\gamma}_n$ that appear in the differentiation formula

$$\sigma(x) y_n'(x) = (\tilde{\alpha}_n x + \tilde{\beta}_n) y_n(x) + \tilde{\gamma}_n y_{n-1}(x).$$

This formula is obtained by substituting the expression for $y_{n+1}(x)$ from the recursion relation (7) into formula (5.7).

Consider a recursion relation of the form (7),

$$z u_n(z) = \alpha_n u_{n+1}(z) + \beta_n u_n(z) + \gamma_n u_{n-1}(z) \tag{7a}$$

for arbitrary complex values of z. One solution of this recursion is the set of polynomials $p_n(z)$ that are orthogonal on (a, b) with weight $\rho(z)$. Another set of solutions for $z \notin [a, b]$ is the set of functions

$$q_n(z) = \int_a^b \frac{p_n(s)\rho(s)}{s - z} ds.$$

Table 2. Coefficients of recursion relations for the Jacobi, Laguerre, and Hermite polynomials

$y_n(x)$	$P_n^{(\alpha,\beta)}(x)$	$L_n^\alpha(x)$	$H_n(x)$
α_n	$\frac{2(n+1)(n+\alpha+\beta+1)}{(2n+\alpha+\beta+1)(2n+\alpha+\beta+2)}$	$-(n+1)$	$\frac{1}{2}$
β_n	$\frac{\beta^2-\alpha^2}{(2n+\alpha+\beta)(2n+\alpha+\beta+2)}$	$2n+\alpha+1$	0
γ_n	$\frac{2(n+\alpha)(n+\beta)}{(2n+\alpha+\beta)(2n+\alpha+\beta+1)}$	$-(n+\alpha)$	n
$\tilde{\alpha}_n$	$-n$	0	0
$\tilde{\beta}_n$	$\frac{(\alpha-\beta)n}{(2n+\alpha+\beta)}$	n	0
$\tilde{\gamma}_n$	$\frac{2(n+\alpha)(n+\beta)}{2n+\alpha+\beta}$	$-(n+\alpha)$	$2n$

For the proof, it is sufficient to integrate the recursion relation for the polynomials $p_n(s)$ on (a,b) after multiplying by $\rho(s)/(s-z)$ and use the equation

$$\int_a^b \frac{sp_n(s)\rho(s)}{s-z}ds = \int_a^b \left(1+\frac{z}{s-z}\right)p_n(s)\rho(s)ds$$

$$= \int_a^b p_n(s)\rho(s)ds + z\int_a^b \frac{p_n(s)\rho(s)}{s-z}ds = zq_n(z) \quad (n\geq 1).$$

The function

$$r_n(z) = \int_a^b \frac{p_n(s)-p_n(z)}{s-z}\rho(s)ds$$

is closely related to $q_n(z)$. It is easily seen to be a polynomial of degree $n-1$, and is called a *polynomial of the second kind*. Since

$$r_n(z) = \int_a^b \frac{p_n(s)\rho(s)}{s-z}ds - p_n(z)\int_a^b \frac{\rho(s)ds}{s-z} = q_n(z) - \frac{1}{a_0}p_n(z)q_0(z)$$

when $z \notin [a,b]$, and both $p_n(z)$ and $q_n(z)$ satisfy the same recursion (7a) for $z \notin [a,b]$, the polynomials $r_n(z)$ satisfy the same recursion. By continuity, they still satisfy (7a) for $z \in [a,b]$.

4. Darboux-Christoffel formula. From the recursion formula (7) there immediately follows a formula that plays an important role in the theory of orthogonal polynomials, namely the *Darboux-Christoffel formula*

$$\sum_{k=0}^{n} \frac{p_k(x)p_k(y)}{d_k^2} = \frac{1}{d_n^2}\frac{a_n}{a_{n+1}}\frac{p_{n+1}(x)p_n(y) - p_n(x)p_{n+1}(y)}{x-y}. \tag{12}$$

To derive this we use (7) and (10):

$$xp_k(x) = \alpha_k p_{k+1}(x) + \beta_k p_k(x) + \alpha_{k-1}\frac{d_k^2}{d_{k-1}^2}p_{k-1}(x),$$

$$yp_k(y) = \alpha_k p_{k+1}(y) + \beta_k p_k(y) + \alpha_{k-1}\frac{d_k^2}{d_{k-1}^2}p_{k-1}(y).$$

These recursions also hold for $k=0$ if we put $\alpha_{-1}/d_{-1}^2 = 0$, $p_{-1}(x) = 0$.

Multiply the first equation by $p_k(y)$, the second by $p_k(x)$, divide each equation by d_k^2, and subtract. This yields

$$(x-y)\frac{p_k(x)p_k(y)}{d_k^2} = A_k(x,y) - A_{k-1}(x,y),$$

where

$$A_k(x,y) = \frac{\alpha_k}{d_k^2}\left[p_{k+1}(x)p_k(y) - p_k(x)p_{k+1}(y)\right].$$

Summing over k from 0 to n, we obtain

$$(x-y)\sum_{k=0}^{n}\frac{p_k(x)p_k(y)}{d_k^2} = A_n(x,y).$$

This formula is evidently equivalent to (12) since $\alpha_n = a_n/a_{n+1}$.

5. Properties of the zeros. Let us show that *all zeros x_j of $p_n(x)$ are simple and on the interval (a,b)*. Let $p_n(x)$ have k changes of sign on (a,b). Evidently $0 \le k \le n$. The property in question will certainly be valid if we show that $k = n$. Put

$$q_k(x) = \begin{cases} 1 & \text{for } k = 0, \\ \prod_{j=1}^{k}(x - x_j) & \text{for } 0 < k \le n. \end{cases}$$

Here x_j are the points where $p_n(x)$ changes sign. The product $p_n(x)q_k(x)$ evidently does not change sign for $x \in (a, b)$. Therefore

$$\int_a^b p_n(x)q_k(x)\rho(x)dx \neq 0.$$

It follows that $k = n$, since if $k < n$

$$\int_a^b p_n(x)q_k(x)\rho(x)dx = 0$$

by (5).

We can show that *the zeros of* $p_n(x)$ *and* $p_{n+1}(x)$ *are interlaced.* For the proof we consider the special case of the Darboux-Christoffel formula obtained from (12) for $y \to x$:

$$\sum_{k=0}^{n} \frac{p_k^2(x)}{d_k^2} = \frac{1}{d_n^2}\frac{a_n}{a_{n+1}}\left[p'_{n+1}(x)p_n(x) - p'_n(x)p_{n+1}(x)\right]. \tag{13}$$

Let $x_j (j = 1, 2, \ldots, n+1)$ be the zeros of $p_{n+1}(x)$. By (13), the sign of the product $p'_{n+1}(x)p_n(x)$ at the zeros x_j of $p_{n+1}(x)$ is independent of j. However, the first factor p'_{n+1} changes sign between x_j and x_{j+1}. Therefore the second factor $p_n(x)$ must also change sign. Consequently $p_n(x)$ has at least one zero in the interval (x_j, x_{j+1}). Since there are n intervals (x_j, x_{j+1}) and each one contains at least one of the n zeros of $p_n(x)$, it is clear that between two successive zeros of $p_{n+1}(x)$ there is exactly one zero of $p_n(x)$.

6. Parity of polynomials from the parity of the weight function. Let $\{p_n(x)\}$ be the polynomials orthogonal on $(-a, a)$ with an even weight function $\rho(x)$. Then if we substitute $-x$ for x in (3), we obtain

$$\int_{-a}^{a} p_n(-x)p_m(-x)\rho(x)dx = 0 \qquad (m \neq n).$$

Since $\rho(x)$ defines the polynomials uniquely up to a normalizing factor, we have $p_n(-x) = c_n p_n(x)$. By comparing the coefficients of the leading terms we obtain $c_n = (-1)^n$. Hence we have, in particular, the following parity properties of the Hermite and Legendre polynomials:

$$H_n(-x) = (-1)^n H_n(x),$$
$$P_n(-x) = (-1)^n P_n(x).$$

The second equation is a special case of

$$P_n^{(\alpha,\beta)}(-x) = (-1)^n P_n^{(\beta,\alpha)}(x),$$

which follows from the Rodrigues formula for the Jacobi polynomials.

We have shown that the polynomials that are orthogonal on $(-a, a)$ with an even weight function $\rho(x)$ have the property

$$p_n(-x) = (-1)^n p_n(x).$$

Equating the coefficients of powers of x, we can easily see that when n is odd, $p_n(x)$ contains only odd powers, and when n is even, only even powers, i.e.

$$p_{2n}(x) = s_n(x^2), \quad p_{2n+1}(x) = x t_n(x^2).$$

Here $s_n(x)$ and $t_n(x)$ are polynomials of degree n. By using the orthogonality of the polynomials $\{p_{2n}(x)\}$, we find that when $m \neq n$

$$\int_{-a}^{a} p_{2n}(x) p_{2m}(x) \rho(x) dx = \int_{-a}^{a} s_n(x^2) s_m(x^2) \rho(x) dx$$

$$= 2\int_0^a s_n(x^2) s_m(x^2) \rho(x) dx = \int_0^{a^2} s_n(\xi) s_m(\xi) \frac{\rho(\sqrt{\xi})}{\sqrt{\xi}} d\xi = 0.$$

Therefore *the polynomials* $s_n(x) = p_{2n}(\sqrt{x})$ *are orthogonal on* $(0, a^2)$ *with weight* $\rho_1(x) = \rho(\sqrt{x})/\sqrt{x}$. Similarly,

$$\int_{-a}^{a} p_{2n+1}(x) p_{2m+1}(x) \rho(x) dx = \int_{-a}^{a} x^2 t_n(x^2) t_m(x^2) \rho(x) dx =$$

$$= \int_0^{a^2} t_n(\xi) t_m(\xi) \sqrt{\xi} \rho(\sqrt{\xi}) d\xi = 0 \quad (m \neq n).$$

Consequently *the polynomials* $t_n(x) = p_{2n+1}(\sqrt{x})/\sqrt{x}$ *are orthogonal on* $(0, a^2)$ *with weight* $\rho_2(x) = \sqrt{x}\rho(\sqrt{x})$. Hence, in particular,

$$H_{2n}(x) = C_n L_n^{-1/2}(x^2), \quad H_{2n+1}(x) = A_n x L_n^{1/2}(x^2).$$

The constants C_n and A_n can be found by equating the coefficients of the highest terms in the respective formulas, and we obtain

$$H_{2n}(x) = (-1)^n 2^{2n} n! L_n^{-1/2}(x^2), \tag{14}$$

$$H_{2n+1}(x) = (-1)^n 2^{2n+1} n! x L_n^{1/2}(x^2). \tag{15}$$

Formulas (14) and (15) let us find, by using (5.9), the values of the Hermite polynomials at $x = 0$:

$$H_{2n}(0) = \frac{(-1)^n(2n)!}{n!}, \qquad H_{2n+1}(0) = 0.$$

7. Relation between two systems of orthogonal polynomials for which the ratio of the weights is a rational function. The evaluation of the orthogonal polynomials by (6), using the moments C_k, is rather tedious when n is large. It is more convenient to calculate the $p_n(x)$ from the recursion relation (7) if the coefficients of the recursion relation are known. These coefficients are easily calculated only for a narrow class of orthogonal polynomials. Hence it would be of great practical value to have simple formulas connecting two systems of polynomials that are orthogonal with respect to different weights. Such formulas can be obtained, for example, when the ratio of the weights is a rational function [U1].

Let $\{p_n(x)\}$ and $\{\bar{p}_n(x)\}$ be the polynomials orthogonal on (a, b) with respective weights $\rho(x)$ and $\bar{\rho}(x)$, and let $\bar{\rho}(x) = R(x)\rho(x)$, where $R(x)$ is a rational function:

$$R(x) = \prod_{j=1}^{k}(x - \alpha_j) \Big/ \prod_{j=1}^{l}(x - \beta_j).$$

We first determine the connection between $\bar{p}_n(x)$ and $p_n(x)$ when $k = 1, l = 0$, i.e. $\bar{\rho}(x) = (x - \alpha_1)\rho(x)$. We expand $\bar{p}_n(x)$ in terms of the $p_m(x)$:

$$\bar{p}_n(x) = \sum_{m=0}^{n} c_m p_m(x).$$

Here

$$c_m = \frac{1}{d_m^2}\int_a^b \bar{p}_n(x)p_m(x)\rho(x)dx =$$

$$= \frac{1}{d_m^2}\left[\int_a^b \bar{p}_n(x)\frac{p_m(x) - p_m(\alpha_1)}{x - \alpha_1}\bar{\rho}(x)dx + p_m(\alpha_1)\int_a^b \bar{p}_n(x)\rho(x)dx\right].$$

Since $(p_m(x) - p_m(\alpha_1))/(x - \alpha_1)$ is a polynomial of degree $m - 1 < n$, the first integral is zero by the orthogonality of the polynomials $\bar{p}_n(x)$. Hence

$$c_m = A_n p_m(\alpha_1)/d_m^2,$$

where A_n is a constant. Consequently

$$\bar{p}_n(x) = A_n \sum_{m=0}^{n} \frac{p_m(\alpha_1)p_m(x)}{d_m^2}.$$

If we now use the Darboux-Christoffel formula, we arrive at the following expression:

$$\bar{p}_n(x) = D_n \frac{p_{n+1}(x)p_n(\alpha_1) - p_{n+1}(\alpha_1)p_n(x)}{x - \alpha_1} \tag{16}$$

(D_n are constants).

Now consider the case $k = 0$, $l = 1$, i.e. $\bar{\rho}(x) = \rho(x)/(x - \beta_1)$. We again use the expansion of $\bar{p}_n(x)$ in terms of the $p_n(x)$:

$$\bar{p}_n = \sum_{m=0}^{n} c_m p_m(x),$$

where

$$c_m = \frac{1}{d_m^2} \int_a^b \bar{p}_n(x)p_m(x)\rho(x)dx$$

$$= \frac{1}{d_m^2} \int_a^b \bar{p}_n(x)(x - \beta_1)p_m(x)\bar{\rho}(x)dx.$$

Since the function $(x - \beta_1)p_m(x)$ is a polynomial of degree $m + 1$, the coefficients c_m are zero for $m < n - 1$ by the orthogonality of the $\bar{p}_n(x)$, i.e.

$$\bar{p}_n(x) = c_{n-1}p_{n-1}(x) + c_n p_n(x).$$

To determine c_{n-1} and c_n, we integrate this equation over (a, b) after multiplying by $\bar{\rho}(x) = \rho(x)/(x - \beta_1)$, i.e.

$$c_{n-1}q_{n-1}(\beta_1) + c_n q_n(\beta_1) = 0,$$

where

$$q_m(z) = \int_a^b \frac{p_m(x)\rho(x)}{x - z} dx.$$

Hence

$$c_n = D_n q_{n-1}(\beta_1), \quad c_{n-1} = -D_n q_n(\beta_1)$$

(D_n are constants).

Therefore for $\bar{\rho}(x) = \rho(x)/(x-\beta_1)$ we have

$$\bar{p}_n(x) = D_n \left[p_n(x) q_{n-1}(\beta_1) - p_{n-1}(x) q_n(\beta_1)\right]. \tag{17}$$

We now consider the general case, where

$$\bar{\rho}(x) = \frac{\prod_{j=1}^{k} (x-\alpha_j)}{\prod_{j=1}^{l} (x-\beta_j)} \rho(x).$$

If we use formulas of the form (16) and (17), we can show inductively, increasing either k or l by 1 at each step, that

$$\bar{p}_n(x) = \frac{D_n}{\prod_{j=1}^{k}(x-\alpha_j)} \begin{vmatrix} q_{n-l}(\beta_1) & \dots & q_{n+k}(\beta_1) \\ \dots & \dots & \dots \\ q_{n-l}(\beta_l) & \dots & q_{n+k}(\beta_l) \\ p_{n-l}(\alpha_1) & \dots & p_{n+k}(\alpha_1) \\ \dots & \dots & \dots \\ p_{n-l}(\alpha_k) & \dots & p_{n+k}(\alpha_k) \\ p_{n-l}(x) & \dots & p_{n+k}(x) \end{vmatrix}$$

(D_n are normalizing constants).

We recall that the functions $q_m(z)$ and the polynomials $p_m(z)$ satisfy the same recursion relations (see part 3).

§ 7 Qualitative behavior and asymptotic properties of the Jacobi, Laguerre and Hermite polynomials.

1. Qualitative behavior. In studying the qualitative behavior of solutions of a differential equation of the form

$$[k(x)y']' + r(x)y = 0 \tag{1}$$

on an interval where $k(x) > 0$ and $R(x) > 0$, it is convenient to use, not the oscillating function $y(x)$, but the function

$$v(x) = y^2(x) + a(x)[k(x)y'(x)]^2, \tag{2}$$

where the multiplier $a(x)$ is chosen so that we know where $v(x)$ is monotonic. For this purpose we calculate $v'(x)$ by using (1):

$$\begin{aligned} v'(x) &= 2yy' + a'(x)[k(x)y']^2 + 2a(x)k(x)[k(x)y']' \\ &= a'(x)[k(x)y']^2 + 2yy'[1 - a(x)k(x)r(x)]. \end{aligned}$$

If we take $a(x) = 1/(k(x)r(x))$ then

$$v'(x) = \left[\frac{1}{k(x)r(x)}\right]' [k(x)y']^2. \tag{3}$$

Since $[k(x)y']^2 \geq 0$, if we choose $a(x)$ in this way the intervals where $v(x)$ is increasing or decreasing are the same as those for $a(x) = 1/(k(x)r(x))$. Notice that the values of $v(x)$ and $y^2(x)$ are equal at the maxima of $y^2(x)$. This lets us find the intervals where the successive maxima of $|y(x)|$ increase or decrease.

Let us apply this transformation to describe the qualitative behavior of the classical orthogonal polynomials on the interval (a, b) of orthogonality, when $\sigma(x) \geq 0$. In this case the polynomials $y = y_n(x)$ satisfy the differential equation (1), where

$$k(x) = \sigma(x)\rho(x), \; r(x) = \lambda\rho(x), \; \lambda = \lambda_n \quad (n \neq 0). \tag{4}$$

Accordingly, we put

$$v(x) = y^2(x) + \lambda^{-1}\sigma(x)[y'(x)]^2. \tag{5}$$

Using the differential equation for $y(x)$, we find

$$v'(x) = \frac{\sigma'(x) - 2\tau(x)}{\lambda}[y'(x)]^2. \tag{6}$$

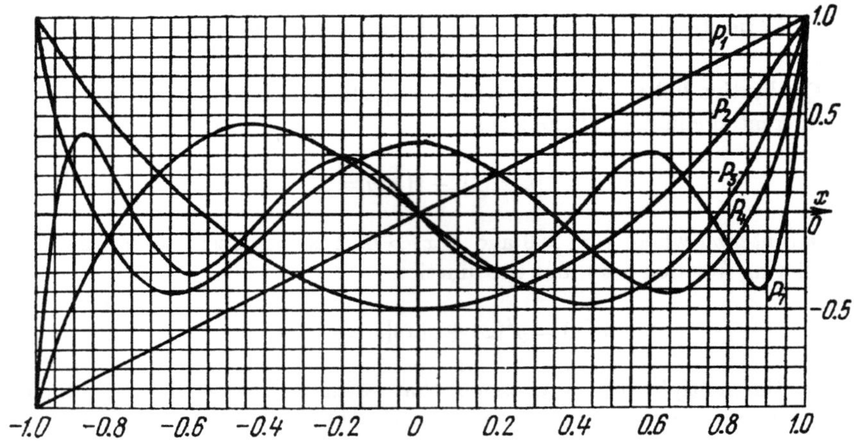

Figure 1.

It is clear from this formula that the sign of $v'(x)$ is the same as the sign of the linear polynomial $(1/\lambda)[\sigma'(x) - 2\tau(x)]$. The values of $v(x)$ and $y^2(x)$ agree at the points where $\sigma(x) = 0$, and also at the maxima of $y^2(x)$ at which $y'(x) = 0$. Hence, in an interval where $v'(x) < 0$, the successive values of $|y(x)|$ at such points will decrease, whereas when $v'(x) > 0$ they will increase.

Examples. 1) For the *Jacobi polynomials*,

$$\sigma(x) = 0 \text{ for } x = \pm 1,\ \sigma'(x) - 2\tau(x) = 2[\alpha - \beta + (\alpha + \beta + 1)x]. \quad (7)$$

Let $\alpha + 1/2 > 0$, $\beta + 1/2 > 0$; then $\lambda_n \geq 1$. When

$$-1 < x < \tilde{x} = \frac{\beta - \alpha}{\alpha + \beta + 1}$$

we have $\sigma'(x) - 2\tau(x) < 0$ and $|P_n^{(\alpha,\beta)}(x)| < |P_n^{(\alpha,\beta)}(-1)|$, and the heights of the maxima of $|P_n^{(\alpha,\beta)}(x)|$ decrease as x increases. Similarly, when $\tilde{x} < x < 1$ the heights of the successive maxima of $|P_n^{(\alpha,\beta)}(x)|$ will increase.

Therefore when $\alpha + 1/2 > 0$ and $\beta + 1/2 > 0, -1 < x < 1$, we have

$$|P_n^{(\alpha,\beta)}(x)| < \max\left[|P_n^{(\alpha,\beta)}(-1)|, |P_n^{(\alpha,\beta)}(1)|\right]. \quad (8)$$

In particular, for the Legendre polynomials, we have (see Figure 1)

$$|P_n(x)| < 1 \text{ for } -1 < x < 1. \quad (9)$$

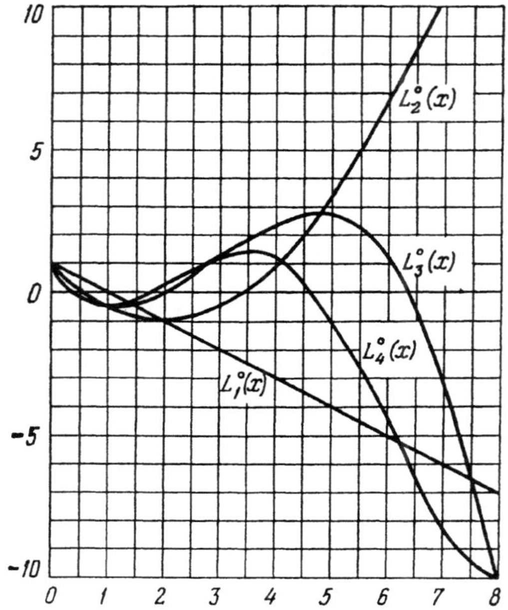

Figure 2.

2) For the *Laguerre polynomials* with $\alpha + 1/2 > 0$ and $0 < x < \tilde{x} = \alpha + 1/2$, we have $|L_n^\alpha(x)| < |L_n^\alpha(0)|$, and the heights of the successive maxima of $|L_n^\alpha(x)|$ decrease. If, however, $x > \tilde{x}$, the heights of successive maxima of $|L_n^\alpha(x)|$ increase (see Figure 2 for $L_n^\alpha(x), \alpha = 0$).

3) For the *Hermite polynomials* $\sigma'(x) - 2\tau(x) = 4x$. Therefore the heights of the successive maxima of $|H_n(x)|$ increase with increasing $|x|$.

2. Asymptotic properties and some inequalities. The preceding inequalities describe the qualitative behavior of $y = y_n(x)$ on the interval of orthogonality. We are now going to obtain some simple quantitative inequalities for the Jacobi and Laguerre polynomials. These describe more precisely how the values of the polynomials depend on n at interior points of (a, b), under the restrictions on the parameters given in part 1. Corresponding inequalities for Hermite polynomials can be obtained by using the connection between the Hermite and Laguerre polynomials, (6.14) and (6.15).*

* The estimates to be obtained in part 2 may be derived more easily (but less rigorously) by means of a quasiclassical approximation (see §19, part 2).

We start from the generalized equation of hypergeometric type

$$u'' + \frac{\tilde{\tau}(x)}{\sigma(x)}u' + \frac{\tilde{\sigma}(x)}{\sigma^2(x)}u = 0, \tag{10}$$

which the transformation $u = \phi(x)y$ discussed in §1 transforms into the equation

$$\sigma(x)y'' + \tau(x)y' + \lambda y = 0. \tag{11}$$

We recall the connection between the coefficients of (10) and (11):

$$\tilde{\tau}(x) = \tau(x) - 2\pi(x),\ \tilde{\sigma}(x) = \lambda\sigma(x) - q(x),$$
$$q(x) = \pi^2(x) + \pi(x)[\tilde{\tau}(x) - \sigma'(x)] + \pi'(x)\sigma(x).$$

Here $\pi(x)$ is the polynomial, at most of degree 1, in the differential equation

$$\phi'(x)/\phi(x) = \pi(x)/\sigma(x)$$

that determines $\phi(x)$. It is convenient to write (10) in the form

$$\sigma(x)u'' + \tilde{\tau}(x)u' + \left[\lambda - \frac{q(x)}{\sigma(x)}\right]u = 0. \tag{12}$$

In order to estimate $u(x)$ we consider the function

$$w(x) = u^2(x) + \lambda^{-1}\sigma(x)[u'(x)]^2,$$

which is similar to $v(x)$. It is evident that on the interval (a, b) we have, for either Jacobi or Laguerre polynomials,

$$|u(x)| \le \sqrt{w(x)}. \tag{13}$$

By using (12), we obtain

$$w'(x) = \frac{\sigma'(x) - 2\tilde{\tau}(x)}{\lambda}[u'(x)]^2 + \frac{2q(x)}{\lambda\sigma(x)}u(x)u'(x).$$

The simplest expression for $w'(x)$ is obtained when $\pi(x)$ is chosen so that $\sigma'(x) - 2\tilde{\tau}(x) = 0$, which leads to

$$\pi(x) = \frac{1}{4}[2\tau(x) - \sigma'(x)].$$

In this case

$$w'(x) = \frac{2q(x)}{\lambda\sigma(x)}u(x)u'(x). \tag{14}$$

It follows from the evident inequality $2ab \le a^2+b^2$ (a and b any real numbers) that

$$2\left(\frac{\sigma(x)}{\lambda}\right)^{1/2} uu' \le u^2 + \frac{\sigma(x)}{\lambda}[u'(x)]^2 = w(x).$$

Consequently by (14)

$$w'(x) \le \frac{|q(x)|}{\sqrt{\lambda}\sigma^{3/2}(x)} w(x).$$

Hence when $x \ge x_0$ we have

$$w(x) = w(x_0)\exp\left[\int_{x_0}^{x} \frac{w'(s)}{w(s)}ds\right] \le w(x_0)\exp\left[\int_{x_0}^{x} \frac{|q(s)|}{\sqrt{\lambda}\sigma^{3/2}(s)}ds\right] \tag{15}$$

and consequently by (13)

$$|u(x)| \le \sqrt{w(x_0)}\exp\left[\int_{x_0}^{x} \frac{|q(s)|}{2\sqrt{\lambda}\sigma^{3/2}(s)}ds\right]. \tag{16}$$

Let us apply (16) to the Jacobi polynomials. By the symmetry relation

$$P_n^{(\alpha,\beta)}(-x) = (-1)^n P_n^{(\alpha,\beta)}(x),$$

it is enough to estimate the Jacobi polynomials $P_n^{(\alpha,\beta)}(x)$ for $-1 < x \le 0$. For these values we have

$$\sigma(x) = 1 - x^2 \ge 1 + x,$$

whence

$$\frac{|q(x)|}{2\sqrt{\lambda}\sigma^{3/2}(x)} \le \frac{A_1}{2\sqrt{\lambda}(1+x)^{3/2}},$$

where $A_1 = \max\limits_{-1\le x\le 0} |q(x)|$. For $-1 < x_0 \le x \le 0$ we then find from (16) that

$$|u(x)| \le \sqrt{w(x_0)}\exp\left[\frac{A_1}{\sqrt{\lambda(1+x_0)}}\right].$$

It is evident that we cannot take $x_0 = -1$ here. Let us determine x_0 from the equation $\sqrt{\lambda(1+x_0)} = C$, where C is a constant independent of n (since

$\lambda = \lambda_n$ does depend on n, the number x_0 also depends on n). Now we have

$$|u(x)| \le A_2 \sqrt{w(x_0)} \tag{17}$$

(A_2 is independent of n).

To estimate $w(x_0)$ we use the connection between $w(x)$ and the Jacobi polynomial $y(x) = P_n^{(\alpha,\beta)}(x)$. We have $u(x) = \phi(x)y(x)$, where $\phi(x)$ is a solution of

$$\frac{\phi'}{\phi} = \frac{\pi(x)}{\sigma(x)}, \quad \pi(x) = \frac{1}{4}[2\tau(x) - \sigma'(x)].$$

Since

$$\frac{\pi(x)}{\sigma(x)} = \frac{1}{2}\frac{\tau(x)}{\sigma(x)} - \frac{1}{4}\frac{\sigma'(x)}{\sigma(x)} = \frac{1}{2}\frac{(\sigma\rho)'}{\sigma\rho} - \frac{1}{4}\frac{\sigma'}{\sigma},$$

we have

$$\phi(x) = \left[\sigma(x)\rho^2(x)\right]^{1/4}, \quad u(x) = \left[\sigma(x)\rho^2(x)\right]^{1/4} y(x),$$

whence

$$\begin{aligned} w(x) &= u^2(x) + \frac{\sigma(x)}{\lambda}[u'(x)]^2 \\ &= \sqrt{\sigma(x)\rho^2(x)}\left[y^2(x) + \left(\frac{2\tau(x) - \sigma'(x)}{4\sqrt{\lambda\sigma(x)}}y + \sqrt{\frac{\sigma(x)}{\lambda}}y'\right)^2\right]. \end{aligned} \tag{18}$$

In estimating $w(x)$ we shall depend on the inequalities already established for $-1 < x < \tilde{x} = (\beta - \alpha)/(\alpha + \beta + 1)$,

$$\begin{gathered} v(x) \le v(-1) = y^2(-1), \\ 0 \le 2\tau(x) - \sigma'(x) < 2\tau(-1) - \sigma'(-1), \end{gathered}$$

and the evident inequalities (see (5))

$$|y(x)| \le \sqrt{v(x)}, \quad \sqrt{\sigma(x)/\lambda}|y'(x)| \le \sqrt{v(x)}.$$

If we use these inequalities and the inequality $\sqrt{\lambda\sigma(x_0)} \ge \sqrt{\lambda(1 + x_0)} = c$, we can deduce from (18) that

$$w(x_0) \le A_3 \sqrt{\sigma(x_0)\rho^2(x_0)}\, y^2(-1)$$

for $x_0 < \tilde{x}$, where

$$A_3 = 1 + \left(\frac{2\tau(-1) - \sigma'(-1)}{4C} + 1\right)^2.$$

The condition $x_0 < \tilde{x}$ will be satisfied automatically for $n \geq 1$ if we take $C < \sqrt{1+\tilde{x}}$, since $\lambda = \lambda_n \geq 1$ for $n \geq 1$.

Using the relation

$$\lim_{z\to\infty} \frac{\Gamma(z+a)}{\Gamma(z)z^a} = 1, \quad |\arg z| \leq \pi - \delta \tag{18a}$$

(see Appendix A, formula (26)), formula (5.9), and the equation

$$\sqrt{\lambda(1+x_0)} = C$$

which determines x_0, we can easily see that the numbers

$$\sqrt{n}[\sigma(x_0)\rho^2(x_0)]^{1/4}|y(-1)|$$

are bounded, uniformly in n. Hence it follows from the inequality for $w(x_0)$ and from (17) that when $\alpha + 1/2 > 0$ and $\beta + 1/2 > 0$,

$$(1-x)^{(\alpha/2)+(1/4)}(1+x)^{(\beta/2)+(1/4)} \left|P_n^{(\alpha,\beta)}(x)\right| \leq \frac{A}{\sqrt{n}}, \tag{19}$$

where the number A is independent of n.

Inequality (19) has been established for $x \geq x_0$. However, it is immediately clear from the behavior of the polynomials $y(x)$ for $-1 \leq x < \tilde{x}$ that (19) remains valid for $x < x_0$, since in this interval

$$[\sigma(x)\rho^2(x)]^{1/4}|y(x)| \leq [\sigma(x_0)\rho^2(x_0)]^{1/4}|y(-1)|,$$

and the numbers $\sqrt{n}\,[\sigma(x_0)\rho^2(x_0)]^{1/4}|y(-1)|$ are bounded, uniformly in n. Consequently (19) holds for $-1 \leq x \leq 0$. From the equation

$$P_n^{(\alpha,\beta)}(-x) = (-1)^n P_n^{(\beta,\alpha)}(x),$$

it is easily seen that (19) is also valid for $0 \leq x \leq 1$.

For the Jacobi polynomials, since

$$d_n^2 = \frac{2^{\alpha+\beta+1}\Gamma(n+\alpha+1)\Gamma(n+\beta+1)}{n!(2n+\alpha+\beta+1)\Gamma(n+\alpha+\beta+1)},$$

we find, by again using (18a), that the numbers nd_n^2 are bounded. Consequently the inequality (19) for the Jacobi polynomials can be rewritten in the form

$$(1-x)^{(\alpha/2)+(1/4)}(1+x)^{(\beta/2)+(1/4)}d_n^{-1}|P_n^{(\alpha,\beta)}(x)| \leq C_1 \tag{20}$$

(the constant C_1 is independent of n).

By the same method as for the Jacobi polynomials, using (15) and putting $\sqrt{\lambda x_0} = C$, we can obtain the following inequality for the Laguerre polynomials $L_n^\alpha(x)$ for $0 \le x \le 1$, $\alpha + 1/2 > 0$:

$$\sqrt{\frac{w(x)}{d_n^2}} \le \frac{C_2}{n^{1/4}}, \tag{21}$$

$$x^{(\alpha/2)+(1/4)} e^{-x/2} d_n^{-1} |L_n^\alpha(x)| \le \{w(x)/d_n^2\}^{1/2} \le C_2/n^{1/4} \tag{22}$$

(the constant C_2 is independent of n).

Inequality (16) is quite poor for the Laguerre polynomials as $x \to +\infty$, since the right-hand side of (15) grows exponentially as $x \to +\infty$, whereas the left-hand side decreases exponentially. The inequality can be improved by using (14) in the following way. Since

$$\sqrt{\lambda^{-1}\sigma(x)}|u'(x)| \le \sqrt{w(x)},$$

we have

$$w'(x) \le \frac{2|q(x)|}{\sqrt{\lambda}\sigma^{3/2}(x)}|u(x)|\sqrt{w(x)},$$

i.e.

$$\frac{d}{dx}\sqrt{w(x)} \le \frac{|q(x)u(x)|}{\sqrt{\lambda}\sigma^{3/2}(x)} = \frac{|q(x)|\sqrt{\rho(x)}|y(x)|}{\sqrt{\lambda}\sigma^{5/4}(x)}. \tag{23}$$

Hence, for $x > 1$,

$$\sqrt{w(x)} = \sqrt{w(1)} + \int_1^x \frac{d}{ds}\left[\sqrt{w(s)}\right] ds$$

$$\le \sqrt{w(1)} + \int_1^x \frac{|q(s)|\sqrt{\rho(s)}|y(s)|}{\sqrt{\lambda}\sigma^{5/4}(s)} ds.$$

Applying Schwarz's inequality,* we obtain

$$\sqrt{w(x)} \le \sqrt{w(1)} + \frac{1}{\sqrt{\lambda}}\sqrt{\int_1^x \frac{q^2(s)ds}{\sigma^{5/2}(s)} \int_0^\infty y^2(s)\rho(s)ds},$$

* In Russian, the Cauchy-Bunyakovsky inequality. Although Bunyakovsky's priority is unquestioned, Western readers are likely to recognize the inequality more easily as Schwarz's inequality.–Translator

whence

$$[\sigma(x)\rho^2(x)]^{1/4}\frac{|y(x)|}{d_n} = \frac{|u(x)|}{d_n} \le \sqrt{\frac{w(x)}{d_n^2}} \le \sqrt{\frac{w(1)}{d_n^2}} + \frac{1}{\sqrt{\lambda}}\sqrt{\int_1^x \frac{q^2(s)ds}{\sigma^{5/2}(s)}}. \tag{24}$$

By (21), we have

$$\sqrt{\frac{w(1)}{d_n^2}} \le \frac{C_2}{n^{1/4}}. \tag{25}$$

Since $q(x)$ is a quadratic polynomial, and $\sigma(x) = x$, there is a constant C_3, independent of n, such that

$$\sqrt{\int_1^x \frac{q^2(s)ds}{\sigma^{5/2}(s)}} \le C_3 x^{5/4}. \tag{26}$$

Combining (24), (25) and (26), we obtain, for $x > 1$ and $\alpha + 1/2 > 0$,

$$x^{(\alpha/2)+(1/4)}e^{-x/2}\frac{|L_n^\alpha(x)|}{d_n} \le \frac{C_2}{n^{1/4}} + \frac{C_3 x^{5/4}}{n^{1/2}}. \tag{27}$$

It is easily verified that this inequality holds for all $x \ge 0$ (see (22)).

It is interesting to observe that (27) is also valid for $\alpha + 1/2 = 0$, since the Laguerre polynomials have

$$q(x) = \frac{1}{4}x^2 - \left(\alpha + \frac{1}{2}\right)x + \frac{1}{4}\left(\alpha^2 - \frac{1}{4}\right)\bigg|_{\alpha=-1/2} = \frac{1}{4}x^2$$

in this case. Consequently the point $x = 0$ is not singular in (23), and in integrating (23) we may take the lower limit at $x = 0$ and immediately obtain (27).

An inequality for Hermite polynomials can be obtained from (27) with $\alpha = \pm 1/2$ by using formulas (6.14) and (6.15):

$$e^{-x^2/2}\frac{|H_n(x)|}{d_n} \le \frac{C_2}{n^{1/4}} + \frac{C_3 x^{5/2}}{n^{1/2}}. \tag{28}$$

Remark. If $x \in [x_1, x_2]$, where $a < x_1 < x_2 < b$, the following simpler inequalities are consequences of (20), (27) and (28):

$$\frac{|P_n^{(\alpha,\beta)}(x)|}{d_n} \leq C_1 \quad \left(\alpha + \frac{1}{2} > 0,\ \beta + \frac{1}{2} > 0\right), \tag{20a}$$

$$|L_n^\alpha(x)|/d_n \leq C_2/n^{1/4} \quad \left(\alpha + \frac{1}{2} > 0\right), \tag{27a}$$

$$|H_n(x)|/d_n \leq C_3/n^{1/4} \tag{28a}$$

(the constants C_1, C_2, C_3 evidently depend on x_1, x_2 and the parameters α, β).

We can show that (20a) and (27a) remain valid for arbitrary real values of α and β. Let us show, for example, that (20a) is valid, or (what amounts to the same thing) that

$$\sqrt{n}|P_n^{(\alpha,\beta)}(x)| \leq c, \tag{19a}$$

where c is a constant.

The proof is by induction. We assume that (19a) holds for $P_n^{(\alpha+1,\beta+1)}(x)$ and $P_n^{(\alpha+2,\beta+2)}(x)$. From the differential equation of the Jacobi polynomials and the differentiation formulas (5.6), we have

$$\begin{aligned}
\sqrt{n}P_n^{(\alpha,\beta)}(x) &= -\frac{\sqrt{n}}{\lambda_n}\left[\tau(x)\frac{d}{dx}P_n^{(\alpha,\beta)}(x) + \sigma(x)\frac{d^2}{dx^2}P_n^{(\alpha,\beta)}(x)\right] \\
&= -\frac{\beta - \alpha - (\alpha+\beta+2)x}{2\sqrt{n}}P_{n-1}^{(\alpha+1,\beta+1)}(x) \\
&\quad - \frac{1-x^2}{4}\left(1 + \frac{\alpha+\beta+2}{n}\right)\sqrt{n}P_{n-2}^{(\alpha+2,\beta+2)}(x).
\end{aligned}$$

Since (19a) holds for $P_{n-1}^{(\alpha+1,\beta+1)}(x)$ and $P_{n-2}^{(\alpha+2,\beta+2)}(x)$, we obtain (19a) for $P_n^{(\alpha,\beta)}(x)$.

Similarly we can prove (27a) for all real values of α.

§ 8 Expansion of functions in series of the classical orthogonal polynomials

1. General considerations. In applications it is important to be able to find numbers a_n, corresponding to a function $f(x)$, that minimize the "mean square deviation"

$$m_N = \int_a^b \left[f(x) - \sum_{n=0}^{N} a_n y_n(x) \right]^2 \rho(x)dx,$$

where $y_n(x)$ are functions that are orthogonal on (a, b) with weight $\rho(x) \geq 0$, and $f(x)$ satisfies

$$\int_a^b f^2(x)\rho(x)dx < \infty.$$

Since the $y_n(x)$ are orthogonal, we have

$$m_N = \int_a^b f^2(x)\rho(x)dx - 2\sum_{n=0}^{N} a_n \int_a^b f(x)y_n(x)\rho(x)dx$$
$$+ \sum_{n=0}^{N} a_n^2 \int_a^b y_n^2(x)\rho(x)dx.$$

Putting

$$d_n^2 = \int_a^b y_n^2(x)\rho(x)dx, \quad c_n = \frac{1}{d_n^2}\int_a^b f(x)y_n(x)\rho(x)dx,$$

we obtain

$$m_N = \int_a^b f^2(x)\rho(x)dx + \sum_{n=0}^{N}(a_n - c_n)^2 d_n^2 - \sum_{n=0}^{N} c_n^2 d_n^2.$$

It is then clear that the minimum of m_N is attained for $a_n = c_n$, i.e.

$$\Delta_N = \min m_N = \int_a^b f^2(x)\rho(x)dx - \sum_{n=0}^{N} c_n^2 d_n^2.$$

Since the constants c_n are independent of N, the sequence $\{\Delta_N\}$ is monotone non-increasing and bounded below ($\Delta_N \geq 0$). Consequently there is a non-negative limit $\lim_{N\to\infty} \Delta_N$, and therefore the series $\sum_{n=0}^{\infty} c_n^2 d_n^2$ converges; moreover,

$$\sum_{n=0}^{\infty} c_n^2 d_n^2 \leq \int_a^b f^2(x)\rho(x)dx. \tag{1}$$

This is *Bessel's inequality.*

If $\lim_{N\to\infty} \Delta_N = 0$, then

$$f(x) = \sum_{n=0}^{\infty} c_n y_n(x), \tag{2}$$

where

$$c_n = \frac{1}{d_n^2} \int_a^b f(x) y_n(x) \rho(x) dx, \tag{3}$$

and the series on the right-hand side of (2) converges in the mean with weight $\rho(x)$ on (a, b), i.e.

$$\lim_{N\to\infty} \int_a^b \left[f(x) - \sum_{n=0}^{N} c_n y_n(x) \right]^2 \rho(x) dx = 0. \tag{4}$$

In this case (1) becomes

$$\sum_{n=0}^{\infty} c_n^2 d_n^2 = \int_a^b f^2(x)\rho(x)dx.$$

This is *Parseval's equation.* The series on the right of (2) is the *Fourier series* of $f(x)$ in terms of the functions $y_n(x)$, and the numbers c_n are its *Fourier coefficients.*

If (4) holds for all $f(x)$ that satisfy the condition of square-integrability,

$$\int_a^b f^2(x)\rho(x)dx < \infty, \tag{5}$$

the system $\{y_n(x)\}$ is called *complete.* A necessary condition for the completeness of $\{y_n(x)\}$ is, as is easily verified, the property of being *closed,* that is, that the equations

$$\int_a^b f(x)y_n(x)\rho(x)dx = 0 \qquad (n = 0, 1, \ldots) \tag{6}$$

imply that $f(x) \equiv 0$ for almost all $x \in (a, b)$, for every $f(x)$ satisfying (5). Consequently a basic question concerning the possibility of expanding $f(x)$ in a series (2) of orthogonal polynomials is whether the system is closed.

2. Closure of systems of orthogonal polynomials. We are going to show that the system $\{p_n(x)\}$ is closed with respect to continuous functions $f(x)$ if $\rho(x)$ is continuous on (a, b) and there is a constant $c_0 > 0$ such that

$$\int_a^b e^{c_0|x|}\rho(x)dx < \infty. \tag{7}$$

To prove this, we consider a function $f(x)$, continuous on (a, b), for which (5) and (6) are satisfied. We also consider the function

$$F(z) = \int_a^b e^{ixz} f(x)\rho(x)dx \tag{8}$$

of a complex variable in a strip $|\mathrm{Im} z| \le C$ with $C < c_0/2$. Let us first show that $F(z)$ is analytic in this strip. It is enough to show that the integral (8) converges uniformly. Since

$$|e^{ixz} f(x)\rho(x)| \le e^{|x \mathrm{Im} z|}|f(x)|\rho(x) \le e^{c_0|x|/2}|f(x)|\rho(x)$$

in the strip in question, the integral for $F(z)$ converges uniformly in the strip if the integral

$$\int_a^b e^{c_0|x|/2}|f(x)|\rho(x)dx$$

converges. The convergence of the latter integral follows from Schwarz's inequality:

$$\int_a^b e^{c_0|x|/2}|f(x)|\rho(x)dx \le \sqrt{\int_a^b e^{c_0|x|}\rho(x)dx \int_a^b f^2(x)\rho(x)dx} < \infty.$$

By the theorem on the analyticity of integrals that depend on a parameter, $F(z)$ will be analytic in $|\mathrm{Im} z| \le C$, and in particular analytic in $|z| \le C$. Hence it can be expanded in a Taylor series,

$$F(z) = \sum_{n=0}^{\infty} F^{(n)}(0)\frac{z^n}{n!}, \quad |z| \le C. \tag{9}$$

By means of similar estimates it is easy to establish the uniform convergence, in the same domain, of the integrals obtained by differentiating the integrand with respect to z. Consequently the derivatives $F^{(n)}(0)$ can be evaluated by differentiation under the integral sign, whence

$$F^{(n)}(0) = \int_a^b (ix)^n f(x)\rho(x)dx \qquad (n = 0, 1, \ldots).$$

Expanding $(ix)^n$ in terms of the polynomials $p_k(x) (k = 0, 1, \ldots, n)$ and using (6), we obtain

$$\begin{aligned} F^{(n)}(0) &= \int_a^b (ix)^n f(x)\rho(x)dx \\ &= \int_a^b \left[\sum_{k=0}^{n} c_{kn} p_k(x)\right] f(x)\rho(x)dx \\ &= \sum_{k=0}^{n} c_{kn} \int_a^b f(x)p_k(x)\rho(x)dx = 0. \end{aligned}$$

Since $F^{(n)}(0) = 0$, it follows from (9) that $F(z)$ is zero in the disk $|z| \le C$. By the principle of analytic continuation, $F(z) = 0$ for all z in the domain of analyticity of $F(z)$. In particular, $F(z) = 0$ for all real z.

Formula (8) for $f(z)$ can also be written in the form

$$F(z) = \int_{-\infty}^{\infty} e^{ixz} f(x)\rho(x)dx \tag{10}$$

if we take $f(x)\rho(x) = 0$ for $x < a$ and $x > b$.

For real z, formula (10) for $F(z)$ is the coefficient in the expansion of $f(x)\rho(x)$ in a Fourier integral. By Parseval's equation for Fourier integrals,

$$\int_{-\infty}^{\infty} [f(x)\rho(x)]^2 \, dx = \frac{1}{2\pi} \int_{-\infty}^{\infty} |F(z)|^2 dz = 0.$$

Since $f(x)$ is continuous and $\rho(x)$ is positive for $x \in (a, b)$, it follows that $f(x) \equiv 0$ on (a, b), i.e. the system of orthogonal polynomials $p_n(x)$ is indeed closed on (a, b).

If we use the explicit forms of $\rho(x)$ for the classical orthogonal polynomials, we can easily verify that the hypotheses imposed on $\rho(x)$ are satisfied for the classical orthogonal polynomials. For the Laguerre polynomials we can take $c_0 < 1$, and for the Jacobi and Hermite polynomials condition (7) is satisfied for any $c_0 > 0$. Hence the systems of classical orthogonal polynomials are closed on the corresponding intervals (a, b) with respect to continuous $f(x)$ that satisfy (5).

3. Expansion theorems. By using the closure of the systems of classical orthogonal polynomials and the estimates obtained in §7, we can find rather simple conditions that guarantee the validity of the expansion (2) of a given $f(x)$. We shall prove the following expansion theorem.

Theorem 1. *Let $f(x)$ be continuous on $a < x < b$ and have a piecewise continuous derivative in this interval; let $\rho(x)$ be the weight function of one of the classical systems of orthogonal polynomials $y_n(x)$. If the integrals*

$$\int_a^b f^2(x)\rho(x)dx \quad \text{and} \quad \int_a^b [f'(x)]^2 \sigma(x)\rho(x)dx$$

converge, the expansion (2) *of $f(x)$ in terms of $y_n(x)$ is valid, and moreover* (2) *converges uniformly on every interval $[x_1, x_2] \subset (a, b)$.*

Proof. We begin by estimating the Fourier coefficients c_n. The derivatives of the classical orthogonal polynomials are, as we proved in §5, orthogonal on (a, b) with weight $\sigma(x)\rho(x)$. Therefore by Bessel's inequality (1) for the coefficients $\tilde{c}_n$ of the expansion of $f'(x)$ in terms of the $y'_n(x)$, we have

$$\sum_{n=0}^{\infty} \tilde{c}_n^2 \tilde{d}_n^2 \leq \int_a^b [f'(x)]^2 \sigma(x)\rho(x)dx < \infty,$$

where

$$\left.\begin{aligned}\tilde{c}_n &= \frac{1}{\tilde{d}_n^2}\int_a^b f'(x)y_n'(x)\sigma(x)\rho(x)dx,\\ \tilde{d}_n^2 &= \int_a^b \left[y_n'(x)\right]^2\sigma(x)\rho(x)dx.\end{aligned}\right\} \tag{11}$$

Let us find the connection between $\tilde{c}_n$ and c_n. By using the equation

$$(\sigma\rho y_n')' + \lambda_n\rho y_n = 0$$

and integration by parts, we obtain

$$\begin{aligned}&\int_{x_1}^{x_2} f'(x)y_n'(x)\sigma(x)\rho(x)dx\\ &= f(x)y_n'(x)\sigma(x)\rho(x)\Big|_{x_1}^{x_2} - \int_{x_1}^{x_2} f(x)\left[\sigma(x)\rho(x)y_n'(x)\right]' dx\\ &= f(x)y_n'(x)\sigma(x)\rho(x)\Big|_{x_1}^{x_2} + \lambda_n\int_{x_1}^{x_2} f(x)y_n(x)\rho(x)dx\\ &\qquad\qquad (a < x_1 < x_2 < b)\,.\end{aligned} \tag{12}$$

By the hypotheses of the theorem and Schwarz's inequality, the integrals in (12) have finite limits as $x_1 \to a$ and $x_2 \to b$. Hence the limits

$$\lim_{x\to a} f(x)y_n'(x)\sigma(x)\rho(x) = A_n,$$
$$\lim_{x\to b} f(x)y_n'(x)\sigma(x)\rho(x) = B_n$$

exist and are finite. Let us show that $B_n = 0$. Suppose that $B_n \neq 0$ for some n. Then as $x \to b$,

$$f(x) \approx \frac{B_n}{y_n'(x)\sigma(x)\rho(x)}. \tag{13}$$

It follows from the explicit form of $\rho(x)$ and from (13) that if $B_n \neq 0$ the function $f(x)$ cannot have

$$\int_a^b f^2(x)\rho(x)dx$$

finite. In fact, if, for example, b is finite, then as $x \to b$ (see §5, Part 1),

$$\sigma(x) \sim b - x, \quad \rho(x) \sim (b-x)^{\alpha} \quad (\alpha > -1),$$

whence

$$f(x) \sim \frac{1}{(b-x)^{\alpha+1}}, \quad f^2(x)\rho(x) \sim \frac{1}{(b-x)^{\alpha+2}}.$$

It is clear from this that the integral $\int_a^b f^2(x)\rho(x)dx$ diverges when $\alpha > -1$.

A similar analysis can be given for the behavior of $f^2(x)\rho(x)$ as $x \to b$ when $b = +\infty$. Hence we have shown that $B_n = 0$ for all n. A similar analysis of the behavior of $f^2(x)\rho(x)$ as $x \to a$ shows that $A_n = 0$.

Consequently if we take limits in (12) as $x_1 \to a$ and $x_2 \to b$, we obtain

$$\int_a^b f'(x)y_n'(x)\sigma(x)\rho(x)dx = \lambda_n \int_a^b f(x)y_n(x)\rho(x)dx.$$

In particular, when $f(x) = y_n(x)$ we have $\tilde{d}_n^2 = \lambda_n d_n^2$, whence $\tilde{c}_n = c_n$. Since the series $\sum_{n=0}^{\infty} \tilde{c}_n^2 \tilde{d}_n^2$ converges, the series $\sum_{n=0}^{\infty} c_n^2 d_n^2 \lambda_n$ must also converge.

Let us now show that our series $\sum_{n=0}^{\infty} c_n y_n(x)$ converges uniformly for $a < x_1 \le x \le x_2 < b$, for arbitrary x_1 and x_2 in (a, b). Using Schwarz's inequality, we have

$$\begin{aligned} \left|\sum_{n=N_1}^{N_2} c_n y_n(x)\right| &\le \sum_{n=N_1}^{N_2} \left|c_n d_n \sqrt{\lambda_n}\right| \frac{|y_n(x)|}{\sqrt{\lambda_n} d_n} \\ &\le \sqrt{\sum_{n=N_1}^{N_2} c_n^2 d_n^2 \lambda_n \sum_{n=N_1}^{N_2} \frac{y_n^2(x)}{\lambda_n d_n^2}} \\ &\le \sqrt{\sum_{n=0}^{\infty} c_n^2 d_n^2 \lambda_n} \sqrt{\sum_{n=N_1}^{N_2} \frac{y_n^2(x)}{\lambda_n d_n^2}}. \end{aligned} \tag{14}$$

By the Cauchy criterion for the uniform convergence of series and inequality (14), it is easy to show that the series $\sum_{n=0}^{\infty} c_n y_n(x)$ will converge uniformly for $a < x_1 \le x \le x_2 < b$ if the series

$$\sum_{n=1}^{\infty} \frac{y_n^2(x)}{\lambda_n d_n^2}$$

converges uniformly in this domain.

In estimating $y_n^2(x)/(\lambda_n d_n^2)$ it is convenient to make a linear change of variable so that the weight function $\rho(x)$ and the polynomials $y_n(x)$ are reduced to a canonical form (see §5). Under this change of variable, $y_n^2(x)/(\lambda_n d_n^2)$ is changed only by a constant multiple independent of n. Hence in studying the convergence of

$$\sum_{n=1}^{\infty} \frac{y_n^2(x)}{\lambda_n d_n^2}$$

it is enough to consider only the cases when $y_n(x)$ are the Jacobi, Laguerre or Hermite polynomials. Using the rough estimates from Part 2 of §7, we can establish the uniform convergence of

$$\sum_{n=1}^{\infty} \frac{y_n^2(x)}{\lambda_n d_n^2}$$

for $a < x_1 \leq x \leq x_2 < b$.

Let us do the calculations for the Laguerre polynomials $y_n(x) = L_n^{\alpha}(x)$. In this case $a = 0$, $b = \infty$, $\rho(x) = x^{\alpha} e^{-x}$. For $0 < x_1 \leq x \leq x_2 < \infty$ we have, by (27a) of §7,

$$\begin{gathered} \frac{1}{d_n} |L_n^{\alpha}(x)| < \frac{c}{n^{1/4}} \quad (c\text{: a positive constant}), \\ \sum_{n=N}^{\infty} \frac{y_n^2(x)}{\lambda_n d_n^2} \leq \sum_{n=N}^{\infty} \frac{c}{n^{3/2}}. \end{gathered} \tag{15}$$

This inequality immediately implies the uniform convergence of

$$\sum_{n=1}^{\infty} \frac{y_n^2(x)}{\lambda_n d_n^2}$$

in the interval $0 < x_1 \leq x \leq x_2 < x_2 < \infty$ when $y_n(x) = L_n^{\alpha}(x)$. The proof is similar in the other cases.

Consequently we see that in virtue of (14) the series $\sum_{n=0}^{\infty} c_n y_n(x)$ converges uniformly for $a < x_1 \leq x \leq x_2 < b$ and therefore represents a continuous function on (a, b), since x_1 and x_2 are arbitrary.

Let us now show that this series converges to $f(x)$. Consider the function

$$\bar{f}(x) = f(x) - \sum_{k=0}^{\infty} c_k y_k(x).$$

Let us find the Fourier coefficients of its expansion in a series of $y_n(x)$. We have

$$\begin{aligned}\int_a^b \bar{f}(x)y_n(x)\rho(x)dx &= \int_a^b y_n(x)\rho(x)\left[f(x) - \sum_{k=0}^{\infty} c_k y_k(x)\right]dx \\ &= c_n d_n^2 - \sum_{k=0}^{N-1} c_k \int_a^b y_n(x)y_k(x)\rho(x)dx - I_N,\end{aligned} \tag{16}$$

where

$$I_N = \int_a^b y_n(x)\rho(x)\left[\sum_{k=N}^{\infty} c_k y_k(x)\right]dx$$

Since

$$\int_a^b y_n(x)y_k(x)\rho(x)dx = 0 \qquad \text{for } k \neq n,$$

we obtain

$$\int_a^b \bar{f}(x)y_n(x)\rho(x)dx = -I_N \tag{17}$$

for $N > n$. We can estimate I_N by (14), taking $N_1 = N$ and $N_2 = \infty$:

$$|I_N| \le \sqrt{\sum_{k=0}^{\infty} c_k^2 d_k^2 \lambda_k} \int_a^b |y_n(x)|\,\rho(x)\sqrt{\sum_{k=N}^{\infty} \frac{y_k^2(x)}{\lambda_k d_k^2}}\,dx.$$

By using an estimate of the same kind as (15), we can show that $I_N \to 0$ as $N \to \infty$. Taking limits in (17), we obtain

$$\int_a^b \bar{f}(x)y_n(x)\rho(x)dx = 0.$$

Since this equation holds for every n, and $\bar{f}(x)$ is continuous for $a < x < b$, the closure of the classical system of orthogonal polynomials implies that $\bar{f}(x) \equiv 0$ for $a < x < b$, which establishes the validity of the expansion

$$f(x) = \sum_{n=0}^{\infty} c_n y_n(x).$$

Remark 1. We have established an expansion theorem for functions $f(x)$ that satisfy some not very restrictive conditions. A more general expansion theorem can be established by means of a more complicated proof. For this purpose it is desirable to transform the differential equation of the classical orthogonal polynomials into a simpler form (see §19, Part 1):

$$u''(s) + [\lambda - q(s)]u(s) = 0 \qquad (0 \le s \le s_0). \tag{18}$$

The eigenfunctions of the original differential equation become eigenfunctions $u = u_n(s)$ of (18). Then the expansion theorem for the classical orthogonal polynomials can be replaced by the following expansion theorem for the function $u_n(s)$ (see [L5]).

Theorem 2. *(Equiconvergence theorem). If $\int_0^{s_0} f^2(s)ds$ is finite, then the expansion of $f(s)$ in terms of the eigenfunctions of* (18) *on* $0 < s < s_0$ *converges or diverges simultaneously with its trigonometric Fourier expansion on this interval (if $s_0 = \infty$, the expansion is a Fourier integral rather than a Fourier series).*

Remark 2. If the expansion of $f(x)$ in a series of classical orthogonal polynomials $y_n(x)$ is needed only in the sense of convergence in the mean, then, as is shown in courses on mathematical analysis, convergence for a function $f(x)$ that satisfies (5) follows immediately from the closure of the system $\{y_n(x)\}$ if all integrals are taken as Lebesgue integrals.

§ 9 Eigenvalue problems that can be solved by means of the classical orthogonal polynomials

1. Statement of the problem. Consider the solution of the equation

$$\sigma(x)y'' + \tau(x)y' + \lambda y = 0 \tag{1}$$

of hypergeometric type for various values of λ, when $\rho(x)$ satisfies the equation $(\sigma\rho)' = \tau\rho$, is bounded on an interval (a, b), and satisfies the conditions imposed on $\rho(x)$ for the classical orthogonal polynomials.

As we have seen, the simplest solutions of (1) are the classical orthogonal polynomials $y_n(x)$, which correspond to

$$\lambda = \lambda_n = -n\tau' - \frac{1}{2}n(n-1)\sigma'' \qquad (n = 0, 1, \ldots).$$

It turns out that the classical orthogonal polynomials are distinguished among the solutions of (1) corresponding to various values of λ not only by their simplicity, but also because they are the only nontrivial solutions of (1) for which $y(x)\sqrt{\rho(x)}$ is both bounded and of integrable square on (a, b).

This property is extensively used in quantum mechanics for solving problems about the energy levels and wave functions of a particle in a potential field. If external forces restrict the particle to a bounded part of space, so that it cannot move off to infinity, one says that the particle is in a bound state. To find the wave functions $\psi(\mathbf{r})$ that describe these states, and the corresponding energy levels E, one solves the time-independent Schrödinger equation

$$-\frac{\hbar^2}{2\mu}\Delta\psi + U\psi = E\psi,$$

where $\hbar$ is Planck's constant, μ is the mass of the particle, $U = U(\mathbf{r})$ is the potential and $\mathbf{r}$ is the radius-vector.

Here the wave function $\psi(\mathbf{r})$ must be bounded for all finite $|\mathbf{r}|$ and be normalized by

$$\int_V |\psi(\mathbf{r})|^2 dV = 1. \tag{2}$$

For many problems of quantum mechanics that can be solved analytically by the method of separation of variables, the Schrödinger equation reduces to a generalized equation of hypergeometric type (§1):

$$u'' + \frac{\tilde{\tau}(x)}{\sigma(x)}u' + \frac{\tilde{\sigma}(x)}{\sigma^2(x)}u = 0 \qquad (a < x < b). \tag{3}$$

The energy E appears as a parameter in the coefficients of equation (3). We assume that $\sigma(x) > 0$ for $x \in (a, b)$ and that $\sigma(x) = 0$ at the endpoints of (a, b), if the endpoints are not at infinity. Since (3) has no singular points at any $x \in (a, b)$, the function $u(x)$ is continuously differentiable on (a, b). Therefore it can have singular points only as $x \to a$ or $x \to b$. In order to state the additional restrictions that should be imposed on $u(x)$ at the endpoints of (a, b), we rewrite (3) in self-adjoint form:

$$(\sigma\tilde{\rho}u')' + (\tilde{\sigma}/\sigma)\tilde{\rho}u = 0. \tag{4}$$

Here $\tilde{\rho}(x) > 0$ and $\tilde{\rho}(x)$ satisfies

$$(\sigma\tilde{\rho})' = \tilde{\tau}\tilde{\rho}. \tag{5}$$

The function $\psi(\mathbf{r})$ will automatically be bounded and satisfy the normalization condition (2) if the problem is formulated in terms of (4) in the following way: *Find all values of E for which* (4) *has a nontrivial solution on* (a, b) *such that* $u(x)\{\tilde{\rho}(x)\}^{1/2}$ *is bounded and of integrable square on* (a, b), i.e. $|u(x)|\{\tilde{\rho}(x)\}^{1/2} < C$ *(some constant) and*

$$\int_a^b |u(x)|^2\tilde{\rho}(x)dx < \infty$$

(if a and b are finite, the last condition can be omitted).

As we showed in §1, equation (3) can be transformed by the substitution $u = \phi(x)y$ into an equation of hypergeometric type

$$\frac{d}{dx}\left[\sigma\rho\frac{dy}{dx}\right] + \lambda\rho y = 0, \tag{6}$$

where $\rho(x)$ satisfies $(\sigma\rho)' = \tau\rho$ and $\tau(x)$ is connected with $\tilde{\tau}(x)$ and $\phi(x)$ by

$$\tau = \tilde{\tau} + 2(\phi'/\phi)\sigma.$$

It follows from this and (5) that $\rho(x) = \tilde{\rho}(x)\phi^2(x)$. Hence the requirements on $u(x)\{\tilde{\rho}(x)\}^{1/2}$ become the requirements listed above on $y(x)\sqrt{\rho(x)}$. The values of λ for which our problem has nontrivial solutions are the *eigenvalues* and the corresponding functions $y(x, \lambda)$ are the *eigenfunctions.*

As we showed in §1, equation (3) can be transformed to the form (6) in various ways. For the majority of the problems of quantum mechanics that admit explicit solutions, this can be done by using a $\rho(x)$ that is bounded on (a, b) and satisfies the conditions imposed on $\rho(x)$ for the classical orthogonal polynomials.

Remark. In order to satisfy the conditions imposed on $\rho(x)$ for the classical orthogonal polynomials, $\tau(x)$ has to vanish at some point of (a, b) and have a negative derivative, $\tau' < 0$.

In fact, as we see from the Rodrigues formula, $y_1(x) = B_1\tau(x)$ and therefore $\tau(x)$ has a zero on (a, b). Moreover, by (5.20),

$$d_{11}^2 = \lambda_1 d_1^2 = -\tau' d_1^2.$$

Since $\rho(x) > 0$ and $\sigma(x) > 0$ for $x \in (a, b)$, we have $d_1^2 > 0$, $d_{11}^2 > 0$, whence $\tau' < 0$.

This remark lets us simplify the selection of a transformation of (3) into (6). We are led to a choice of the constant k and of the sign in formula (1.11) for $\pi(x)$ that will make the function

$$\tau(x) = \tilde{\tau}(x) + 2\pi(x)$$

satisfy the necessary conditions.

2. Classical orthogonal polynomials as eigenfunctions of some eigenvalue problems. Let us consider the eigenvalue problem stated in Part 1.

Theorem. *Let $y = y(x)$ be a solution of the equation*

$$\sigma(x)y'' + \tau(x)y' + \lambda y = 0$$

of hypergeometric type, and let $\rho(x)$, a solution of $(\sigma\rho)' = \tau\rho$, be bounded on (a, b) and satisfy the conditions imposed on $\rho(x)$ for the classical orthogonal polynomials. Then nontrivial solutions of the equation of hypergeometric type for which $y(x)\sqrt{\rho(x)}$ is bounded and of integrable square on (a, b) exist only when

$$\lambda = \lambda_n = -n\tau' - \frac{1}{2}n(n-1)\sigma'' \qquad (n = 0, 1, \ldots) \tag{7}$$

and they have the form

$$y(x, \lambda_n) = y_n(x) = \frac{B_n}{\rho(x)} \frac{d^n}{dx^n}[\sigma^n(x)\rho(x)], \tag{8}$$

i.e. they are the classical polynomials that are orthogonal with weight $\rho(x)$ on (a, b) (if a and b are finite, the condition of quadratic integrability can be omitted).

Proof. That the classical orthogonal polynomials $y_n(x)$ with $\lambda = \lambda_n$ are nontrivial solutions can be verified immediately.

Let us show that the problem has no other solutions. Suppose the contrary, i.e. that for some λ there is a nontrivial solution $y = y(x, \lambda)$ which is not a classical orthogonal polynomial. We have

$$(\sigma\rho y')' + \lambda\rho y = 0, \qquad (\sigma\rho y_n')' + \lambda_n\rho y_n = 0.$$

Multiply the first equation by $y_n(x)$ and the second by $y(x,\lambda)$; subtract the second equation from the first and integrate over $(x_1, x_2), a < x_1 < x_2 < b$ (note that the equations for $y(x,\lambda)$ and $y_n(x)$ have no singular points on $[x_1, x_2]$). We obtain

$$(\lambda - \lambda_n)\int_{x_1}^{x_2} y(x,\lambda)y_n(x)\rho(x)dx + \sigma(x)\rho(x)W(y_n, y)\Big|_{x_1}^{x_2} = 0, \tag{9}$$

where

$$W(y_n, y) = y_n(x)y'(x,\lambda) - y_n'(x)y(x,\lambda)$$

is the Wronskian. When $\lambda \neq \lambda_n$ $(n = 0, 1, \ldots)$ it follows from (9) that

$$\lim_{x\to a} \sigma(x)\rho(x)W(y_n, y) = c_1, \tag{10}$$

$$\lim_{x\to b} \sigma(x)\rho(x)W(y_n, y) = c_2 \tag{11}$$

(c_1 and c_2 are constants).

For the proof, it is enough to take the limit in (9) as $x_1 \to a$ (or $x_2 \to b$) and use the convergence of the integral

$$\int_a^b y(x,\lambda)y_n(x)\rho(x)dx.$$

The convergence follows from the Schwarz inequality

$$\left|\int_{x_1}^{x_2} y(x,\lambda)y_n(x)\rho(x)dx\right| \leq \left\{\int_{x_1}^{x_2} y^2(x,\lambda)\rho(x)dx \int_{x_1}^{x_2} y_n^2(x)\rho(x)dx\right\}^{1/2}$$

and the convergence of the integrals

$$\int_a^b y^2(x,\lambda)\rho(x)dx, \qquad \int_a^b y_n^2(x)\rho(x)dx.$$

Equations (10) and (11) also hold for $\lambda = \lambda_n$ if we put $c_1 = c_2 = c$, since it is clear from (9) that when $\lambda = \lambda_n$ we have

$$\sigma(x)\rho(x)W[y_n(x), y(x,\lambda)] = \text{const.}$$

Let us show that the constant c_2 in (11) must be zero. Since

$$\frac{d}{dx}\left[\frac{y(x,\lambda)}{y_n(x)}\right] = \frac{W[y_n(x), y(x,\lambda)]}{y_n^2(x)},$$

we have

$$y(x,\lambda) = y_n(x)\left[\frac{y(x_0,\lambda)}{y_n(x_0)} + \int_{x_0}^{x} \frac{W[y_n(s), y(s,\lambda)]}{y_n^2(s)}\,ds\right]. \tag{12}$$

In (12) we choose the point $x_0 < b$ so that it lies to the right of all the zeros of $y_n(x)$. To discuss the behavior of $y(x,\lambda)$ as $x \to b$, we use the explicit forms of $\rho(x)$ (see §5, Part 1). There are three cases:

1) b is finite and

$$\sigma(x) \sim b - x, \quad \rho(x) \sim (b-x)^\alpha \qquad (\alpha \geq 0)$$

as $x \to b$;

2) $b = +\infty$ and

$$\sigma(x) \sim x, \quad \rho(x) \sim x^\alpha e^{\beta x} \qquad (\alpha \geq 0, \beta < 0)$$

as $x \to +\infty$;

3) $b = +\infty$ and

$$\sigma(x) = 1, \quad \rho(x) \sim e^{\alpha x^2 + \beta x} \qquad (\alpha < 0)$$

as $x \to +\infty$.

As follows from (11) with $c_2 \neq 0$, in the first case the integrand in (12) has the following behavior as $s \to b$:

$$\frac{W[y_n(s), y(s,\lambda)]}{y_n^2(s)} \approx \frac{c_2}{\sigma(s)\rho(s)y_n^2(s)}.$$

Hence as $x \to b$, in the first case we have

$$\int_{x_0}^{x} \frac{W[y_n(s), y(s,\lambda)]}{y_n^2(s)} ds \sim \begin{cases} (b-x)^{-\alpha} & (\alpha > 0), \\ \ln(b-x) & (\alpha = 0), \end{cases}$$

$$\sqrt{\rho(x)}y(x,\lambda) \sim \begin{cases} (b-x)^{-\alpha/2} & \text{if } \alpha > 0, \\ \ln(b-x) & \text{if } \alpha = 0, \end{cases}$$

i.e. the function $\sqrt{\rho(x)}y(x,\lambda)$ is unbounded. Hence, in the first case, we must have $c_2 = 0$.

In the other two cases we use the asymptotic behavior of the functions in (12) as $s \to +\infty$:

$$y_n(s) \sim s^n;$$

$$\sigma(s)\rho(s)y_n^2(s) \sim \begin{cases} s^{\alpha+2n+1}e^{\beta s} & \text{if } \rho(s) = s^{\alpha}e^{\beta s} & (\beta < 0), \\ s^{2n}e^{\alpha s^2+\beta s} & \text{if } \rho(s) = e^{\alpha s^2+\beta s} & (\alpha < 0). \end{cases}$$

If we use L'Hospital's rule, we obtain, as $x \to -\infty, c_2 \neq 0$,

$$\int_{x_0}^{x} \frac{W[y_n(s), y(s,\lambda)]}{y_n^2(s)} ds \sim \begin{cases} x^{-(\alpha+2n+1)}e^{-\beta x}, \\ x^{-(2n+1)}e^{-\alpha x^2-\beta x}; \end{cases}$$

$$\sqrt{\rho(x)}y(x,\lambda) \sim \begin{cases} x^{-((\alpha/2)+n+1)}e^{-\beta x/2}, \\ x^{-(n+1)}e^{-(\alpha x^2+\beta x)/2}. \end{cases}$$

In either case, $\sqrt{\rho(x)}y(x,\lambda)$ is not of integrable square on (a,b). Consequently $c_2 = 0$ in these cases also. A similar study of $\sqrt{\rho(x)}y(x,\lambda)$ as $x \to a$ shows that $c_1 = 0$.

Hence we have shown that

$$\lim_{x\to a} \sigma(x)\rho(x)W(y_n, y) = 0,$$
$$\lim_{x\to b} \sigma(x)\rho(x)W(y_n, y) = 0,$$

for all n. These equations are possible only if $y(x,\lambda) \equiv 0$. In fact, if $\lambda \neq \lambda_n (n = 0, 1, \ldots)$, then from relation (9) we obtain

$$\int_a^b y(x,\lambda)y_n(x)\rho(x)dx = 0 \qquad (n = 0, 1, \ldots)$$

as $x_1 \to a$ and as $x_2 \to b$. Because of the closure of the classical orthogonal polynomials, this equation is possible only when $y(x,\lambda) = 0$ for $x \in (a,b)$.

On the other hand, if $\lambda = \lambda_n$ then (10) implies that $W[y_n(x), y(x,\lambda)] = 0$, i.e. the solutions $y_n(x)$ and $y(x,\lambda)$ are linearly dependent, contrary to hypothesis. This completes the proof of the theorem.

3. Quantum mechanics problems that lead to classical orthogonal polynomials. We illustrate the applicability of the theorem of Part 2 by means of some quantum mechanics problems in which the Schrödinger equation reduces to a generalized equation of hypergeometric type.

Example 1. We consider the problem of finding the eigenvalues and eigenfunctions for the linear harmonic oscillator, i.e. for a particle in a field with potential $U = m\omega^2 x^2/2$ (m is the mass, x the displacement from equilibrium, ω the angular frequency). The problem of the harmonic oscillator plays an important role in the foundations of quantum electrodynamics, and has applications to various types of oscillations in crystals and molecules.

The Schrödinger equation for the wave function $\psi(x)$ of the harmonic oscillator has the form

$$-\frac{\hbar^2}{2m}\frac{d^2\psi}{dx^2} + \frac{1}{2}m\omega^2 x^2\psi = E\psi \quad (-\infty < x < \infty).$$

Here $\psi(x)$ must be bounded and satisfy the normalization condition

$$\int_{-\infty}^{\infty} \psi^2(x)dx = 1.$$

In solving the problem, it is convenient to replace x and E by dimensionless variables ξ and ϵ:

$$x = \xi\sqrt{\frac{\hbar}{m\omega}} = \alpha\xi, \quad E = \hbar\omega\epsilon.$$

Then we obtain the equation

$$\psi'' + (2\epsilon - \xi^2)\psi = 0$$

(primes denote differentiation with respect to ξ). This is a generalized equation of hypergeometric type for which

$$\sigma(\xi) = 1, \quad \tilde{\tau}(\xi) = 0, \quad \tilde{\sigma}(\xi) = 2\epsilon - \xi^2.$$

We now have a problem of a kind that we have solved before. In the present case $\tilde{\rho}(\xi) = 1$. Hence the requirement that $\sqrt{\tilde{\rho}(\xi)}\psi(\xi)$ is of integrable square follows from the normalization condition. Following the method used above, we transform the equation for ψ to an equation of hypergeometric type,

$$\sigma(\xi)y'' + \tau(\xi)y' + \lambda y = 0,$$

by putting $\psi(\xi) = \phi(\xi)y(\xi)$, where $\phi(\xi)$ satisfies the equation

$$\phi'/\phi = \pi(\xi)/\sigma(\xi).$$

In the present case the polynomial $\pi(\xi)$ is

$$\pi(\xi) = \pm\sqrt{k - 2\epsilon + \xi^2}.$$

The constant k can be determined from the condition that the function under the square root sign has a double zero, i.e. $k = 2\epsilon$. There are two possible polynomials $\pi(\xi) = \pm\xi$; we select the one for which

$$\tau(\xi) = \tilde{\tau}(\xi) + 2\pi(\xi)$$

has a negative derivative. The conditions on $\tau(\xi)$ are satisfied if we take $\tau(\xi) = -2\xi$, in which case

$$\pi(\xi) = -\xi, \quad \phi(\xi) = e^{-\xi^2/2},$$
$$\lambda = 2\epsilon - 1, \quad \rho(\xi) = e^{-\xi^2}.$$

The energy eigenvalues are determined by

$$\lambda + n\tau' + \frac{n(n-1)}{2}\sigma'' = 0,$$

which yields

$$\epsilon = \epsilon_n = n + \frac{1}{2}, \quad \text{i.e.} \quad E = E_n = \hbar\omega\left(n + \frac{1}{2}\right) \quad (n = 0, 1, \ldots).$$

We obtain the eigenfunctions in the form

$$y_n(\xi) = B_n e^{\xi^2}\frac{d^n}{d\xi^n}(e^{-\xi^2}).$$

These are, up to numerical factors, the Hermite polynomials $H_n(\xi)$. The wave functions $\psi(x)$ are

$$\psi_n(x) = C_n e^{-\xi^2/2}H_n(\xi), \quad x = \alpha\xi, \quad \alpha = (\hbar/(m\omega))^{\frac{1}{2}}$$

The constants C_n can be found from the normalization condition

$$\int_{-\infty}^{\infty} \psi_n^2(x)dx = 1.$$

Example 2. Consider the problem of finding the eigenvalues and eigenfunctions for the one-dimensional Schrödinger equation

$$-\frac{\hbar^2}{2m}\psi'' + U(x)\psi = E\psi \qquad (-\infty < x < \infty)$$

for a particle in the field

$$U(x) = -\frac{U_0}{\cosh^2 \alpha x}, \qquad \text{where } U_0 > 0$$

(the Pöschl-Teller potential; see [19], problems 38, 39.) Here $\psi(x)$ is to be bounded, and normalized by

$$\int_{-\infty}^{\infty} \psi^2(x)dx = 1.$$

Since $U(x) < 0$, only values of $E < 0$ are admissible. To simplify the form of the equation we make the change of independent variable $s = \tanh \alpha x$.*

We then obtain the generalized equation of hypergeometric type

$$\Phi'' + \frac{\tilde{\tau}(s)}{\sigma(s)}\Phi' + \frac{\tilde{\sigma}(s)}{\sigma^2(s)}\Phi = 0, \quad \Phi(s) = \Psi(x),$$

for which $a = -1, b = 1$,

$$\sigma(s) = 1 - s^2, \quad \tilde{\tau}(s) = -2s, \quad \tilde{\sigma}(s) = -\beta^2 + \gamma^2(1 - s^2),$$
$$\beta^2 = -\frac{2mE}{\hbar^2\alpha^2}, \quad \gamma^2 = \frac{2mU_0}{\hbar^2\alpha^2} \qquad (\beta > 0, \gamma > 0).$$

* In many quantum mechanics problems that can be solved explicitly, the Schrödinger equation can be reduced to an equation with rational coefficients by a natural change of variable suggested by the form of $U(x)$, where the transformation must be one-to-one. In the present case the potential has a simple expression in terms of hyperbolic functions, so it is natural to try $\sinh \alpha x$, $\tanh \alpha x$, or $\exp(\pm\alpha x)$ as a new variable. We chose the substitution $s = \tanh \alpha x$.

This is again a problem of the kind we have discussed. Here $\tilde{\rho}(s) = 1$. Hence the integrability of the square of $\sqrt{\tilde{\rho}(s)}\Phi(s)$ follows from the normalization condition $\int_{-\infty}^{\infty} \psi^2(x)dx = 1$. In fact,

$$\int_{-1}^{1} \Phi^2(s)ds = \alpha \int_{-\infty}^{\infty} \frac{\psi^2(x)}{\cosh^2 \alpha x} dx < \alpha \int_{-\infty}^{\infty} \psi^2(x)dx = \alpha.$$

The solution is obtained by the previous method. We transform the equation for $\Phi(s)$ to the equation of hypergeometric type

$$\sigma(s)y'' + \tau(s)y' + \lambda y = 0$$

by putting $\Phi(s) = \phi(s)y(s)$, where $\phi(s)$ satisfies

$$\phi'/\phi = \pi(s)/\sigma(s).$$

The polynomial $\pi(s)$ is now given by

$$\pi(s) = \pm\sqrt{\beta^2 - \gamma^2(1 - s^2) + k(1 - s^2)}.$$

The constant k is determined by the condition that the expression under the square root sign has a double zero, that is, that $k = \gamma^2$ or $k = \gamma^2 - \beta^2$. In the first case $\pi(s) = \pm\beta$; in the second, $\pi(s) = \pm\beta s$. We choose the one for which $\tau(s) = \tilde{\tau}(s) + 2\pi(s)$ has a negative derivative and a zero on $(-1, +1)$. These conditions are satisfied by

$$\tau(s) = -2(1 + \beta)s,$$

which corresponds to

$$\pi(s) = -\beta s, \quad \phi(s) = (1 - s^2)^{\beta/2},$$
$$\lambda = \gamma^2 - \beta^2 - \beta, \quad \rho(s) = (1 - s^2)^{\beta}.$$

The energy eigenvalues are determined by

$$\lambda + n\tau' + \frac{1}{2}n(n-1)\sigma'' = 0 \qquad (n = 0, 1, \ldots),$$

which reduces to

$$\gamma^2 - \beta^2 - \beta = 2n(1 + \beta) + n(n - 1).$$

Hence the eigenvalues are

$$E_n = -\frac{\hbar^2\alpha^2}{2m}\beta_n^2 \qquad \text{where } \beta_n = -n - \frac{1}{2} + \sqrt{\gamma^2 + \frac{1}{4}} \quad (\beta_n > 0).$$

The condition $\beta_n > 0$ can be satisfied only for

$$n < \sqrt{\gamma^2 + \frac{1}{4}} - \frac{1}{2},$$

i.e. there are only finitely many eigenvalues. In this case the eigenfunctions $y_n(s)$ have the form $y_n(s) = P_n^{(\beta,\beta)}(s)$ with $\beta = \beta_n$. The wave functions $\psi_n(x)$ are

$$\psi_n(x) = C_n(1-s^2)^{\beta/2}P_n^{(\beta,\beta)}(s),$$

with $\beta = \beta_n$, $s = \tanh \alpha x$. Here C_n is a normalizing constant determined by

$$\int_{-\infty}^{\infty} \psi_n^2(x)dx = 1.$$

Other examples of the process of solving quantum mechanics problems are discussed in §26.

§ 10 Spherical harmonics

1. Solution of Laplace's equation in spherical coordinates. Spherical harmonics are an important class of special functions that are closely related to the classical orthogonal polynomials. They arise, for example, when Laplace's equation is solved in spherical coordinates. Since continuous solutions of Laplace's equation are *harmonic functions*, these solutions are called *spherical harmonics*; the term *spherical functions* is also used.

Let us find the bounded solutions of Laplace's equation $\Delta u = 0$ in spherical coordinates r, θ, ϕ. We have

$$\Delta u = \Delta_r u + \frac{1}{r^2}\Delta_{\theta,\phi} u,$$

where

$$\Delta_r u = \frac{1}{r^2}\frac{\partial}{\partial r}\left(r^2\frac{\partial u}{\partial r}\right),$$

$$\Delta_{\theta,\phi} u = \frac{1}{\sin\theta}\frac{\partial}{\partial\theta}\left(\sin\theta\frac{\partial u}{\partial\theta}\right) + \frac{1}{\sin^2\theta}\frac{\partial^2 u}{\partial\phi^2}.$$

We look for particular solutions by assuming $u = R(r)Y(\theta, \phi)$. Substituting this into Laplace's equation, we obtain

$$\frac{r^2\Delta_r R(r)}{R(r)} = -\frac{\Delta_{\theta,\phi}Y(\theta,\phi)}{Y(\theta,\phi)}.$$

Since the left-hand side is independent of θ and ϕ, and the right-hand side is independent of r, we have

$$\frac{r^2\Delta_r R}{R} = -\frac{\Delta_{\theta,\phi}Y(\theta,\phi)}{Y(\theta,\phi)} = \mu,$$

where μ is a constant. Hence we have the equations

$$(r^2R')' = \mu R, \tag{1}$$

$$\Delta_{\theta,\phi}Y + \mu Y = 0. \tag{2}$$

We can also solve (2) by separating variables, by putting

$$Y(\theta,\phi) = \Theta(\theta)\Phi(\phi).$$

This yields

$$\frac{\sin\theta \dfrac{d}{d\theta}\left(\sin\theta \dfrac{d\Theta}{d\theta}\right)}{\Theta(\theta)} + \mu \sin^2\theta = -\frac{\Phi''(\phi)}{\Phi(\phi)} = \nu,$$

where ν is a constant. Therefore we obtain the following equations for $\Phi(\phi)$ and $\Theta(\theta)$:

$$\Phi'' + \nu\Phi = 0, \tag{3}$$

$$\sin\theta \frac{d}{d\theta}\left(\sin\theta \frac{d\Theta}{d\theta}\right) + (\mu \sin^2\theta - \nu)\Theta = 0. \tag{4}$$

The requirement that $\Phi(\phi)$ is single-valued yields the periodicity $\Phi(\phi+2\pi) = \Phi(\phi)$. Under this condition, (3) can be solved only when $\nu = m^2$ with m an integer. Thus we obtain the linearly independent solutions of (3):

$$\Phi_m(\phi) = C_m e^{im\phi},$$
$$\Phi_{-m}(\phi) = C_{-m} e^{-im\phi}$$

(C_m is a normalizing constant).

The functions $\Phi_m(\phi) = C_m e^{im\phi}$ $(m = 0, \pm 1, \ldots)$ satisfy the orthogonality condition

$$\int_0^{2\pi} \Phi_m^*(\phi)\Phi_{m'}(\phi)d\phi = A_m \delta_{mm'},$$

where

$$A_m = 2\pi|C_m|^2, \quad \delta_{mm'} = \begin{cases} 1, & m' = m, \\ 0, & m' \neq m. \end{cases}$$

It is convenient to take $A_m = 1$, so that $C_m = 1/\sqrt{2\pi}$, i.e.

$$\Phi_m(\phi) = \frac{1}{\sqrt{2\pi}} e^{im\phi} \qquad (m = 0, \pm 1, \pm 2, \ldots).$$

Now let us solve equation (4) with $\nu = m^2$. If we put $\cos\theta = x$, equation (4) becomes the generalized equation of hypergeometric type (see §1)

$$\frac{d}{dx}\left[(1-x^2)\frac{d\Theta}{dx}\right] + \left(\mu - \frac{m^2}{1-x^2}\right)\Theta = 0, \tag{5}$$

with $\sigma(x) = 1 - x^2$, $\tilde{\tau}(x) = -2x$, $\tilde{\sigma}(x) = \mu(1-x^2) - m^2$.

The problem of finding bounded solutions of (5) on $(-1,1)$ leads to the same kind of eigenvalue problem as the one discussed in §9, since in the present case $\sigma(x)|_{x=\pm 1} = 0$ and $\tilde{\rho}(x) = 1$. We shall therefore use the method of §9.

We transform (5) to an equation of hypergeometric type by putting $\Theta(x) = \phi(x)y(x)$, where $\phi(x)$ is a solution of $\phi'/\phi = \pi(x)/\sigma(x)$ ($\pi(x)$ is a polynomial of degree at most 1). In the present case

$$\pi(x) = \pm\sqrt{k(1-x^2) + m^2 - \mu(1-x^2)},$$

where k is to be determined by the condition that the expression under the square root sign has a double zero. We obtain the following possibilities for $\pi(x)$:

$$\pi(x) = \begin{cases} \pm m & \text{for } k = \mu, \\ \pm mx & \text{for } k = \mu - m^2. \end{cases}$$

We must choose the form of $\pi(x)$ for which

$$\tau(x) = \tilde{\tau}(x) + 2\pi(x)$$

has a negative derivative and a zero on $(-1,1)$. For $m \geq 0$ these conditions are satisfied by

$$\tau(x) = -2(m+1)x,$$

which corresponds to

$$\pi(x) = -mx, \qquad \phi(x) = (1-x^2)^{m/2},$$
$$\lambda = \mu - m(m+1), \qquad \rho(x) = (1-x^2)^m.$$

The eigenvalues μ are determined by

$$\lambda + n\tau' + \frac{n(n-1)}{2}\sigma'' = 0,$$

which yields $\mu = \mu_n = l(l+1)$, where $l = m+n$ $(n = 0, 1, \ldots)$. The functions $y_n(x)$ have the form

$$y_n(x) = \frac{B_{nm}}{(1-x^2)^m} \frac{d^n}{dx^n}[(1-x^2)^{n+m}]$$

and are, up to constant factors, the Jacobi polynomials $P_n^{(m,m)}(x)$. Since $n = l - m$, where l is an integer such that $l \geq m$, we have, for $m \geq 0$,

$$\Theta(x) \equiv \Theta_{lm}(x) = C_{lm}(1-x^2)^{m/2} P_{l-m}^{(m,m)}(x). \tag{6}$$

Here C_{lm} are normalizing constants. The functions $\Theta_{lm}(x)$ evidently satisfy an orthogonality condition which follows from the orthogonality of the Jacobi polynomials:

$$\int_{-1}^{1} \Theta_{lm}(x)\Theta_{l'm}(x)dx = A_{lm}\delta_{ll'},$$

where

$$A_{lm} = C_{lm}^2 \int_{-1}^{1} \left[P_{l-m}^{(m,m)}(x)\right]^2 (1-x^2)^m dx.$$

It is convenient to take* $A_{lm} = 1$, which yields

$$C_{lm} = \frac{1}{2^m l!}\sqrt{\frac{2l+1}{2}(l-m)!(l+m)!}.$$

A different expression for the functions $\Theta_{lm}(x), m \geq 0$, follows from the properties of the Jacobi polynomials. From the differentiation formula (5.6) for Jacobi polynomials it follows that

$$P_{l-m}^{(m,m)}(x) = \frac{2^m l!}{(l+m)!}\frac{d^m}{dx^m}P_l(x),$$

where $P_l(x) = P_l^{(0,0)}(x)$ are the Legendre polynomials.

Hence for $m \geq 0$

$$\Theta_{lm}(x) = \sqrt{\frac{2l+1}{2}\frac{(l-m)!}{(l+m)!}}P_l^m(x),$$

where

$$P_l^m(x) = (1-x^2)^{m/2}\frac{d^m P_l(x)}{dx^m}$$

are the *associated Legendre functions of the first kind.*

* This choice of the sign of C_{lm} is not always used. We follow the notation of [B3].

We can obtain explicit expressions for the $\Theta_{lm}(x)$ by using the Rodrigues formulas for $P_l(x)$ and $P_{l-m}^{(m,m)}(x)$:

$$\Theta_{lm}(x) = \frac{(-1)^l}{2^l l!}\sqrt{\frac{2l+1}{2}\frac{(l-m)!}{(l+m)!}}(1-x^2)^{m/2}\frac{d^{l+m}}{dx^{l+m}}(1-x^2)^l, \tag{7}$$

$$\Theta_{lm}(x) = \frac{(-1)^{l-m}}{2^l l!}\sqrt{\frac{2l+1}{2}\frac{(l+m)!}{(l-m)!}}(1-x^2)^{-m/2}\frac{d^{l-m}}{dx^{l-m}}(1-x^2)^l. \tag{8}$$

We define $\Theta_{lm}(x)$ for $m < 0$ by using (7) and (8). This yields

$$\Theta_{l,-m}(x) = (-1)^m \Theta_{lm}(x). \tag{9}$$

It is then clear that, when $m < 0$, $\Theta_{lm}(x)$ is again a solution of (5). Therefore when $\mu = l(l+1)$ equation (2) has the bounded single-valued solutions

$$Y_{lm}(\theta,\phi) = \frac{1}{\sqrt{2\pi}}e^{im\phi}\Theta_{lm}(\cos\theta) \qquad (-l \le m \le l). \tag{10}$$

The functions $Y_{lm}(\theta,\phi)$ are the *spherical harmonics of order l.*

We give the spherical harmonics explicitly for the simplest cases:

$$Y_{l0}(\theta,\phi) = \sqrt{\frac{2l+1}{4\pi}}P_l(\cos\theta), \tag{11}$$

$$\begin{aligned} Y_{10}(\theta,\phi) &= \sqrt{\frac{3}{4\pi}}\cos\theta, \\ Y_{1,\pm1}(\theta,\phi) &= \pm\sqrt{\frac{3}{8\pi}}\sin\theta e^{\pm i\phi}. \end{aligned} \tag{12}$$

It is easily verified that the $Y_{lm}(\theta,\phi)$ satisfy the *orthogonality condition*

$$\int_\Omega Y_{lm}(\theta,\phi)Y^*_{l'm'}(\theta,\phi)d\Omega = \delta_{ll'}\delta_{mm'}. \tag{13}$$

The integration in (13) is with respect to solid angle,

$$d\Omega = \sin\theta\, d\theta d\phi \qquad (0 \le \theta \le \pi, 0 \le \phi \le 2\pi).$$

It is clear from (9) and (10) that

$$Y^*_{lm}(\theta,\phi) = \Theta_{lm}(\cos\theta)\Phi_{-m}(\phi) = (-1)^m Y_{l,-m}(\theta,\phi). \tag{14}$$

Hence we have explicitly obtained the functions $Y(\theta,\phi)$ that give the angle dependence of the bounded solutions $u = R(r)Y(\theta,\phi)$ of Laplace's equation.

To determine the functions $R(r)$ we reduce (1) to the Euler equation

$$r^2R'' + 2rR' - l(l+1)R = 0,$$

whose general solution is

$$R(r) = C_1 r^l + C_2 r^{-l-1}$$

(C_1 and C_2 are constants). Hence the functions $r^l Y_{lm}(\theta,\phi)$ and $r^{-l-1}Y_{lm}(\theta,\phi)$ are particular solutions of Laplace's equation; the former are used in solving interior boundary value problems for spherical regions, and the latter, for exterior problems. They are known as *solid spherical harmonics.*

Remark. A different approach to spherical harmonics, based on the representations of the rotation group, is discussed, for example, in [G2]. This approach is useful in the general theory of angular momentum in quantum mechanics.

2. Properties of spherical harmonics. We now obtain the basic properties of the spherical harmonics $Y_{lm}(\theta,\phi)$.

1) From the recursion relation for the Jacobi polynomials and the connection of $\Theta_{lm}(x)$ with $P_{l-m}^{(m,m)}(x)$, it is easy to derive the recursion relation (on l) for $Y_{lm}(\theta,\phi)$:

$$\cos\theta \cdot Y_{lm} = \left(\frac{(l+1)^2 - m^2}{4(l+1)^2 - 1}\right)^{1/2} Y_{l+1,m} + \left(\frac{l^2 - m^2}{4l^2 - 1}\right)^{1/2} Y_{l-1,m}.$$

This formula remains valid for $m < 0$, as is easily seen by using (14).

2) Differentiating (7), we obtain the differentiation formula

$$\frac{d\Theta_{lm}}{dx} = -\frac{mx}{1-x^2}\Theta_{lm} + \left(\frac{l(l+1) - m(m+1)}{1-x^2}\right)^{1/2} \Theta_{l,m+1}.$$

Replacing m by $-m$ and using (9), we can obtain another differentiation formula

$$\frac{d\Theta_{lm}}{dx} = -\frac{mx}{1-x^2}\Theta_{lm} - \left(\frac{l(l+1) - m(m-1)}{1-x^2}\right)^{1/2} \Theta_{l,m-1}.$$

In these formulas we take $\Theta_{lm}(x) = 0$ for $m = \pm(l+1)$.

By eliminating $d\Theta_{lm}/dx$ from the differentiation formulas, we obtain a recursion on m for $\Theta_{lm}(x)$:

$$\frac{2mx}{\sqrt{1-x^2}}\Theta_{lm} = \Big[\sqrt{l(l+1)-m(m+1)}\Theta_{l,m+1} + \sqrt{l(l+1)-m(m-1)}\Theta_{l,m-1}\Big].$$

By using (10), we can obtain a *differentiation formula for spherical harmonics*. Since

$$\frac{\partial Y_{lm}(\theta,\phi)}{\partial\theta} = -\sin\theta \frac{e^{im\phi}}{\sqrt{2\pi}} \frac{d\Theta_{lm}(x)}{dx}\bigg|_{x=\cos\theta},$$

the differentiation formulas for $\Theta_{lm}(x)$ can be written in the form

$$e^{\pm i\phi}\left(\pm\frac{\partial Y_{lm}}{\partial\theta} + m\cot\theta\cdot Y_{lm}\right) = \sqrt{l(l+1)-m(m\pm1)}Y_{l,m\pm1}. \tag{15}$$

Here we are to put $Y_{lm}(\theta,\phi) = 0$ if $m = \pm(l+1)$.

The following *differentiation formula* can be derived from the explicit form of the spherical harmonics:

$$\frac{\partial Y_{lm}(\theta,\phi)}{\partial\phi} = imY_{lm}(\theta,\phi). \tag{16}$$

3. Integral representation. Let us obtain an *integral representation* for the $Y_{lm}(\theta,\phi)$. Starting from the expression (7) for $\Theta_{lm}(x)$, let us represent $(d^{l+m}/dx^{l+m})\,(1-x^2)^l$ by Cauchy's integral formula:

$$\frac{d^{l+m}}{dx^{l+m}}(1-x^2)^l = \frac{(l+m)!}{2\pi i}\int_C \frac{(1-s^2)^l}{(s-x)^{l+m+1}}ds$$

(C is a contour surrounding the point $s = x$). Let us take C to be a circumference with center at $s = x$ and radius $\sqrt{1-x^2}$. Then, putting $s = x + \sqrt{1-x^2}e^{i\alpha}$, we obtain

$$\frac{d^{l+m}}{dx^{l+m}}(1-x^2)^l = \frac{(-2)^l(l+m)!}{2\pi}(1-x^2)^{-m/2}\int_0^{2\pi} e^{-im\alpha}(x+i\sqrt{1-x^2}\sin\alpha)^l d\alpha.$$

Substituting this into (7) and using (10), we obtain the integral representation for $Y_{lm}(\theta, \phi)$:

$$Y_{lm}(\theta,\phi) = B_{lm} \int_0^{2\pi} e^{-im(\alpha-\phi)}(\cos\theta + i\sin\theta\sin\alpha)^l d\alpha$$
$$= B_{lm} \int_{-\phi}^{2\pi-\phi} e^{-im\alpha}[\cos\theta + i\sin\theta\sin(\alpha+\phi)]^l d\alpha.$$

where

$$B_{lm} = \frac{1}{4\pi l!}\left(\frac{2l+1}{\pi}(l-m)!(l+m)!\right)^{1/2}.$$

Since the integral of a periodic function is the same over any interval whose length is a period, we have

$$Y_{lm}(\theta,\phi) = B_{lm} \int_0^{2\pi} e^{-im\alpha}[\cos\theta + i\sin\theta\sin(\alpha+\phi)]^l d\alpha. \tag{17}$$

4. Connection between homogeneous harmonic polynomials and spherical harmonics. When we solved Laplace's equation $\Delta u = 0$ in spherical coordinates, we found the solutions that are bounded as $r \to 0$:

$$u_{lm}(r,\theta,\phi) = r^l Y_{lm}(\theta,\phi).$$

We can write the integral representation (17) for $u_{lm}(r,\theta,\phi)$ in Cartesian coordinates $x = r\sin\theta\cos\phi$, $y = r\sin\theta\sin\phi$, $z = r\cos\theta$:

$$u_{lm}(r,\theta,\phi) = B_{lm} \int_0^{2\pi} e^{-im\alpha}[r\cos\theta + ir\sin\theta\sin(\alpha+\phi)]^l d\alpha$$
$$= B_{lm} \int_0^{2\pi} e^{-im\alpha}(z + ix\sin\alpha + iy\cos\alpha)^l d\alpha.$$

It is then clear that the functions $u_{lm}(r,\theta,\phi)$ are the homogeneous polynomials of degree l in x, y, z.

Recall that a homogeneous polynomial of degree l has the form

$$u_l(x,y,z) = \sum_{l_1,l_2,l_3} C_{l_1 l_2 l_3} x^{l_1} y^{l_2} z^{l_3},$$

where the summation is over all nonnegative indices $l_1 \geq 0, l_2 \geq 0, l_3 \geq 0$ whose sum is l. For example, $r^2 = x^2 + y^2 + z^2$ is a homogeneous polynomial.

Let us find the number of linearly independent homogeneous polynomials of degree l. It is enough to find all combinations of l_1 and l_2, since for a given l the value of l_3 is then determined: $l_3 = l - l_1 - l_2$. For a given l_1 the value of l_2 varies from $l_2 = 0$ to $l_2 = l - l_1$, i.e. it takes $l - l_1 + 1$ values. Hence the number of linearly independent homogeneous polynomials of degree l is

$$N_1 = \sum_{l_1=0}^{l} (l - l_1 + 1) = \frac{(l+1)(l+2)}{2}.$$

A homogeneous polynomial that satisfies Laplace's equation is a *homogeneous harmonic polynomial.* The function $r^l Y_{lm}(\theta,\phi)$ is an example.

From the homogeneous polynomials r^2 and $r^{l-2n} Y_{l-2n,m}(\theta,\phi)$ we can construct homogeneous polynomials of degree l,

$$u_{lmn}(x,y,z) = (r^2)^n r^{l-2n} Y_{l-2n,m}(\theta,\phi) = r^l Y_{l-2n,m}(\theta,\phi).$$

Here the indices m and n can take integral values satisfying

$$0 \leq 2n \leq l, -(l-2n) \leq m \leq l-2n.$$

Since the spherical harmonics $Y_{l-2n,m}(\theta,\phi)$ are linearly independent (as follows from their orthogonality), the homogeneous polynomials $u_{lmn}(x,y,z)$ are linearly independent. For a given $l - 2n$ there are $2(l-2n)+1$ possible values of m. Hence the total number of these homogeneous polynomials is

$$\sum_n [2(l-2n)+1] = \frac{(l+1)(l+2)}{2}.$$

Since we have constructed as many homogeneous polynomials as the total number of linearly independent homogeneous polynomials of degree l, every homogeneous polynomial of degree l can be represented as a linear combination of $r^l Y_{l-2n,m}(\theta,\phi)$, i.e.

$$u_l(x,y,z) = r^l \sum_{m,n} C_{mn} Y_{l-2n,m}(\theta,\phi). \tag{18}$$

We have thus obtained an expansion of an arbitrary homogeneous polynomial in terms of spherical harmonics. It is easy to show by using (18) that *every homogeneous harmonic polynomial of degree l is a linear combination of the homogeneous harmonic polynomials $r^l Y_{lm}(\theta, \phi)$.*

In fact, let $u_l(x, y, z)$ be a homogeneous harmonic polynomial, $\Delta u_l = 0$. If we apply the Laplacian $\Delta_r + (1/r^2)\Delta_{\theta,\phi}$ to (18), we obtain

$$\begin{aligned}\Delta u_l &= r^{l-2} \sum_{m,n} [l(l+1) - (l-2n)(l-2n+1)] C_{mn} Y_{l-2n,m}(\theta, \phi) \\ &= r^{l-2} \sum_{m,n} 2n(2l-2n+1) C_{mn} Y_{l-2n,m}(\theta, \phi) = 0.\end{aligned}$$

Since the spherical harmonics $Y_{l-2n,m}(\theta, \phi)$ are linearly independent, we obtain

$$2n(2l-2n+1)C_{mn} = 0,$$

that is, $C_{mn} = 0$ for $n > 0$, as required.

5. Generalized spherical harmonics. Under rotations of the coordinate system, a homogeneous polynomial becomes a homogeneous polynomial of the same degree. On the other hand, the Laplacian is invariant under rotations, $\Delta_{xyz} = \Delta_{x'y'z'}$. Therefore every homogeneous harmonic polynomial becomes a homogeneous harmonic polynomial of the same degree under a rotation. Hence

$$u_{lm}(x, y, z) = \sum_{m'} D^l_{mm'} u_{lm'}(x', y', z'),$$

where

$$u_{lm}(x, y, z) = r^l Y_{lm}(\theta, \phi).$$

Consequently

$$Y_{lm}(\theta, \phi) = \sum_{m'} D^l_{mm'} Y_{lm'}(\theta', \phi'). \tag{19}$$

Thus the linear combinations of the $Y_{lm}(\theta, \phi)$ with a given l form a $(2l+1)$-dimensional space that is invariant under rotation.

The coefficients $D^l_{mm'}$ evidently depend on the parameters that determine the rotation. Every rotation of the coordinate system about the origin is completely determined by three real parameters. In fact, every rotation can be described by giving the direction of the rotation axis (two parameters) and the angle of rotation (one parameter). A more commonly used set of parameters that determine the rotation are the *Euler angles* α, β, γ. Every rotation can be realized by three successive rotations about the coordinate axes:

a) a rotation about the z axis through the angle α;

b) a rotation about the new y axis through the angle β;

c) a rotation about the new z axis through the angle* γ.

Consequently

$$D^l_{mm'} = D^l_{mm'}(\alpha, \beta, \gamma).$$

We shall denote the matrix with elements $D^l_{mm'}(\alpha, \beta, \gamma)$ by $D(\alpha, \beta, \gamma)$ and call it a *finite rotation matrix*.

Every rotation is uniquely specified by the Euler angles if they are taken so that $0 \le \alpha < 2\pi, 0 \le \beta \le \pi, 0 \le \gamma < 2\pi$. If the Euler angles are not in these intervals, one must keep in mind that a rotation with angles $(\alpha+2\pi n_1, \beta+2\pi n_2, \gamma+2\pi n_3)$ coincides with the rotation (α, β, γ) if n_1, n_2, n_3 are integers. Therefore

$$D(\alpha + 2\pi n_1, \beta + 2\pi n_2, \gamma + 2\pi n_3) = D(\alpha, \beta, \gamma).$$

Moreover, we observe that the rotation (α, β, γ) is equivalent to the rotation $(\pi + \alpha, -\beta, \pi + \gamma)$.

The inverse rotation is described by the angles

$$\alpha_1 = -\gamma, \quad \beta_1 = -\beta, \quad \gamma_1 = -\alpha,$$

and is equivalent to the rotation

$$(\pi + \alpha_1, -\beta_1, \pi + \gamma_1) = (\pi - \gamma,\ \beta,\ \pi - \alpha).$$

Consequently the matrix of the inverse rotation is the matrix of the rotation $(\pi - \gamma, \beta, \pi - \alpha)$, i.e.

$$D^{-1}(\alpha, \beta, \gamma) = D(\pi - \gamma, \beta, \pi - \alpha).$$

The functions $D^l_{mm'}(\alpha, \beta, \gamma)$ are known as *generalized spherical harmonics*, since in many special cases they coincide with ordinary spherical harmonics. They are also known as *Wigner's D-functions*. They are extensively used in quantum mechanics. (See, for example, [B2], [D1], [E1], [F1], [R2], [V1]. Note that in [B2] the functions $D^l_{mm'}(\alpha, \beta, \gamma)$ are denoted by $D^l_{m',m}(\alpha, \beta, \gamma)$.)

* Sometimes the rotation through the angle β is taken about the new x axis instead of about the new y axis. The Euler angles α', β', γ' defined in this way are connected with α, β, γ by $\alpha' = \alpha + \pi/2$, $\beta' = \beta$, $\gamma' = \gamma - \pi/2$.

We shall present a number of basic properties of generalized spherical harmonics. Since the element of solid angle is invariant under a rotation of coordinates, i.e. $d\Omega = d\Omega'$, the orthogonality conditions

$$\int Y_{lm}(\theta, \phi) Y^*_{lm_1}(\theta, \phi) d\Omega = \delta_{mm_1},$$

$$\int Y_{lm'}(\theta', \phi') Y^*_{lm'_1}(\theta', \phi') d\Omega' = \delta_{m'm'_1}$$

(the superscript * means "complex conjugate") imply the formula

$$\sum_{m'} D^l_{mm'}(\alpha, \beta, \gamma)[D^l_{m_1 m'}(\alpha, \beta, \gamma)]^* = \delta_{mm_1},$$

i.e. the matrix $D^\dagger(\alpha, \beta, \gamma)$, the transpose conjugate of $D(\alpha, \beta, \gamma)$, is equal to $D^{-1}(\alpha, \beta, \gamma)$. This means that $D(\alpha, \beta, \gamma)$ is a unitary matrix (see [G1] or [S6]). From (19) we obtain

$$Y_{lm'}(\theta', \phi') = \sum_m [D^l_{mm'}(\alpha, \beta, \gamma)]^* Y_{lm}(\theta, \phi). \tag{20}$$

If we use the equations $D^{-1}(\alpha, \beta, \gamma) = D(\pi - \gamma, \beta, \pi - \alpha)$ and $D^{-1}(\alpha, \beta, \gamma) = D^\dagger(\alpha, \beta, \gamma)$ we obtain the following formula:

$$D^l_{mm'}(\pi - \gamma, \beta, \pi - \alpha) = [D^l_{m'm}(\alpha, \beta, \gamma)]^*. \tag{21}$$

Another elementary property of the generalized spherical harmonics is easily obtained from property (14) of the $Y_{lm}(\theta, \phi)$:

$$D^l_{mm'}(\alpha, \beta, \gamma) = (-1)^{m-m'}[D^l_{-m,-m'}(\alpha, \beta, \gamma)]^*. \tag{22}$$

6. Addition theorem. We present a useful relation for spherical harmonics, which is known as the addition theorem. To obtain it, we consider two arbitrary vectors $\mathbf{r}_1$ and $\mathbf{r}_2$ whose directions are specified by the spherical coordinates (θ_1, ϕ_1) and (θ_2, ϕ_2). Let the angle between the vectors be ω. According to (20) and (19) we have

$$Y_{lm'}(\theta'_1, \phi'_1) = \sum_m (D^l_{mm'})^* Y_{lm}(\theta_1, \phi_1), \tag{23}$$

$$Y_{lm}(\theta_2, \phi_2) = \sum_{m'} D^l_{mm'} Y_{lm'}(\theta'_2, \phi'_2). \tag{24}$$

As the result of a rotation of the coordinate system let the direction of the z' axis coincide with that of the radius vector $\mathbf{r}_2$. Then $\theta'_1 = \omega$, $\theta'_2 = 0$.

We set $m' = 0$ in (23). Since by virtue of (11)

$$Y_{l0}(\theta_1', \phi_1') = \sqrt{\frac{2l+1}{4\pi}} P_l(\cos\omega),$$

we have

$$\sqrt{\frac{2l+1}{4\pi}} P_l(\cos\omega) = \sum_m (D^l_{m0})^* Y_{lm}(\theta_1, \phi_1). \tag{23a}$$

According to (7), at $\theta_2' = 0$ we have $Y_{lm'}(0, \phi_2') = 0$ if $m' \neq 0$. Therefore, by (11) and (5.12), formula (24) takes the form

$$Y_{lm}(\theta_2, \phi_2) = D^l_{m0} Y_{l0}(0, \phi_2) = D^l_{m0}\sqrt{\frac{2l+1}{4\pi}}. \tag{24a}$$

Comparing (23a) and (24a) yields

$$P_l(\cos\omega) = \frac{4\pi}{2l+1} \sum_{m=-l}^{l} Y_{lm}(\theta_1, \phi_1) Y^*_{lm}(\theta_2, \phi_2). \tag{25}$$

Relation (25) is known as the *addition theorem for spherical harmonics.* It has numerous applications, for example in the theory of atomic spectra. Formula (25) is very often applied in order to expand $1/|\mathbf{r}_1 - \mathbf{r}_2|$ in a series of the spherical harmonics $Y_{lm}(\theta_1, \phi_1)$ and $Y_{lm}(\theta_2, \phi_2)$. Since (see (5.14))

$$\frac{1}{|\mathbf{r}_1 - \mathbf{r}_2|} = \sum_{l=0}^{\infty} \frac{r_<^l}{r_>^{l+1}} P_l(\cos\omega),$$

we find from the addition theorem that

$$\frac{1}{|\mathbf{r}_1 - \mathbf{r}_2|} = \sum_{l=0}^{\infty} \sum_{m=-l}^{l} \frac{4\pi}{2l+1} \frac{r_<^l}{r_>^{l+1}} Y_{lm}(\theta_1, \phi_1) Y^*_{lm}(\theta_1, \phi_2). \tag{26}$$

Here $r_< = \min(r_1, r_2)$ and $r_> = \max(r_1, r_2)$.

Example 1. Consider the potential

$$u(\mathbf{r}) = \int_V \frac{\rho(\mathbf{r}')}{|\mathbf{r} - \mathbf{r}'|} d\tau', \tag{27}$$

of a distribution of electric charge of density $\rho(\mathbf{r})$ inside a volume V. To calculate the potential $u(\mathbf{r})$ at great distances from V we need its expansion in powers of $1/r$, taking the origin inside V. Using (26) with $\mathbf{r}_1 = r$, $\mathbf{r}_2 = r'$ and $r > r'$, we can represent (24) in the form

$$u(\mathbf{r}) = \sum_{l=0}^{\infty} \sum_{m=-l}^{l} \frac{Q_{lm}}{r^{l+1}} Y_{lm}(\theta, \phi), \tag{28}$$

where

$$Q_{lm} = \frac{4\pi}{2l+1} \int_V (r')^l \rho(\mathbf{r}') Y_{lm}(\theta', \phi') d\tau'. \tag{29}$$

Formula (28) is known as the *multipole expansion of the potential.*

If V is a sphere $0 < r' < a$ and $\rho(\mathbf{r}') = \rho(r')$, the integral (29) is easily evaluated: $Q_{lm} = \sqrt{4\pi} Q \delta_{l0} \delta_{m0}$, where Q is the total charge. In this case, as one would expect, $u(\mathbf{r}) = Q/r$.

Example 2. Let us use (25) to solve the *first interior boundary value problem (Dirichlet problem) for Laplace's equation in a spherical region:*

$$\Delta u = 0, \quad u(r, \theta, \phi)|_{r=a} = f(\theta, \phi).$$

We look for a solution by separation of variables in the form of a series of spherical harmonics $r^l Y_{lm}(\theta, \phi)$:

$$u(r, \theta, \phi) = \sum_{l,m} C_{lm} \left(\frac{r}{a}\right)^l Y_{lm}(\theta, \phi). \tag{30}$$

The coefficients C_{lm} are obtained from the boundary conditions on $r = a$ and the orthogonality of the functions $Y_{lm}(\theta, \phi)$. Then

$$C_{lm} = \int f(\theta', \phi') Y^*_{lm}(\theta', \phi') d\Omega'.$$

The solution (30) can also be represented as an integral. To obtain this, we substitute the expression for C_{lm} into (30), interchange summation and integration, and then carry out the summation on m by means of the addition formula:

$$\begin{aligned} u(r, \theta, \phi) &= \int d\Omega' f(\theta', \phi') \left[\sum_{l,m} \left(\frac{r}{a}\right)^l Y_{lm}(\theta, \phi) Y^*_{lm}(\theta', \phi') \right] \\ &= \int d\Omega' f(\theta', \phi') \left[\sum_{l} \frac{2l+1}{4\pi} \left(\frac{r}{a}\right)^l P_l(\mu) \right]. \end{aligned}$$

Here μ is the cosine of the angle between the directions (θ, ϕ) and (θ', ϕ'):

$$\mu = \cos\theta\cos\theta' + \sin\theta\sin\theta'\cos(\phi - \phi').$$

To carry out the summation on l, we use the generating function of the Legendre polynomials:

$$\sum_{l=0}^{\infty} t^l P_l(\mu) = \frac{1}{\sqrt{1 - 2t\mu + t^2}}.$$

Since

$$\begin{aligned}\sum_l (2l+1)t^l P_l(\mu) &= 2\sqrt{t}\sum_l \left(l + \frac{1}{2}\right) t^{l-1/2} P_l(\mu) \\ &= 2\sqrt{t}\frac{d}{dt}\left[\sqrt{t}\sum_l t^l P_l(\mu)\right] \\ &= 2\sqrt{t}\frac{d}{dt}\left(\frac{\sqrt{t}}{\sqrt{1 - 2t\mu + t^2}}\right) = \frac{1 - t^2}{(1 - 2t\mu + t^2)^{3/2}}\end{aligned}$$

we have

$$\sum_l (2l+1)\left(\frac{r}{a}\right)^l P_l(\mu) = \frac{1 - (r/a)^2}{[1 - 2\mu r/a + (r/a)^2]^{3/2}}$$

and consequently the solution of the first boundary value problem for Laplace's equation in a spherical region can be represented in the form

$$u(r, \theta, \phi) = \frac{1}{4\pi}\int d\Omega' f(\theta', \phi') \frac{1 - (r/a)^2}{[1 - 2\mu r/a + (r/a)^2]^{3/2}}.$$

7. Explicit expressions for generalized spherical harmonics. We now obtain explicit expressions for the functions $D^l_{mm'}(\alpha, \beta, \gamma)$. Let us perform two successive rotations determined by the parameters $\alpha_1, \beta_1, \gamma_1$ and $\alpha_2, \beta_2, \gamma_2$; let the equivalent single rotation be determined by the parameters α, β, γ. Furthermore let the first rotation transform the coordinates of some given vector from (θ, ϕ) to (θ_1, ϕ_1), and let the second rotation transform these coordinates to (θ', ϕ'). Then

$$Y_{lm}(\theta, \phi) = \sum_{m_1} D^l_{mm_1}(\alpha_1, \beta_1, \gamma_1) Y_{lm_1}(\theta_1, \phi_1),$$

$$Y_{lm_1}(\theta_1, \phi_1) = \sum_{m'} D^l_{m_1 m'}(\alpha_2, \beta_2, \gamma_2) Y_{lm'}(\theta', \phi').$$

On the other hand,

$$Y_{lm}(\theta,\phi)=\sum_{m'} D^l_{mm'}(\alpha,\beta,\gamma)Y_{lm'}(\theta',\phi').$$

If we combine these expansions, we obtain, by the linear independence of the spherical harmonics,

$$D^l_{mm'}(\alpha,\beta,\gamma)=\sum_{m_1} D^l_{mm_1}(\alpha_1,\beta_1,\gamma_1)D^l_{m_1m'}(\alpha_2,\beta_2,\gamma_2),$$

i.e.

$$D(\alpha,\beta,\gamma)=D(\alpha_1,\beta_1,\gamma_1)D(\alpha_2,\beta_2,\gamma_2),$$

so that when two rotations are performed successively their matrices are multiplied in inverse order. There is a similar result for the effect of several rotations of the coordinate system. It follows from the definition of the Euler angles and the preceding considerations that in order to calculate the generalized spherical functions $D^l_{mm'}(\alpha,\beta,\gamma)$ we need only obtain expressions for them when the rotations are about the z and y axes. Let $C^l_{mm'}(\alpha)$ and $d^l_{mm'}(\beta)$ be the generalized spherical harmonics corresponding to a rotation through the angle α about the z axis and a rotation through the angle β about the y axis. Then we have

$$D^l_{mm'}(\alpha,\beta,\gamma)=\sum_{m_1m_2} C^l_{mm_1}(\alpha)d^l_{m_1m_2}(\beta)C^l_{m_2m'}(\gamma).$$

We now look for explicit expressions for the functions $C^l_{mm'}(\alpha)$. Under a rotation through the angle α around the z axis, the spherical coordinates of a given vector become $\theta'=\theta$, $\phi'=\phi-\alpha$. Therefore

$$Y_{lm}(\theta,\phi)=Y_{lm}(\theta',\phi'+\alpha)=e^{im\alpha}Y_{lm}(\theta',\phi').$$

On the other hand,

$$Y_{lm}(\theta,\phi)=\sum_{m'} C^l_{mm'}(\alpha)Y_{lm'}(\theta',\phi').$$

Hence

$$C^l_{mm'}(\alpha)=e^{im\alpha}\delta_{mm'}$$

and consequently

$$D^l_{mm'}(\alpha,\beta,\gamma)=e^{i(m\alpha+m'\gamma)}d^l_{mm'}(\beta). \tag{31}$$

We now find the generalized spherical harmonics $d^l_{mm'}(\beta)$ corresponding to a rotation of the coordinate system through the angle β about the y axis. In this case

$$Y_{lm}(\theta,\phi)=\sum_{m'} d^l_{mm'}(\beta)Y_{lm'}(\theta',\phi'). \tag{32}$$

The new coordinates (x', y', z') are connected with the old coordinates (x, y, z) by

$$\begin{aligned} x &= x'\cos\beta + z'\sin\beta,\\ y &= y',\\ z &= z'\cos\beta - x'\sin\beta. \end{aligned}$$

Changing to spherical coordinates, we find the connection between (θ,ϕ) and (θ',ϕ'):

$$\begin{aligned} \sin\theta\cos\phi &= \sin\theta'\cos\phi'\cos\beta + \cos\theta'\sin\beta,\\ \sin\theta\sin\phi &= \sin\theta'\sin\phi',\\ \cos\theta &= \cos\theta'\cos\beta - \sin\theta'\cos\phi'\sin\beta. \end{aligned} \tag{33}$$

To determine $d^l_{mm'}(\beta)$ we find a differential relation between these functions. Since it is simplest to differentiate with respect to β and ϕ' on the right-hand side of (32), we consider this equation for a fixed θ', taking θ and ϕ as functions of β and ϕ'. Consequently

$$\begin{aligned} \frac{\partial Y_{lm}(\theta,\phi)}{\partial\beta} &= \frac{\partial Y_{lm}}{\partial\theta}\frac{\partial\theta}{\partial\beta}+\frac{\partial Y_{lm}}{\partial\phi}\frac{\partial\phi}{\partial\beta},\\ \frac{\partial Y_{lm}(\theta,\phi)}{\partial\phi'} &= \frac{\partial Y_{lm}}{\partial\theta}\frac{\partial\theta}{\partial\phi'}+\frac{\partial Y_{lm}}{\partial\phi}\frac{\partial\phi}{\partial\phi'}. \end{aligned}$$

The derivatives $\partial\theta/\partial\beta, \partial\theta/\partial\phi', \partial\phi/\partial\beta$, and $\partial\phi/\partial\phi'$ are calculated by using (33). Differentiating the last of these equations yields

$$\frac{\partial\theta}{\partial\beta}=\cos\phi,\qquad \frac{\partial\theta}{\partial\phi'}=-\sin\beta\sin\phi.$$

The derivatives $\partial\phi/\partial\beta$ and $\partial\phi/\partial\phi'$ are easily found by differentiating the second and first equations in (33):

$$\begin{aligned} \frac{\partial\phi}{\partial\beta} &= -\cot\theta\sin\phi,\\ \frac{\partial\phi}{\partial\phi'} &= -\sin\beta\cot\theta\cos\phi+\cos\beta. \end{aligned}$$

Hence

$$\frac{\partial Y_{lm}(\theta,\phi)}{\partial\beta} = \cos\phi\frac{\partial Y_{lm}(\theta,\phi)}{\partial\theta} - im\cot\theta\sin\phi\, Y_{lm}(\theta,\phi),$$

$$\frac{\partial Y_{lm}(\theta,\phi)}{\partial\phi'} = -\sin\beta\left[\sin\phi\frac{\partial Y_{lm}(\theta,\phi)}{\partial\theta} + im\cot\theta\cos\phi\, Y_{lm}(\theta,\phi)\right] + im\cos\beta\, Y_{lm}(\theta,\phi).$$

To calculate $\partial Y_{lm}(\theta,\phi)/\partial\theta$ we use the differentiation formula (15). Since (15) involves $e^{\pm i\phi}\partial Y_{lm}/\partial\theta$, and the expressions for $\partial Y_{lm}/\partial\beta$ and $\partial Y_{lm}/\partial\phi'$ involve $\cos\phi\cdot\partial Y_{lm}/\partial\theta$ and $\sin\phi\cdot\partial Y_{lm}/\partial\theta$, in order to use (15) we must first form the corresponding linear combinations of $\partial Y_{lm}/\partial\beta$ and $\partial Y_{lm}/\partial\phi'$. We have

$$\begin{aligned}\frac{\partial Y_{lm}(\theta,\phi)}{\partial\beta} &\mp \frac{i}{\sin\beta}\frac{\partial Y_{lm}(\theta,\phi)}{\partial\phi'}\\ &= e^{\pm i\phi}\left[\frac{\partial Y_{lm}}{\partial\theta} \mp m\cot\theta\, Y_{lm}\right] \pm m\cot\beta\, Y_{lm}\\ &= \mp\sqrt{l(l+1)-m(m\pm 1)}\, Y_{l,m\pm 1}(\theta,\phi) \pm m\cot\beta\, Y_{lm}(\theta,\phi).\end{aligned}$$

If we use the expansion (32) for $Y_{lm}(\theta,\phi)$ and $Y_{l,m\pm1}(\theta,\phi)$ and equate the coefficients of $Y_{lm'}(\theta',\phi')$ on the left-hand and right-hand sides of the equation, we obtain the required differential relation for $d^l_{mm'}(\beta)$:

$$\frac{d}{d\beta}d^l_{mm'} \pm \frac{m'-m\cos\beta}{\sin\beta}d^l_{mm'} = \mp\sqrt{l(l+1)-m(m\pm1)}\, d^l_{m\pm1,m'}. \tag{34}$$

Here we are to take $d^l_{\pm(l+1),m'}(\beta) = 0$. By using (34) and the condition $d^l_{mm'}(0) = \delta_{mm'}$, which follows from (32) for $\beta = 0$, we can determine the functions $d^l_{mm'}(\beta)$.

Let us rewrite (34) in a more compact form. If we multiply it by

$$\exp\left(\pm\int\frac{m'-m\cos\beta}{\sin\beta}d\beta\right) = (1-\cos\beta)^{\pm(m'-m)/2}(1+\cos\beta)^{\mp(m'+m)/2},$$

we obtain

$$\begin{aligned}\frac{d}{d\beta}&\left[(1-\cos\beta)^{\pm(m'-m)/2}(1+\cos\beta)^{\mp(m'+m)/2}d^l_{mm'}(\beta)\right]\\ &= \mp\sqrt{l(l+1)-m(m\pm1)}(1-\cos\beta)^{\pm(m'-m)/2}\\ &\qquad\times(1+\cos\beta)^{\mp(m'+m)/2}d^l_{m\pm1,m'}(\beta).\end{aligned} \tag{35}$$

If we use the upper signs and take $m = l$, we find

$$(1-\cos\beta)^{(m'-l)/2}(1+\cos\beta)^{-(m'+l)/2} d^l_{lm'}(\beta) = \text{ const.}$$

Hence

$$d^l_{lm'}(\beta) = C_{lm'}(1-\cos\beta)^{(l-m')/2}(1+\cos\beta)^{(l+m')/2}$$

($C_{lm'}$ are constants). When $m < l$ the functions $d^l_{mm'}(\beta)$ can be expressed recursively in terms of $d^l_{lm'}(\beta)$ by taking the lower signs in (35). After the change of variable

$$x = \cos\beta, \quad v_{mm'}(x) = (1-x)^{(m-m')/2}(1+x)^{(m+m')/2} d^l_{mm'}(\beta),$$

we obtain

$$v_{m-1,m'} = -\frac{1}{\sqrt{l(l+1)-m(m-1)}}\frac{dv_{mm'}}{dx},$$

whence

$$v_{mm'} = (-1)^{l-m}\prod_{s=m+1}^{l}\frac{1}{\sqrt{l(l+1)-s(s-1)}}\frac{d^{l-m}}{dx^{l-m}}v_{lm'},$$

i.e.

$$\begin{aligned} d^l_{mm'}(\beta) = C_{lm'}&\frac{(-1)^{l-m}(1-x)^{(m'-m)/2}(1+x)^{-(m'+m)/2}}{\prod\limits_{s=m+1}^{l}\sqrt{l(l+1)-s(s-1)}} \\ &\times\frac{d^{l-m}}{dx^{l-m}}\left[(1-x)^{l-m'}(1+x)^{l+m'}\right]. \end{aligned} \tag{36}$$

To determine $C_{lm'}$ we use the equation $d^l_{m'm'}(0) = 1$. Carrying out the differentiations in (36) by Leibniz's rule, we obtain

$$d^l_{m'm'}(0) = C_{lm'}\frac{2^{-m'}2^{l+m'}(l-m')!}{\prod\limits_{s=m'+1}^{l}\sqrt{l(l+1)-s(s-1)}} = 1.$$

This yields

$$C_{lm'} = \frac{\prod\limits_{s=m'+1}^{l}\sqrt{l(l+1)-s(s-1)}}{2^l(l-m')!}.$$

Since

$$\prod_{s=m+1}^{l} [l(l+1) - s(s-1)] = \prod_{s=m+1}^{l} (l+s)(l-s+1) = \frac{(2l)!(l-m)!}{(l+m)!},$$

we finally obtain

$$\begin{aligned} d^l_{mm'}(\beta) &= \frac{(-1)^{l-m}}{2^l(l-m)!}\sqrt{\frac{(l+m)!(l-m)!}{(l+m')!(l-m')!}} \\ &\times (1-x)^{(m'-m)/2}(1+x)^{-(m'+m)/2}\frac{d^{l-m}}{dx^{l-m}}\left[(1-x)^{l-m'}(1+x)^{l+m'}\right]. \end{aligned} \tag{37}$$

Observe that the $d^l_{mm'}(\beta)$ are real. They can be expressed in terms of Jacobi polynomials:

$$\begin{aligned} d^l_{mm'}(\beta) &= \frac{1}{2^m}\sqrt{\frac{(l+m)!(l-m)!}{(l+m')!(l-m')!}} \\ &\times (1-x)^{(m-m')/2}(1+x)^{(m+m')/2}P^{(m-m',m+m')}_{l-m}(x), \end{aligned}$$

where $x = \cos\beta$. The $d^l_{mm'}(\beta)$ can be written in a different form by using the symmetry relations that follow from (21), (22), (31), and the fact that $d^l_{mm'}$ is real:

$$d^l_{mm'} = (-1)^{m-m'} d^l_{m'm}, \quad d^l_{mm'} = (-1)^{m-m'} d^l_{-m,-m'}. \tag{38}$$

By using (38) we can always ensure that

$$m - m' \geq 0, \quad m + m' \geq 0.$$

Comparing (37) for $m' = 0$ with (8), we have

$$d^l_{m0}(\beta) = \left(\frac{2}{2l+1}\right)^{1/2} \Theta_{lm}(x),$$

whence

$$\begin{aligned} D^l_{m0}(\alpha,\beta,\gamma) &= \left(\frac{4\pi}{2l+1}\right)^{1/2} Y_{lm}(\beta,\alpha), \\ D^l_{00}(\alpha,\beta,\gamma) &= P_l(\cos\beta). \end{aligned} \tag{39}$$

By using (21) we can obtain a similar equation

$$D^l_{0m}(\alpha,\beta,\gamma) = (-1)^m \left(\frac{4\pi}{2l+1}\right)^{1/2} Y_{lm}(\beta,\gamma).$$

§ 11 Functions of the second kind

1. Integral representations. As we showed in §3, the differential equation of the classical orthogonal polynomials has solutions of the form

$$y(z) = \frac{C_n}{\rho(z)} \int_C \frac{\sigma^n(s)\rho(s)}{(s-z)^{n+1}} ds, \tag{1}$$

where the contour C is chosen so that

$$\left. \frac{\sigma^{n+1}(s)\rho(s)}{(s-z)^{n+2}} \right|_{s_1}^{s_2} = 0 \tag{2}$$

(s_1 and s_2 are the endpoints of the contour). A closed contour surrounding the point $s = z$, with $C_n = B_n n!/(2\pi i)$, yields the classical orthogonal polynomials $y_n(z)$ that are orthogonal on (a, b). When $z \notin [a, b]$, another possibility is to take C to be the line segment joining $s_1 = a$ and $s_2 = b$. In this case (2) is satisfied, by (5.17). With $C_n = B_n n!$, the corresponding solution is a *function of the second kind*, and is denoted by $Q_n(z)$:

$$Q_n(z) = \frac{B_n n!}{\rho(z)} \int_a^b \frac{\sigma^n(s)\rho(s)}{(s-z)^{n+1}} ds. \tag{3}$$

It is easy to see from the explicit form of $\rho(z)$ (see §5) that this function can have branch points at $z = a$ and $z = b$. In order to have a single-valued function $Q_n(z)$ it is necessary to introduce a cut in the complex plane, for example along $(a, +\infty)$, that is, from $z = a$ to the right along the real axis. Then we may suppose that $\rho(x + i0) = \rho(x)$ for $x \in (a, b)$.

If we integrate by parts n times in (3), we obtain an integral representation that connects $Q_n(z)$ with the polynomials $y_n(z)$:

$$Q_n(z) = \frac{B_n(n-1)!}{\rho(z)} \left\{ -\left. \frac{\sigma^n(s)\rho(s)}{(s-z)^n} \right|_a^b + \int_a^b \frac{d/ds[\sigma^n(s)\rho(s)]}{(s-z)^n} ds \right\}$$

$$= \dots = \frac{B_n}{\rho(z)} \int_a^b \frac{(d^n/ds^n)[\sigma^n \rho(s)]}{s-z} ds.$$

Here the integrated terms are zero by (5.17), since

$$[\sigma^n(z)\rho(z)]^{(n-m)} = \frac{1}{A_{mn}B_n} \sigma^m(z)\rho(z) y_n^{(m)}(z).$$

By using the Rodrigues formula for the $y_n(z)$, we can write the preceding equation in the form

$$Q_n(z) = \frac{1}{\rho(z)} \int_a^b \frac{y_n(s)\rho(s)}{s-z} ds. \tag{4}$$

It is sometimes convenient to write (4) in the form

$$Q_n(z) = \frac{1}{\rho(z)} \left[\int_a^b \frac{y_n(s) - y_n(z)}{s-z} \rho(s) ds + y_n(z) \int_a^b \frac{\rho(s) ds}{s-z} \right].$$

The first integral on the right is a polynomial $r_n(z)$ of the second kind (see §6, Part 3); the second integral can be expressed in terms of $Q_0(z)$. We obtain

$$Q_n(z) = \frac{1}{\rho(z)} r_n(z) + \frac{y_n(z)}{y_0(z)} Q_0(z). \tag{5}$$

Consequently the singularities of $Q_n(z)$ are determined by the behavior of $Q_0(z)$ and $1/\rho(z)$.

2. Asymptotic formula. By using (4), we can obtain an *asymptotic formula* for $Q_n(z)$ for large $|z|$. We use the equation

$$\begin{aligned} \frac{1}{s-z} &= -\frac{1}{z} \frac{1}{1-s/z} = -\frac{1}{z} \left[\sum_{k=0}^{p} \left(\frac{s}{z}\right)^k + \frac{(s/z)^{p+1}}{1-s/z} \right] \\ &= -\frac{1}{z} \sum_{k=0}^{p} \left(\frac{s}{z}\right)^k + \frac{s^{p+1}}{(s-z)z^{p+1}}. \end{aligned}$$

If we multiply by $y_n(s)\rho(s)$ and integrate from a to b, we obtain

$$\rho(z) Q_n(z) = -\sum_{k=n}^{p} \frac{1}{z^{k+1}} \int_a^b s^k y_n(s) \rho(s) ds + \frac{R_p(z)}{z^{p+1}} \tag{6}$$

where

$$R_p(z) = \int_a^b \frac{s^{p+1} y_n(s) \rho(s)}{s-z} ds.$$

Here we used the orthogonality property (6.5).

If $z \to \infty$ and the shortest distance from z to (a, b) is bounded away from $0, |R_p(z)|$ is bounded and (6) provides an asymptotic formula for $Q_n(z)$. In particular, when $p = n + 1$ we see from (6) that

$$Q_n(z) = -\frac{d_n^2}{a_n} \frac{1}{\rho(z) z^{n+1}} \left[1 + O\left(\frac{1}{z}\right)\right]. \tag{7}$$

Here (and later) we use the notation $f_1(z) = O[f_2(z)]$ as $z \to z_0$ to mean

$$|f_1(z)| \leq C|f_2(z)|$$

in some neighborhood of $z = z_0$, where C is a constant.

It is clear from (7) that $Q_n(z)$ and $y_n(z)$ have different asymptotic behavior and therefore are linearly independent solutions of the differential equation of the classical orthogonal polynomials (except in the case $n = 0, \alpha + \beta + 1 = 0$ for the Jacobi polynomials $P_n^{(\alpha,\beta)}(z)$).

3. Recursion relations and differentiation formulas. Since the integral representation (3) for $Q_n(z)$ differs from the integral representation of $y_n(z)$ only by the constant factor $1/(2\pi i)$ and the choice of contour, $Q_n(z)$ and $y_n(z)$ must satisfy the same recursion relations and differentiation formulas (see §§5 and 6):

$$zQ_n(z) = \alpha_n Q_{n+1}(z) + \beta_n Q_n(z) + \gamma_n Q_{n-1}(z) \quad (n \geq 1), \tag{8}$$

$$\sigma(z) Q_n'(z) = \frac{\lambda_n}{n\tau_n'} \left[\tau_n(z) Q_n(z) - \frac{B_n}{B_{n+1}} Q_{n+1}(z)\right]. \tag{9}$$

For the derivative of the function of the second kind we can use (4.7) to obtain the integral representation

$$Q_n'(z) = \frac{\chi_n B_n n!}{\sigma(z)\rho(z)} \int_a^b \frac{\sigma^n(s)\rho(s)}{(s-z)^n} ds,$$

where $\chi_n = \tau' + (n-1)\sigma''/2$. When $n = 0$, this leads to a differential equation for $Q_0(z)$:

$$\sigma(z)\rho(z) Q_0'(z) = C, \tag{10}$$

where

$$C = \chi_0 B_0 \int_a^b \rho(s) ds = \frac{\chi_0 d_0^2}{a_0}.$$

We have used the fact that, by the Rodrigues formula, $y_0 = B_0 = a_0$.

From (10), it is easy to obtain a convenient formula for $Q_0(z)$:

$$Q_0(z) = Q_0(z_0) - C \int_z^{z_0} \frac{ds}{\sigma(s)\rho(s)}. \tag{11}$$

It is convenient to take z_0 to be a value of z for which $Q_0(z_0) = 0$. It is clear from the asymptotic formula (7) that for the Laguerre function of the second kind, $Q_0(z)$, we may take $z_0 = -\infty$, and for the Hermite function of the second kind we may take $z_0 = \pm i\infty$. For the Jacobi function $Q_0^{(\alpha,\beta)}(z)$ we may take $z_0 = \infty$ if $\alpha + \beta > -1$.

It follows from (11) that when $z \to x$, $x \in (a, b)$, the limits $Q_0(x \pm i0)$ exist, and therefore by (5) both of the limits $Q_n(x \pm i0)$ necessarily exist. Since $\overline{\rho(z)Q_n(z)} = \rho(\bar{z})Q_n(\bar{z})$ by (4) (the bar denotes the complex conjugate), for $z = x$ the second solution of the equation for the classical orthogonal polynomials can be taken to be, instead of $Q_n(x \pm i0)$, the real combination

$$\rho(x)Q_n(x) = \frac{1}{2}[\rho(x + i0)Q_n(x + i0) + \rho(x - i0)Q_n(x - i0)]$$

(recall that $\rho(x + i0) = \rho(x)$).

It can be shown that, with this definition, $Q_n(x)$ will satisfy the same equations as $y_n(x)$ for $x \in (a, b)$. The integral representation (4) will also remain valid if the integral is interpreted as a principal value, since the integral in (4) is an integral of Cauchy type (see, for example, [L3]).

4. Some special functions related to $Q_0(z)$: Incomplete beta and gamma functions, exponential integrals, exponential integral function, integral sine and cosine, error function, Fresnel integrals.

It follows from (11) that $Q_0(z)$ for the Jacobi polynomials reduces to the *incomplete beta function* $B_z(p, q)$ by the substitution $t = 2/(1 + s)$; for the Laguerre polynomials, to the *incomplete gamma function* $\Gamma(a, z)$ by the substitution $t = -s$; and for the Hermite polynomials, to the *error function* $\operatorname{erf}(z)$ by the substitution $t = \pm is$. These functions are defined as follows:

$$B_z(p, q) = \int_0^z t^{p-1}(1 - t)^{q-1}\,dt,$$

$$\Gamma(a, z) = \int_z^\infty e^{-t}t^{a-1}\,dt,$$

$$\operatorname{erf}(z) = \frac{2}{\sqrt{\pi}} \int_0^z e^{-t^2}\, dt.$$

To make $\Gamma(a, z)$ single-valued, we need to introduce a cut from $z = 0$ to $z = -\infty$ along the real axis. Accordingly, in the formula for $\Gamma(a, z)$, we should take $|\arg t| < \pi$ in t^{a-1}. Similarly, in the formula for $B_z(p, q)$ we should take $0 < \arg z < 2\pi$, $0 < \arg t < 2\pi$, $|\arg(1-t)| < \pi$.

As an illustration we consider (11) for the Laguerre and Hermite functions of the second kind.

1) For the *Laguerre function of the second kind*, if we put $z_0 = -\infty$ in (11) we obtain

$$\begin{aligned} Q_0(z) \equiv Q_0^\alpha(z) &= \Gamma(\alpha+1) \int_z^{-\infty} \frac{ds}{\sigma(s)\rho(s)} \\ &= \Gamma(\alpha+1) \int_z^{-\infty} \frac{e^s}{s^{\alpha+1}} ds = \Gamma(\alpha+1) e^{-i\pi\alpha} \Gamma(-\alpha, -z). \end{aligned} \tag{12}$$

To make $Q_n(z)$ single-valued we introduced a cut from $z = 0$ to $z = +\infty$ along the real axis. Correspondingly, in evaluating $s^{\alpha+1}$ in (12) we must suppose that $0 < \arg s < 2\pi$.

For integral values $\alpha = m$ $(m = 0, 1, 2, \ldots)$ the function $Q_0^\alpha(z)$ can be expressed in terms of the functions

$$E_m(z) = z^{m-1} \int_z^{+\infty} \frac{e^{-s}}{s^m} ds = \int_1^{\infty} \frac{e^{-zt}}{t^m} dt, \tag{13}$$

which have many applications to problems of the transmission of radiation through matter, in the theory of nuclear reactors. The functions $E_m(z)$ are *exponential integrals*. If we put $\alpha = m$ in (12) for $z > 0$, we obtain

$$Q_0^m(-z) = \frac{(-1)^m m!}{z^m} E_{m+1}(z) \quad (z > 0). \tag{14}$$

By using (7), it is easy to obtain from (14) an *asymptotic representation* for $E_m(z)$:

$$E_m(z) = z^{-1} e^{-z} \left[1 + O\left(\frac{1}{z}\right) \right]. \tag{15}$$

Differentiating (13), we obtain the differentiation formula

$$E'_m(z) = \frac{m-1}{z}E_m(z) - \frac{e^{-z}}{z} = -E_{m-1}(z),$$

which leads to the following *recursion relation:*

$$E_m(z) = \frac{1}{m-1}[e^{-z} - zE_{m-1}(z)]. \tag{16}$$

Let us now investigate the behavior of $E_m(z)$ as $z \to 0$. By (16) it is enough to study $E_1(z)$, which has a singular point at $z = 0$. We isolate it by using the identity

$$\begin{aligned} E_1(z) &= \int_z^\infty \frac{e^{-s}}{s}ds = \int_1^\infty \frac{e^{-s}}{s}ds + \int_z^1 \frac{e^{-s}-1}{s}ds + \int_z^1 \frac{ds}{s} \\ &= C - \ln z - \int_0^z \frac{e^{-s}-1}{s}ds, \end{aligned}$$

where

$$C = \int_1^\infty \frac{e^{-s}}{s}ds + \int_0^1 \frac{e^{-s}-1}{s}ds.$$

If we expand e^{-s} in powers of s and integrate term by term, we obtain an expansion of $E_1(z)$:

$$E_1(z) = C - \ln z + \sum_{k=1}^\infty \frac{(-1)^{k-1}}{k}\frac{z^k}{k!}. \tag{17}$$

We evaluate C by integration by parts:

$$\begin{aligned} C &= e^{-s}\ln s|_1^\infty + \int_1^\infty e^{-s}\ln s\, ds + (e^{-s}-1)\ln s|_0^1 + \int_0^1 e^{-s}\ln s\, ds \\ &= \int_0^\infty e^{-s}\ln s\, ds = \Gamma'(1) = -\gamma \end{aligned}$$

(γ is Euler's constant; see Appendix A).

In practice, the related *exponential integral function* $\mathrm{Ei}(z)$ is often used instead of $E_1(z)$; these are connected by

$$E_1(z) = -\mathrm{Ei}(-z).$$

Other related functions are

$$\mathrm{Si}(z) = \int_0^z \frac{\sin s}{s} ds, \quad \mathrm{Ci}(z) = \int_\infty^z \frac{\cos s}{s} ds,$$

the *integral sine* and *integral cosine*. When $z > 0$ we can use Jordan's lemma (see [L3] or [W3]) to obtain

$$\begin{aligned} E_1(iz) &= \int_{iz}^{\infty} \frac{e^{-s}}{s} ds = \int_{iz}^{+i\infty} \frac{e^{-s}}{s} ds = \int_z^\infty \frac{e^{-it}}{t} dt \\ &= \int_z^\infty \frac{\cos t}{t} dt - i \int_z^\infty \frac{\sin t}{t} dt = -\left\{ \mathrm{Ci}(z) + i \left[\frac{1}{2}\pi - \mathrm{Si}(z) \right] \right\}, \end{aligned}$$

since

$$\int_0^\infty \frac{\sin t}{t} dt = \frac{\pi}{2}.$$

Hence, for $z > 0$,

$$\begin{aligned} \mathrm{Ci}(z) &= -\frac{1}{2}[E_1(iz) + E_1(-iz)], \\ \mathrm{Si}(z) &= \frac{\pi}{2} + \frac{1}{2i}[E_1(iz) - E_1(-iz). \end{aligned} \tag{18}$$

By the principle of analytic continuation, (18) remains valid in a larger set in the z plane.

From (15), (17) and (18) we can easily obtain asymptotic representations and expansions in powers of z for $\mathrm{Si}(z)$ and $\mathrm{Ci}(z)$. For example

$$\begin{aligned} \mathrm{Si}(z) &= \sum_{k=0}^{\infty} \frac{(-1)^k z^{2k+1}}{(2k+1)(2k+1)!}, \\ \mathrm{Ci}(z) &= \gamma + \ln z - \sum_{k=1}^{\infty} \frac{(-1)^{k+1} z^{2k}}{2k(2k)!}. \end{aligned}$$

Since these power series converge in the whole complex plane, $\mathrm{Si}(z)$ is analytic in the whole plane, and $\mathrm{Ci}(z)$ is analytic in the plane cut along $(-\infty, 0)$.

2) For the *Hermite function of the second kind*, if we take $z_0 = \pm i\infty$ (the sign is the sign of Imz), we obtain

$$Q_0(z) = 2\sqrt{\pi} \int_z^{\pm i\infty} e^{s^2}\, ds.$$

For $z > 0$ we have

$$Q_0(iz) = 2\sqrt{\pi} \int_{iz}^{i\infty} e^{s^2}\, ds = 2\sqrt{\pi}\, i \int_z^{\infty} e^{-s^2}\, ds = \pi i[1 - \mathrm{erf}(z)],$$

where

$$\mathrm{erf}(z) = \frac{2}{\sqrt{\pi}} \int_0^z e^{-s^2}\, ds \tag{19}$$

is the error function.

From the asymptotic formula (7) for $Q_0(z)$ it is easy to obtain an *asymptotic formula for the error function*:

$$\mathrm{erf}(z) = 1 - \frac{1}{\sqrt{\pi}} \frac{e^{-z^2}}{z} \left[1 + O\left(\frac{1}{z}\right)\right].$$

We can obtain an *expansion of* erf(z) *in powers of* z by expanding e^{-s^2} in (19) in powers of z and integrating term by term:

$$\mathrm{erf}(z) = \frac{2}{\sqrt{\pi}} \sum_{k=0}^{\infty} \frac{(-1)^k z^{2k+1}}{k!(2k+1)}.$$

The Fresnel integrals

$$S(z) = \int_0^z \sin \frac{\pi t^2}{2} dt, \quad C(z) = \int_0^z \cos \frac{\pi t^2}{2} dt$$

are closely related to the error function. In fact, when $z > 0$ we have

$$C(z) - iS(z) = \int_0^z e^{-i\pi t^2/2} dt = \frac{1}{\sqrt{2i}} \mathrm{erf}\left(\sqrt{\frac{i\pi}{2}}\, z\right).$$

This connection lets one obtain the asymptotic behavior and series expansions for the Fresnel integrals.

Graphs of $E_1(x)$, Si(x), Ci(x), erf(x), $S(x)$ and $C(x)$ are given in Figures 3–5.

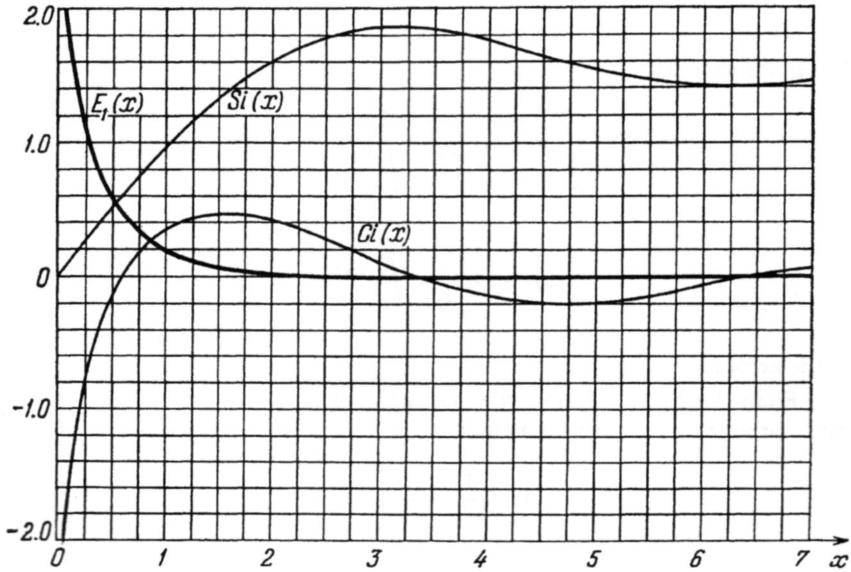

Figure 3.

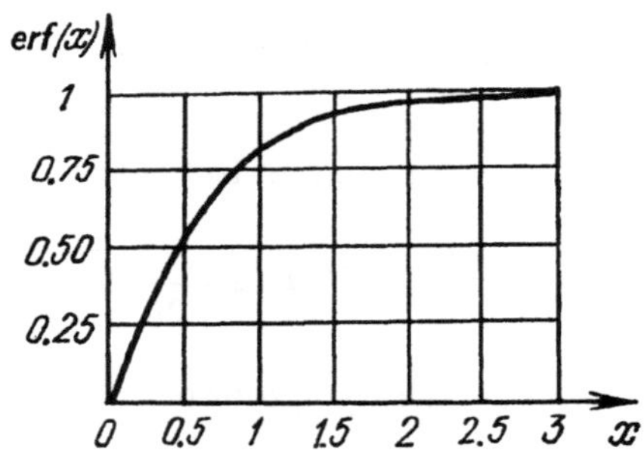

Figure 4.

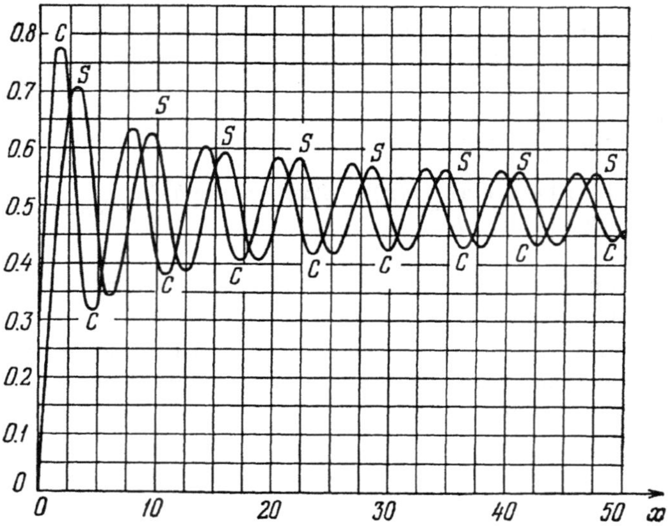

Figure 5. $S(x)$ and $C(x)$

§ 12 Classical Orthogonal Polynomials of a Discrete Variable

1. The difference equation of hypergeometric type. The theory that we have developed of polynomial solutions of the differential equation of hypergeometric type,

$$\tilde{\sigma}(x)y'' + \tilde{\tau}(x)y' + \lambda y = 0, \tag{1}$$

where $\tilde{\sigma}(x)$ and $\tilde{\tau}(x)$ are polynomials of at most second and first degree*, and λ is a constant, admits a natural generalization to the case when the differential equation is replaced by a difference equation. Let us consider the simplest case, when (1) is replaced by the difference equation

$$\begin{aligned}\tilde{\sigma}(x)\frac{1}{h}\left[\frac{y(x+h)-y(x)}{h} - \frac{y(x)-y(x-h)}{h}\right]& \\ + \frac{\tilde{\tau}(x)}{2}\left[\frac{y(x+h)-y(x)}{h} + \frac{y(x)-y(x-h)}{h}\right] &+ \lambda y(x) = 0,\end{aligned} \tag{2}$$

which approximates (1) on a lattice with the constant mesh $\Delta x = h$ up to second order in h.**

By the linear change of variable $x \to hx$, which preserves the type of the equation, we can always arrange that the mesh in (2) is equal to unity: $\Delta x = h = 1$. We then have

$$\begin{aligned}\tilde{\sigma}(x)[y(x+1) - 2y(x) + y(x-1)] + \frac{1}{2}\tilde{\tau}(x)\{[y(x+1) - y(x)]& \\ + [y(x) - y(x-1)]\} + \lambda y(x) = 0.&\end{aligned}$$

This equation can be written in the form

$$\tilde{\sigma}(x)\Delta\nabla y(x) + \frac{1}{2}\tilde{\tau}(x)(\Delta + \nabla)y(x) + \lambda y(x) = 0, \tag{2a}$$

where

$$\Delta f(x) = f(x+1) - f(x), \quad \nabla f(x) = f(x) - f(x-1).$$

Since $\nabla f(x) = \Delta f(x) - \Delta\nabla f(x)$, equation (2a) is equivalent to

$$\sigma(x)\Delta\nabla y(x) + \tau(x)\Delta y(x) + \lambda y(x) = 0, \tag{3}$$

---

* It will be convenient to denote the coefficients in (1) by $\tilde{\sigma}(x)$ and $\tilde{\tau}(x)$ instead of $\sigma(x)$ and $\tau(x)$ as in Chapter 1.

** We say that a difference operator L_h approximates the differential operator L at the point x to order m in h if $Ly(x) - L_h y(x) = O(h^m)$, $h \to 0$.

where

$$\sigma(x) = \tilde{\sigma}(x) - \frac{1}{2}\tilde{\tau}(x), \qquad \tilde{\tau}(x) = \tau(x).$$

Evidently $\sigma(x)$ is a polynomial of at most degree 2.

Before we turn to studying the solutions of (3), we consider a number of properties of the operators Δ and ∇. We have

$$\Delta f(x) = \nabla f(x+1), \tag{4}$$

$$\Delta\nabla f(x) = \nabla\Delta f(x) = f(x+1) - 2f(x) + f(x-1), \tag{5}$$

$$\Delta[f(x)g(x)] = f(x)\Delta g(x) + g(x+1)\Delta f(x). \tag{6}$$

From (6) we obtain the formula for summation by parts:

$$\sum_i f(x_i)\Delta g(x_i) = f(x_i)g(x_i)\Big|_a^b - \sum_i g(x_{i+1})\Delta f(x_i). \tag{7}$$

Here $x_{i+1} = x_i + 1$, and the summation is over indices i for which $a \le x_i \le b-1$. We observe that for a polynomial $q_m(x)$ of degree m the expressions $\Delta q_m(x)$ and $\nabla q_m(x)$ are polynomials of degree $m-1$; and that $\Delta^m q_m(x) = \nabla^m q_m(x) = q_m^{(m)}(x)$.

We can establish a number of properties of the solutions of (3) that are analogous to those of solutions of (1).

Let us show that *the function* $v_1(x) = \Delta y(x)$ *satisfies a difference equation of the form* (3). For the proof, we apply the operator Δ to both sides of equation (3):

$$\Delta[\sigma(x)\nabla v_1(x)] + \Delta[\tau(x)v_1(x)] + \lambda v_1(x) = 0.$$

By using (6) and (4), we can write this equation in the form

$$\sigma(x)\Delta\nabla v_1 + \tau_1(x)\Delta v_1 + \mu_1 v_1 = 0, \tag{8}$$

where $\tau_1(x) = \tau(x+1) + \Delta\sigma(x)$, $\mu_1 = \lambda + \Delta\tau(x)$. Since $\tau_1(x)$ is a polynomial of at most the first degree, and μ_1 is independent of x, equation (8) for $v_1(x)$ is of the same form as (3).

It is also easy to verify the converse: *every solution of* (8) *with* $\lambda \neq 0$ *can be represented in the form* $v_1(x) = \Delta y(x)$, where $y(x)$ is a solution of (3) that can be expressed in terms of $v_1(x)$ by

$$y(x) = -(1/\lambda)[\sigma(x)\nabla v_1 + \tau(x)v_1].$$

In a similar way we can obtain a difference equation of hypergeometric type,

$$\sigma(x)\Delta\nabla v_n + \tau_n(x)\Delta v_n + \mu_n v_n = 0, \tag{9}$$

for $v_n(x) = \Delta^n y(x)$. Here

$$\tau_n(x) = \tau_{n-1}(x+1) + \Delta\sigma(x), \quad \tau_0(x) = \tau(x), \tag{10}$$

$$\mu_n = \mu_{n-1} + \Delta\tau_{n-1}(x), \quad \mu_0 = \lambda. \tag{11}$$

The converse is also valid: every solution of (9) with $\mu_k \neq 0 \quad (k = 0, 1, \ldots, n-1)$ can be represented as $v_n(x) = \Delta^n y(x)$, where $y(x)$ is a solution of (3).

If we rewrite (10) in the form

$$\tau_n(x) + \sigma(x) = \tau_{n-1}(x+1) + \sigma(x+1), \tag{10a}$$

we easily obtain an explicit expression for $\tau_n(x)$:

$$\tau_n(x) = \tau(x+n) + \sigma(x+n) - \sigma(x).$$

To obtain an explicit formula for μ_n we have only to observe that $\Delta\tau_n(x)$ and $\Delta^2\sigma(x)$ are independent of x. Therefore

$$\Delta\tau_n = \Delta\tau_{n-1} + \Delta^2\sigma = \ldots = \Delta\tau + n\Delta^2\sigma,$$

and consequently $\mu_n = \mu_{n-1} + \Delta\tau + (n-1)\Delta^2\sigma$. Hence

$$\begin{aligned} \mu_n = \mu_0 + \sum_{k=1}^{n}(\mu_k - \mu_{k-1}) &= \lambda + n\Delta\tau + \frac{1}{2}n(n-1)\Delta^2\sigma \\ &= \lambda + n\tau' + \frac{1}{2}n(n-1)\sigma''. \end{aligned} \tag{12}$$

2. Finite difference analogs of polynomials of hypergeometric type and of their derivatives. A Rodrigues formula. The properties of the higher differences $\Delta^n y(x)$ established in part 1 allow us to construct a theory of classical orthogonal polynomials of a discrete variable along the same lines as the discussion in the preceding chapter. It is clear that $v_n(x) = \text{const}$ is a solution of (9) for $\mu_n = 0$. Since $v_n(x) = \Delta^n y(x)$, this means that if

$$\lambda = \lambda_n = -n\tau' - \frac{1}{2}n(n-1)\sigma''$$

there is a particular solution $y = y_n(x)$ of (3) which is a polynomial of degree n, provided that $\mu_k \neq 0$ for $k = 0, 1, \ldots, n-1$. In fact, the equation

$$\sigma(x)\Delta\nabla v_k + \tau_k(x)\Delta v_k + \mu_k v_k = 0$$

for $v_k(x)$ can be rewritten in the form

$$v_k(x) = -(1/\mu_k)[\sigma(x)\nabla v_{k+1}(x) + \tau_k(x)v_{k+1}(x)].$$

It is clear from this that if $v_{k+1}(x)$ is a polynomial, then $v_k(x)$ is also a polynomial if $\mu_k \neq 0$.

To obtain an explicit expression for $y_n(x)$, we write (3) and (9) in self-adjoint form:

$$\Delta(\sigma\rho\nabla y) + \lambda\rho y = 0, \tag{13}$$

$$\Delta(\sigma\rho_n\nabla v_n) + \mu_n\rho_n v_n = 0. \tag{14}$$

Here $\rho(x)$ and $\rho_n(x)$ satisfy difference equations,

$$\Delta(\sigma\rho) = \tau\rho, \tag{15}$$

$$\Delta(\sigma\rho_n) = \tau_n\rho_n. \tag{16}$$

We can determine the connection between $\rho_n(x)$ and $\rho(x)$ by writing (16) in the form

$$\frac{\sigma(x+1)\rho_n(x+1)}{\rho_n(x)} = \tau_n(x) + \sigma(x).$$

It is now clear that (10a) is equivalent to

$$\frac{\sigma(x+1)\rho_n(x+1)}{\rho_n(x)} = \frac{\sigma(x+2)\rho_{n-1}(x+2)}{\rho_{n-1}(x+1)},$$

i.e.

$$\frac{\rho_n(x+1)}{\sigma(x+2)\rho_{n-1}(x+2)} = \frac{\rho_n(x)}{\sigma(x+1)\rho_{n-1}(x+1)} = C_n(x),$$

where $C_n(x)$ is any function of period 1. We only need to find a particular solution of (16), so we may take $C_n(x) = 1$. We then obtain

$$\rho_n(x) = \sigma(x+1)\rho_{n-1}(x+1).$$

Since $\rho_0(x) = \rho(x)$, we have

$$\rho_n(x) = \rho(x+n)\prod_{k=1}^{n}\sigma(x+k). \tag{17}$$

By using the connection between $\rho_n(x)$ and $\rho_{n+1}(x)$ we can present (14) as a simple relation between $v_n(x)$ and $v_{n+1}(x)$. In fact,

$$\begin{aligned}\rho_n(x)v_n(x) &= -(1/\mu_n)\Delta[\sigma(x)\rho_n(x)\nabla v_n(x)]\\ &= -(1/\mu_n)\nabla[\sigma(x+1)\rho_n(x+1)\Delta v_n(x)],\end{aligned}$$

i.e.

$$\rho_n(x)v_n(x) = -(1/\mu_n)\nabla[\rho_{n+1}(x)v_{n+1}(x)].$$

For $m < n$ we now obtain successively

$$\begin{aligned}\rho_m v_m &= -\frac{1}{\mu_m}\nabla(\rho_{m+1}v_{m+1})\\ &= \left(-\frac{1}{\mu_m}\right)\left(-\frac{1}{\mu_{m+1}}\right)\nabla^2(\rho_{m+2}v_{m+2}) = \ldots = \frac{A_m}{A_n}\nabla^{n-m}(\rho_n v_n),\end{aligned} \tag{18}$$

where

$$A_n = (-1)^n \prod_{k=0}^{n-1}\mu_k, \qquad A_0 = 1. \tag{19}$$

We now proceed to obtain an explicit form for the polynomials of degree n, i.e. $y = y_n(x)$.

If $y = y_n(x)$ we have $v_n(x) = y_n(x) = \text{const}$, and by using (18) we arrive at the following expression for $v_{mn}(x) = \Delta^m y_n(x)$:

$$v_{mn}(x) = \Delta^m y_n(x) = \frac{A_{mn}B_n}{\rho_m(x)}\nabla^{n-m}[\rho_n(x)], \tag{20}$$

where

$$\begin{aligned}A_{mn} &= A_m(\lambda)|_{\lambda=\lambda_n} = \frac{n!}{(n-m)!}\prod_{k=0}^{m-1}\left(\tau' + \frac{n+k-1}{2}\sigma''\right),\\ A_{0n} &= 1 \qquad\qquad (m \le n),\\ B_n &= \frac{\Delta^n y_n(x)}{A_{nn}} = \frac{1}{A_{nn}}y_n^{(n)}(x).\end{aligned} \tag{21}$$

Hence, in particular, for $m = 0$ we obtain an explicit expression for $y_n(x)$:

$$y_n(x) = \frac{B_n}{\rho(x)}\nabla^n[\rho_n(x)]. \tag{22}$$

Thus the polynomial solutions of (3) are determined by (22) up to the normalizing factors B_n. These solutions correspond to the values

$$\lambda = \lambda_n = -n\tau' - \frac{1}{2}n(n-1)\sigma''.$$

By using (4) and (17), we can also write (22) in the form

$$y_n(x) = \frac{B_n}{\rho(x)}\Delta^n[\rho_n(x-n)] = \frac{B_n}{\rho(x)}\Delta^n\left[\rho(x)\prod_{k=0}^{n-1}\sigma(x-k)\right]. \tag{22a}$$

Equation (20) is the finite-difference analog of the Rodrigues formula for the classical orthogonal polynomials and their derivatives (compare formula (2.10)). The Rodrigues formula for the polynomials $y_n(x)$ and their differences $\Delta y_n(x)$ leads to a relation between $\Delta y_n(x)$ and the polynomials themselves. To find it, it is enough to observe that if $m = 1$ in (20) we have $A_{1n} = -\lambda_n$ and according to (17) we have $[\rho_1(x)]_{n-1} = \rho_n(x)$. In fact,

$$\rho_1(x) = \sigma(x+1)\rho(x+1),$$
$$[\rho_1(x)]_{n-1} = \rho_1(x+n-1)\prod_{k=1}^{n-1}\sigma(x+k) = \rho(x+n)\prod_{k=1}^{n}\sigma(x+k) = \rho_n(x).$$

Hence

$$\begin{aligned}\Delta y_n(x) &= -\lambda_n\frac{B_n}{\rho_1(x)}\nabla^{n-1}[\rho_n(x)] \\ &= -\lambda_n\frac{B_n}{\bar{B}_{n-1}}\frac{\bar{B}_{n-1}}{\rho_1(x)}\nabla^{n-1}\left\{[\rho_1(x)]_{n-1}\right\} = -\lambda_n\frac{B_n}{\bar{B}_{n-1}}\bar{y}_{n-1}(x).\end{aligned} \tag{23}$$

Here $\bar{y}_n(x)$ is the polynomial obtained by replacing $\rho(x)$ by $\rho_1(x)$ in the formula for $y_n(x)$, and $\bar{B}_n$ is the normalizing constant in the Rodrigues formula for $\bar{y}_n(x)$.

By using (20) with $m = n-1$ we can easily calculate the coefficients a_n and b_n of the highest powers of x in the expansion

$$y_n(x) = a_n x^n + b_n x^{n-1} + \dots.$$

For this purpose we first calculate the $(n-1)$th difference $\Delta^{n-1}(x^n)$, which is a polynomial of the first degree. We have

$$\Delta^{n-1}(x^n) = \alpha_n(x+\beta_n),$$

where α_n and β_n are constants. To determine α_n and β_n, we observe that

$$\Delta^n(x^n) = \Delta[\alpha_n(x+\beta_n)] = \alpha_n.$$

Hence

$$\begin{aligned}\alpha_{n+1}(x+\beta_{n+1}) &= \Delta^n(x^{n+1}) = \Delta^{n-1}(\Delta x^{n+1}) = \Delta^{n-1}[(x+1)^{n+1} - x^{n+1}] \\ &= \Delta^{n-1}\left[(n+1)x^n + \frac{1}{2}n(n+1)x^{n-1} + \dots\right] \\ &= (n+1)\alpha_n(x+\beta_n) + \frac{1}{2}n(n+1)\alpha_{n-1}.\end{aligned}$$

Equating coefficients of the powers of x on the two sides of this equation, we obtain

$$\begin{aligned}\alpha_{n+1} &= (n+1)\alpha_n, \\ \alpha_{n+1}\beta_{n+1} &= (n+1)\alpha_n\beta_n + \frac{1}{2}n(n+1)\alpha_{n-1}.\end{aligned}$$

Since $\alpha_1 = 1$, the first equation yields $\alpha_n = n!$, whence $\beta_{n+1} = \beta_n + \frac{1}{2}$. Since $\beta_1 = 0$, we have $\beta_n = (n-1)/2$. Therefore

$$\Delta^{n-1}(x^n) = n!\left(x + \frac{n-1}{2}\right).$$

Consequently

$$\begin{aligned}\Delta^{n-1}y_n(x) &= \Delta^{n-1}(a_n x^n + b_n x^{n-1} + \dots) \\ = a_n\alpha_n(x+\beta_n) + b_n\alpha_{n-1} &= n!\, a_n\left(x + \frac{1}{2}(n-1)\right) + (n-1)!\, b_n.\end{aligned}$$

On the other hand,

$$\nabla\rho_n(x) = \Delta\rho_n(x-1) = \Delta[\sigma(x)\rho_{n-1}(x)] = \tau_{n-1}(x)\rho_{n-1}(x).$$

Consequently if we take $m = n-1$ in (20) we obtain

$$A_{n-1,n}B_n\tau_{n-1}(x) = n!\, a_n\left(x + \frac{1}{2}(n-1)\right) + (n-1)!\, b_n,$$

whence

$$a_n = \frac{A_{n-1,n}B_n}{n!}\tau'_{n-1} = B_n\prod_{k=0}^{n-1}\left(\tau' + \frac{n+k-1}{2}\sigma''\right), \quad a_0 = B_0; \tag{24}$$

$$\frac{b_n}{a_n} = n\frac{\tau_{n-1}(0)}{\tau'_{n-1}} - \frac{n(n-1)}{2} = n\frac{\tilde{\tau}(0) + (n-1)\tilde{\sigma}'(0)}{\tilde{\tau}' + (n-1)\tilde{\sigma}''}. \tag{25}$$

3. The orthogonality property. We now discuss the orthogonality of the polynomial solutions of (3). To do this, we write the equations for $y_n(x)$ and $y_m(x)$ in self-adjoint form,

$$\Delta[\sigma(x)\rho(x)\nabla y_n(x)] + \lambda_n \rho(x) y_n(x) = 0,$$
$$\Delta[\sigma(x)\rho(x)\nabla y_m(x)] + \lambda_m \rho(x) y_m(x) = 0.$$

Multiply the first equation by y_m and the second by y_n, and subtract the second from the first. We obtain

$$(\lambda_m - \lambda_n)\rho(x) y_m(x) y_n(x) = \Delta\left\{\sigma(x)\rho(x)[y_m(x)\nabla y_n(x) - y_n(x)\nabla y_m(x)]\right\}.$$

If we now put $x = x_i$ and $x_{i+1} = x_i + 1$, and sum over the indices i for which $a \le x_i \le b-1$, we obtain

$$(\lambda_m - \lambda_n)\sum_i y_m(x_i) y_n(x_i)\rho(x_i)$$
$$= \sigma(x)\rho(x)\left[y_m(x)\nabla y_n(x) - y_n(x)\nabla y_m(x)\right]\Big|_a^b.$$

The expression $y_m\nabla y_n - y_n\nabla y_m$ is a polynomial in x. Hence under the boundary conditions

$$\sigma(x)\rho(x)x^k\Big|_{x=a,b} = 0 \qquad (k = 0, 1, \ldots) \tag{26}$$

the polynomial solutions of (3) are orthogonal on $[a, b-1]$ with weight $\rho(x)$:

$$\sum_{x_i=a}^{b-1} y_m(x_i) y_n(x_i)\rho(x_i) = \delta_{mn} d_n^2. \tag{27}$$

We call the polynomials $y_n(x)$ *classical orthogonal polynomials of a discrete variable* provided that (a, b) is an interval on the real axis and $\rho(x)$ satisfies (15) and (26). They are usually considered under the additional condition $\rho(x_i) > 0$ for $a \le x_i \le b-1$.

The polynomials $\Delta y_n(x)$ satisfy the equation obtained from the equation for $y_n(x)$ by replacing $\rho(x)$ by $\rho_1(x) = \sigma(x+1)\rho(x+1) = [\tau(x) + \sigma(x)]\rho(x)$ and λ by $\mu_1 = \lambda + \tau'$. The function $\rho_1(x)$ evidently satisfies a condition similar to (26):

$$\sigma(x)\rho_1(x)x^k\Big|_{x=a,b-1} = 0 \qquad (k = 0, 1, \ldots).$$

Hence the polynomials $\Delta y_n(x)$ have the orthogonality property

$$\sum_{x_i=a}^{b-2} \Delta y_m(x_i)\Delta y_n(x_i)\rho_1(x_i) = \delta_{mn} d_{1n}^2.$$

Proceeding similarly, we can easily show that the polynomials $\Delta^k y_n(x)$ satisfy

$$\sum_{x_i=a}^{b-k-1} \Delta^k y_m(x_i)\Delta^k y_n(x_i)\rho_k(x_i) = \delta_{mn}d_{kn}^2. \tag{28}$$

If we take $\rho(a) > 0$, and

$$\begin{aligned} &\sigma(x_i) > 0 \text{ for } a+1 \le x_i \le b-1, \\ &\sigma(x_i) + \tau(x_i) > 0 \text{ for } a \le x_i \le b-2, \end{aligned} \tag{29}$$

it follows from (15), written in the form

$$\frac{\rho(x+1)}{\rho(x)} = \frac{\sigma(x)+\tau(x)}{\sigma(x+1)},$$

and the explicit form of $\rho_k(x)$ that

$$\rho_k(x_i) > 0 \text{ for } a \le x \le b-k-1 \quad (k = 0, 1, \ldots).$$

We now discuss the choice of a and b to satisfy the boundary conditions (26), and the positivity condition on $\rho(x_i)$ on the interval $[a, b-1]$ of orthogonality. If a is finite, then by hypothesis $\rho(a) > 0$, i.e. a is a root of $\sigma(x)$. Since a linear change of variable $x \to x + a$ does not change the type of the equation, it is always possible, if $\sigma(x) \not\equiv$ const, to take $\sigma(0) = 0$. That is, we may suppose that $a = 0$. If b is finite, we have by (15)

$$\sigma(b)\rho(b) = [\sigma(b-1) + \tau(b-1)]\rho(b-1).$$

Since $\rho(b-1) > 0$ we have

$$\sigma(b-1) + \tau(b-1) = 0. \tag{30}$$

When $b = +\infty$ the boundary condition (26) will be satified if

$$\lim_{x\to+\infty} x^k\rho(x) = 0 \qquad (k = 0, 1, \ldots).$$

A similar remark applies when $a = -\infty$.

To calculate the squared norms d_n^2 we first establish the connection between the squared norms d_{kn}^2 and $d_{k+1,n}^2$, where

$$d_{kn}^2 = \sum_{x_i=a}^{b-k-1} v_{kn}^2(x_i)\rho_k(x_i), \quad d_{0n}^2 = d_n^2, \quad v_{kn}(x) = \Delta^k y_n(x).$$

To do this, we write the difference equation for $v_{kn}(x)$,

$$\Delta[\sigma(x)\rho_k(x)\nabla v_{kn}(x)] + \mu_{kn}\rho_k(x)v_{kn}(x) = 0.$$

where $\mu_{kn} = \mu_k(\lambda)|_{\lambda=\lambda_n} = \lambda_n - \lambda_k$, multiply by $v_{kn}(x)$, and sum over the values $x = x_i$ for which $a \le x_i \le b-k-1$:

$$\sum_i v_{kn}(x_i)\Delta[\sigma(x_i)\rho_k(x_i)\nabla v_{kn}(x_i)] + \mu_{kn}d_{kn}^2 = 0.$$

We use summation by parts (7) and the equations

$$\Delta v_{kn}(x) = v_{k+1,n}(x), \quad \sigma(x+1)\rho_k(x+1) = \rho_{k+1}(x),$$

to find that

$$\sum_i v_{kn}(x_i)\Delta[\sigma(x_i)\rho_k(x_i)\nabla v_{kn}(x_i)]$$
$$= \sigma(x)\rho_k(x)\nabla v_{kn}(x)v_{kn}(x)\Big|_a^{b-k} - d_{k+1,n}^2.$$

Since the first part of the right-hand side is zero because of the boundary conditions (26), we have

$$d_{kn}^2 = \frac{1}{\mu_{kn}}d_{k+1,n}^2.$$

Hence we obtain successively

$$\begin{aligned} d_n^2 \equiv d_{0n}^2 &= \frac{1}{\mu_{0n}}d_{1n}^2 = \frac{1}{\mu_{0n}}\frac{1}{\mu_{1n}}d_{2n}^2 = \ldots = \frac{d_{nn}^2}{\prod_{k=0}^{n-1}\mu_{kn}} \\ &= \frac{v_{nn}^2(x)}{\prod_{k=0}^{n-1}\mu_{kn}}S_n = (-1)^n A_{nn}B_n^2 S_n, \end{aligned} \tag{31}$$

where

$$S_n = \sum_{x_i=a}^{b-n-1} \rho_n(x_i). \tag{32}$$

For $n = b - a - 1$ (in the case when $b - a = N$ is finite) the sum S_n contains only one term and hence is easily calculated:

$$S_{N-1} = \rho_{N-1}(a). \tag{33}$$

To calculate S_n for $n < N - 1$ it is enough to be able to calculate the ratio S_{n-1}/S_n. To do this, we transform the expression (32) for S_n by using the connection between $\rho_n(x)$ and $\rho_{n-1}(x)$:

$$S_n = \sum_i \rho_n(x_i) = \sum_i \sigma(x_i + 1)\rho_{n-1}(x_i + 1) = \sum_i \sigma(x_i)\rho_{n-1}(x).$$

We expand $\sigma(x)$ in powers of the first-degree polynomial $\tau_{n-1}(x)$:

$$\sigma(x) = A\tau_{n-1}^2(x) + B\tau_{n-1}(x) + C.$$

Then, by using the equation for $\rho_{n-1}(x)$ and summing by parts, we obtain

$$\begin{aligned}
S_n &= \sum_i [A\tau_{n-1}(x_i) + B]\tau_{n-1}(x_i)\rho_{n-1}(x_i) + CS_{n-1} \\
&= \sum_i [A\tau_{n-1}(x_i) + B]\Delta[\sigma(x_i)\rho_{n-1}(x_i)] + CS_{n-1} \\
&= -\sum_i \sigma(x_i + 1)\rho_{n-1}(x_i + 1)\Delta[A\tau_{n-1}(x_i) + B] + CS_{n-1} \\
&= -A\tau_{n-1}' S_n + CS_{n-1}.
\end{aligned}$$

Hence

$$\frac{S_{n-1}}{S_n} = \frac{1 + A\tau_{n-1}'}{C} = \frac{1 + \sigma''/(2\tau_{n-1}')}{\sigma(x_n^*)}, \tag{34}$$

where x_n^* is the root of the equation $\tau_{n-1}(x) = 0$. We use the equations $\sigma(x_n^*) = C$ and $\sigma'' = 2A(\tau_{n-1}')^2$. With the aid of formulas (31)–(34), we finally obtain

$$d_n^2 = (-1)^n A_{nn} B_n^2 \rho_{N-1}(a) \prod_{k=n+1}^{N-1} \left[\frac{1 + \sigma''/(2\tau_{k-1}')}{\sigma(x_k^*)}\right], \tag{31a}$$

where $N = b - a$, $\tau_{k-1}' = \tau' + (k-1)\sigma''$, and x_k^* is the root of the equation

$$\tau(x) + (k-1)\sigma'(x) + (k-1)\tau' + \frac{1}{2}(k-1)^2\sigma'' = 0.$$

4. The Hahn, Chebyshev, Meixner, Kravchuk, and Charlier polynomials. Let us find explicit expressions for the weight functions $\rho(x)$ with respect to which the classical orthogonal polynomials of a discrete variable are orthogonal. For this purpose we rewrite the difference equation (15) in the form

$$\frac{\rho(x+1)}{\rho(x)} = \frac{\sigma(x)+\tau(x)}{\sigma(x+1)}. \tag{35}$$

It is easily verified that the solution of a difference equation

$$\frac{\rho(x+1)}{\rho(x)} = f(x),$$

whose right-hand side is expressible as a product or quotient of two functions, has the following simple property:

If $\rho_1(x)$ and $\rho_2(x)$ are solutions of the equations

$$\frac{\rho_1(x+1)}{\rho_1(x)} = f_1(x), \quad \frac{\rho_2(x+1)}{\rho_2(x)} = f_2(x),$$

then the solution of the equation

$$\frac{\rho(x+1)}{\rho(x)} = f(x)$$

with $f(x) = f_1(x)f_2(x)$ is $\rho(x) = \rho_1(x)\rho_2(x)$, and with $f(x) = f_1(x)/f_2(x)$ it is $\rho(x) = \rho_1(x)/\rho_2(x)$.

Since the right-hand side of (35) is a rational function, its solution can be expressed in terms of the solutions of the difference equations

$$\rho(x+1)/\rho(x) = \gamma + x, \tag{36}$$

$$\rho(x+1)/\rho(x) = \gamma - x, \tag{37}$$

$$\rho(x+1)/\rho(x) = \gamma, \tag{38}$$

where γ is a constant. Since

$$\gamma + x = \Gamma(\gamma+x+1)/\Gamma(\gamma+x),$$

a particular solution of (36) has the form $\rho(x) = \Gamma(\gamma+x)$. Similarly, by using the equation

$$\gamma - x = \frac{\Gamma(\gamma-x+1)}{\Gamma(\gamma-x)} = \frac{1}{\Gamma[(\gamma+1)-(x+1)]} : \frac{1}{\Gamma(\gamma+1-x)},$$

we obtain a particular solution of (37):

$$\rho(x) = 1/\Gamma(\gamma + 1 - x).$$

It is easily verified that a particular solution of (38) is $\rho(x) = \gamma^x$.

Let us now find solutions of (35) corresponding to the different possible degrees of $\sigma(x)$.

1) Let $\sigma(x) = x(\gamma_1 - x)$, $\sigma(x) + \tau(x) = (x + \gamma_2)(\gamma_3 - x)$.

Here γ_1, γ_2 and γ_3 are constants. With $a = 0$, $b = N$, conditions (29) and (30), namely

$$\begin{aligned} &\sigma(x_i) > 0 \qquad \text{for } 1 \le x_i \le N - 1, \\ &\sigma(x_i) + \tau(x_i) > 0 \text{ for } 0 \le x_i \le N - 2, \\ &\sigma(N - 1) + \tau(N - 1) = 0 \end{aligned} \tag{39}$$

will be satisfied if we take

$$\gamma_1 = N + \alpha, \quad \gamma_2 = \beta + 1 \quad (\alpha > -1,\ \beta > -1), \quad \gamma_3 = N - 1.$$

In this case (35) assumes the form

$$\frac{\rho(x+1)}{\rho(x)} = \frac{(x + \beta + 1)(N - 1 - x)}{(x + 1)(N + \alpha - 1 - x)}. \tag{40}$$

A solution of this equation is

$$\rho(x) = \frac{\Gamma(N + \alpha - x)\Gamma(x + \beta + 1)}{\Gamma(x + 1)\Gamma(N - x)} \quad (\alpha > -1,\ \beta > -1). \tag{41}$$

Let us discuss the reasons for choosing γ_1 and γ_2 in the forms $\gamma_1 = N + \alpha, \gamma_2 = \beta + 1$. It is natural to expect that a polynomial solution $y_n(x)$, after the linear change of variable $x = \frac{1}{2}N(1 + s)$, which carries the interval $(0, N)$ to $(-1, 1)$, will tend to the Jacobi polynomial $P_n^{(\alpha,\beta)}(s)$ when $N \to \infty$ (that is, when $\Delta s = h = 2/N \to 0$), and that the weight function $\rho(x)$ will tend, except for a constant multiple, to the weight function $(1 - s)^\alpha(1 + s)^\beta$ for the Jacobi polynomials. A solution of (35) for

$$\sigma(x) = x(\gamma_1 - x), \quad \sigma(x) + \tau(x) = (x + \gamma_2)(N - 1 - x)$$

is given by

$$\rho(x) = \frac{\Gamma(\gamma_1 - x)\Gamma(x + \gamma_2)}{\Gamma(x + 1)\Gamma(N - x)} = \frac{\Gamma\left[\frac{1}{2}N(1 - s) + \gamma_1 - N\right]}{\Gamma\left[\frac{1}{2}N(1 - s)\right]} \frac{\Gamma\left[\frac{1}{2}N(1 + s) + \gamma_2\right]}{\Gamma\left[\frac{1}{2}N(1 + s) + 1\right]}.$$

Since

$$\lim_{z\to\infty}\frac{\Gamma(z+a)}{\Gamma(z)z^a}=1, \tag{42}$$

we have

$$\rho(x)\approx\left[\frac{1}{2}N(1-s)\right]^{\gamma_1-N}\left[\frac{1}{2}N(1+s)\right]^{\gamma_2-1}$$

as $N\to\infty$. Consequently it is natural to take $\gamma_1-N=\alpha$, $\gamma_2-1=\beta$.

The polynomials $y_n(x)$ obtained by the Rodrigues formula (22) when $B_n=(-1)^n/n!$, with weight function $\rho(x)$ defined by (41), are called the *Hahn polynomials* and denoted by $h_n^{(\alpha,\beta)}(x,N)$. We shall also use the notation $h_n^{(\alpha,\beta)}(x)$ when N is fixed in the corresponding formulas.

An important special case of the Hahn polynomials, called the *Chebyshev polynomials of a discrete variable*, denoted by

$$t_n(x)=h_n^{(0,0)}(x,N),$$

arises when $\rho(x)=1$.

The Hahn polynomials $h_n^{(\alpha,\beta)}(x,N)$ and their differences are orthogonal on $[0,N-1]$ when $\alpha>-1$ and $\beta>-1$.

It is interesting that a simple relation was recently discovered connecting the Hahn polynomials with the Clebsch-Gordan coefficients, which are extensively used in quantum mechanics and in the theory of group representations. This has stimulated further study of the properties of these coefficients. The connection will be discussed in §26, part 5.

2) Let $\sigma(x)=x(x+\gamma_1)$, $\sigma(x)+\tau(x)=(\gamma_2-x)(\gamma_3-x)$. Condition (39) will be satisfied if

$$\gamma_1>-1,\quad \gamma_2>N-2,\quad \gamma_3=N-1.$$

In this case a solution of (35) is

$$\rho(x)=\frac{1}{\Gamma(x+1)\Gamma(x+\mu+1)\Gamma(N+\nu-x)\Gamma(N-x)}(\mu>-1,\nu>-1). \tag{43}$$

Here $\mu=\gamma_1,\nu=\gamma_2-N+1$.

The polynomials $y_n(x)$ obtained by the Rodrigues formula with $B_n=1/n!$, when $\rho(x)$ is defined by (43), are also called Hahn polynomials; they are denoted by $\tilde{h}_n^{(\mu,\nu)}(x,N)$.

There is a simple connection between the polynomials $\tilde{h}_n^{(\mu,\nu)}(x)$ and $h_n^{(\alpha,\beta)}(x)$. If we formally set $\mu=-N-\alpha$ and $\nu=-N-\beta$, the expressions for $\sigma(x)$ and $\sigma(x)+\tau(x)$ corresponding to $\tilde{h}_n^{(\mu,\nu)}(x)$ and $h_n^{(\alpha,\beta)}(x)$ differ only in sign. Consequently the polynomials $\tilde{h}_n^{(-N-\alpha,-N-\beta)}(x)$ and $h_n^{(\alpha,\beta)}(x)$ satisfy the same difference equation. It is easily verified that, with our normalization, these polynomials are the same. In fact, by using the gamma-function formula

$$\Gamma(z)\Gamma(1-z)=\frac{\pi}{\sin\pi z},$$

we can show that the expressions (41) and (43) for $\rho(x)$ differ, when $\mu = -N-\alpha$, $\nu=-N-\beta$, only by the periodic factor

$$C(x)=\frac{\pi^2}{\sin\pi(N+\alpha-x)\sin\pi(\beta+1+x)},$$

which does not affect the explicit formulas obtained by using the Rodrigues formula. Hence the formulas for $\tilde{h}_n^{(-N-\alpha,-N-\beta)}(x)$ and $h_n^{(\alpha,\beta)}(x)$ agree if the normalizing constants B_n differ by the factor $(-1)^n$, since the corresponding expressions for $\sigma(x)$ differ in sign.

Consequently the polynomials $\tilde{h}_n^{(-N-\alpha,-N-\beta)}(x)$ provide analytic continuations of $h_n^{(\alpha,\beta)}(x)$ with respect to the parameters α and β from the domain $\alpha>-1$, $\beta>-1$ into the domain $\alpha<1-N$, $\beta<1-N$.

Remark 1. When $\sigma(x)$ is a polynomial of degree 2, the solution of (35) for $\rho(x)$ behaves like a power as $x\to\pm\infty$, as we can see by using (42). Therefore the choice $a=-\infty$ or $b=+\infty$ has the effect that the moments $\sum_i x_i^k\rho(x_i)$ $(k=0,1,\ldots)$ of the weight function will not exist from some k onward, i.e. in this case there is only a finite set of orthogonal polynomials $\{y_n(x)\}$, although the number of points over which we sum in the orthogonality relation is infinite.

Remark 2. The polynomials $\tilde{h}_n^{(\mu,\nu)}(x,N)$ and $h_n^{(\alpha,\beta)}(x,N)$ can be expressed in terms of the polynomials $p_n(x,\beta,\gamma,\delta)$, discussed in [E2], for which

$$B_n=\frac{1}{n!},\quad \sigma(x)=x(\delta-1+x),\quad \rho(x)=\frac{(\beta)_x(\gamma)_x}{x!(\delta)_x},$$

where $(a)_x=\Gamma(a+x)/\Gamma(a)$. Since the functions $\sigma(x)$ for the polynomials $\tilde{h}_n(x,N)$ and $p_n(x,\beta,\gamma,\delta)$ coincide for $\mu=\delta-1$, $\nu=\gamma-\beta$, and $N=1-\gamma$, and the weight functions $\rho(x)$ differ only by a constant factor, we have

$$\tilde{h}_n^{(\mu,\nu)}(x,N)=p_n(x,1-N-\nu,1-N,1+\mu).$$

By the relation

$$h_n^{(\alpha,\beta)}(x,N)=\tilde{h}_n^{(-N-\alpha,-N-\beta)}(x,N)$$

that we had above, we also obtain

$$h_n^{(\alpha,\beta)}(x,N) = p_n(x,\beta+1,1-N,1-N-\alpha).$$

3) Let $\sigma(x) = x$. We consider the three cases

$$\sigma(x)+\tau(x) = \begin{cases} \mu(\gamma+x), \\ \mu(\gamma-x), \\ \mu. \end{cases}$$

Here μ and γ are constants. Then (35) has the following solutions:

$$\rho(x) = \begin{cases} C\dfrac{\mu^x\Gamma(\gamma+x)}{\Gamma(x+1)}, \\ C\dfrac{\mu^x}{\Gamma(x+1)\Gamma(\gamma+1-x)}, \\ C\dfrac{\mu^x}{\Gamma(x+1)}. \end{cases}$$

In the first case we can satisfy the boundary conditions (26) and the positivity of the weight function $\rho_k(x_i)$ by taking

$$a=0, \quad b=+\infty, \quad 0<\mu<1, \quad \gamma>0.$$

It is convenient to take C to be $1/\Gamma(\gamma)$. We then obtain the *Pascal distribution* from probability theory,

$$\rho(x) = \frac{\mu^x(\gamma)_x}{\Gamma(x+1)}. \tag{44}$$

With $B_n = \mu^{-n}$, the corresponding polynomials are the *Meixner polynomials* $m_n^{(\gamma,\mu)}(x)$.

Arguing similarly in the second case, we take

$$a=0,\ b=N+1,\ \gamma=N,\ \mu=p/q \quad (p>0,\ q>0,\ p+q=1),\ C=q^N N!.$$

The numbers $\rho(x_i)$ become the familiar *binomial distribution* from probability theory,

$$\rho(x_i) = C_N^i p^i q^{N-i}, \quad C_N^i = \frac{N!}{i!(N-i)!}. \tag{45}$$

With $B_n = (-1)^n q^n/n!$ the corresponding polynomials are the *Kravchuk polynomials* $k_n^{(p)}(x,N)$.

In the third case, with $a = 0$, $b = +\infty$, $C = e^{-\mu}$, we have the *Poisson distribution*

$$\rho(x_i) = \frac{e^{-\mu}\mu^i}{i!}. \tag{46}$$

The corresponding orthogonal polynomials of a discrete variable, with $B_n = \mu^{-n}$, are the *Charlier polynomials* $c_n^{(\mu)}(x)$.

4) The case $\sigma(x) = 1$ is not of interest, since it does not lead to any new polynomials.

From (23) we obtain the following *difference formulas* for the Hahn, Meixner, Kravchuk, and Charlier polynomials:

$$\Delta h_n^{(\alpha,\beta)}(x,N) = (\alpha+\beta+n+1)h_{n-1}^{(\alpha+1,\beta+1)}(x,N-1); \tag{47}$$

$$\Delta \tilde{h}_n^{(\mu,\nu)}(x,N) = -(\mu+\nu+2N-n-1)\tilde{h}_{n-1}^{(\mu,\nu)}(x,N-1); \tag{48}$$

$$\Delta k_n^{(p)}(x,N) = k_{n-1}^{(p)}(x,N-1); \tag{49}$$

$$\Delta m_n^{(\gamma,\mu)}(x) = -\frac{n(1-\mu)}{\mu} m_{n-1}^{(\gamma+1,\mu)}(x); \tag{50}$$

$$\Delta c_n^{(\mu)}(x) = -\frac{n}{\mu} c_{n-1}^{(\mu)}(x). \tag{51}$$

Let us consider the *symmetry properties* of the orthogonal polynomials of a discrete variable that follow from the symmetry of $\rho(x)$. For the Hahn polynomials $h_n^{(\alpha,\beta)}(x,N)$, the weight function $\rho(x)$ has the following symmetries:

$$\rho(x) \equiv \rho(x,\alpha,\beta) = \rho(N-1-x,\beta,\alpha).$$

Hence by replacing i by $N-1-i$ we can rewrite the orthogonality relation

$$\sum_{i=0}^{N-1} h_n^{(\alpha,\beta)}(x_i)h_m^{(\alpha,\beta)}(x_i)\rho(x_i,\alpha,\beta) = 0, \quad n \neq m,$$

in the form

$$\sum_{i=0}^{N-1} h_n^{(\alpha,\beta)}(N-1-x_i)h_m^{(\alpha,\beta)}(N-1-x_i)\rho(x_i,\beta,\alpha) = 0, \quad n \neq m.$$

Since the weight function and the interval of orthogonality determine the polynomials uniquely, up to a constant multiple, we have

$$h_n^{(\alpha,\beta)}(N-1-x) = C_n h_n^{(\beta,\alpha)}(x),$$

where C_n is a constant. Equating the coefficients of x^n on both sides by using (24), we obtain $C_n = (-1)^n$, i.e.

$$h_n^{(\alpha,\beta)}(N-1-x) = (-1)^n h_n^{(\beta,\alpha)}(x). \tag{52a}$$

Similarly, the Kravchuk polynomials satisfy

$$k_n^{(p)}(x) = (-1)^n k_n^{(q)}(N-x), \quad p+q=1. \tag{52b}$$

The relation (52a) remains valid for all complex values of x, α, β, N. For the proof it is sufficient to use the difference equation for the Hahn polynomials $y_n(x) = h_n^{(\alpha,\beta)}(x, N)$:

$$x(N+\alpha-x)\Delta\nabla y_n(x) + [(\beta+1)(N-1)-(\alpha+\beta+2)x]\Delta y_n(x) \\ + n(n+\alpha+\beta+1)y_n(x) = 0.$$

It is easy to verify that if we replace x by $N-1-x$, α by β, and β by α, this equation retains its form. Since, under this replacement, the polynomial $y_n(x)$ remains a polynomial of the same degree, then because of the uniqueness of polynomial solutions of difference equations of hypergeometric type we obtain the relation

$$h_n^{(\alpha,\beta)}(x, N) = C_n h_n^{(\beta,\alpha)}(N-1-x, N),$$

where C_n is a constant which may be found by equating the coefficients of x^n. The resulting relation is obviously equivalent to (52a). Similarly we may obtain (52b) for any complex values of x, p, N as well as the following relations:

$$h_n^{(\alpha,\beta)}(x, N) = h_n^{(-N,\alpha+\beta+N)}(x-\alpha-N, -\alpha), \tag{52c}$$

$$m_n^{(\gamma,\mu)}(x) = \mu^{-n} m_n^{(\gamma,1/\mu)}(-\gamma-x), \tag{52d}$$

$$k_n^{(p)}(x, N) = \frac{p^n}{n!} m_n^{(-N,-p/q)}(x). \tag{52e}$$

By using the Rodrigues formula, it is easy to find *the values of the Hahn, Meixner, Kravchuk, and Charlier polynomials at the endpoints of the interval of orthogonality*. Let us, as an example, find $h_n^{(\alpha,\beta)}(0)$. We use the formula

$$\nabla^n f(x) = \sum_{k=0}^{n} (-1)^k \frac{n!}{k!(n-k)!} f(x-k),$$

which can be proved by induction. Since the function $\rho_n(x)$ for the Hahn polynomials $h_n^{(\alpha,\beta)}(x)$ is zero at $x = -1, -2, \ldots$, we have $\nabla^n \rho_n(0) = \rho_n(0)$

and by the Rodrigues formula (22)

$$h_n^{(\alpha,\beta)}(0) = (-1)^n \frac{(N-1)!}{n!(N-n-1)!} \cdot \frac{\Gamma(n+\beta+1)}{\Gamma(\beta+1)}.$$

Similarly, for the Meixner, Kravchuk, and Charlier polynomials we obtain

$$\begin{aligned} m_n^{(\gamma,\mu)}(0) &= \frac{\Gamma(n+\gamma)}{\Gamma(\gamma)}, \\ k_n^{(p)}(0) &= (-1)^n \frac{N!}{n!(N-n)!} p^n, \\ c_n^{(\mu)}(0) &= 1. \end{aligned}$$

By using their symmetry properties, we can easily find expressions for $h_n^{(\alpha,\beta)}(N-1)$ and $k_n^{(p)}(N)$:

$$\begin{aligned} h_n^{(\alpha,\beta)}(N-1) &= \frac{(N-1)!}{n!(N-n-1)!} \cdot \frac{\Gamma(n+\alpha+1)}{\Gamma(\alpha+1)}. \\ k_n^{(p)}(N) &= \frac{N!}{n!(N-n)!} \cdot q^n. \end{aligned}$$

Finally we observe that *for any polynomials $p_n(x)$ that satisfy an orthogonality relation of the form*

$$\sum_{i=0}^{N-1} p_n(x_i) p_m(x_i) \rho_i = d_n^2 \cdot \delta_{mn}$$

there is another orthogonality relation. In fact, if we consider the matrix C whose elements are

$$c_{ni} = \frac{p_n(x_i)\sqrt{\rho_i}}{d_n},$$

the orthogonality property for the polynomials $p_n(x)$ is equivalent to the unitary property of the matrix C:

$$\sum_{i=0}^{N-1} c_{ni} \cdot c_{mi} = \delta_{mn}.$$

Hence C also satisfies

$$\sum_{n=0}^{N-1} c_{nk} \cdot c_{nl} = \delta_{kl},$$

which is equivalent to the *"dual" orthogonality relation for the* $p_n(x)$:

$$\sum_{n=0}^{N-1} p_n(x_k) \cdot p_n(x_l)\bar{\rho}_n = \frac{1}{\rho_k}\delta_{kl},$$

where $\bar{\rho}_n = 1/d_n^2$.

Let us show that the dual orthogonality relation for the Hahn polynomials $h_n^{(\alpha,\beta)}(x)$ leads to a new system of orthogonal polynomials. We need to know the dependence on n of the values of $h_n^{(\alpha,\beta)}(x)$ at $x = i$ $(i = 0, 1, \ldots)$.

For this purpose we rewrite the difference equation for the Hahn polynomials in the form

$$y_{i+1} = (-A_i\lambda_n + B_i)y_i + C_i y_{i-1} \quad (C_0 = 0),$$

where

$$y_i = h_n^{(\alpha,\beta)}(i), \quad y_0 = (-1)^n \frac{(N-1)!\Gamma(n+\beta+1)}{n!(N-n-1)!\Gamma(\beta+1)},$$

$$A_i = \frac{1}{\sigma(i)+\tau(i)} = [(N-i-1)(i+\beta+1)]^{-1},$$

B_i and C_i are constants independent of n.

Hence by induction we find that y_i/y_0 is a polynomial of degree i in λ_n with leading coefficient equal to

$$(-1)^i \prod_{k=0}^{i-1} A_k.$$

Since

$$\lambda_n = t_n - \frac{1}{4}(\alpha+\beta)(\alpha+\beta+2),$$

where $t_n = s_n(s_n+1)$, $s_n = n + (\alpha+\beta)/2$, we obtain

$$\begin{aligned} h_n^{(\alpha,\beta)}(i) &= (-1)^i h_n^{(\alpha,\beta)}(0) i! w_i^{(\alpha,\beta)}(t_n) \prod_{k=0}^{i-1} A_k \\ &= (-1)^{n+i} \frac{i!(N-i-1)!}{n!(N-n-1)!} \cdot \frac{\Gamma(n+\beta+1)}{\Gamma(i+\beta+1)} w_i^{(\alpha,\beta)}(t_n), \end{aligned}$$

where $w_i^{(\alpha,\beta)}(t)$ is a polynomial of degree i in t, with leading coefficient $1/i!$. Hence the dual orthogonality relation for the Hahn polynomials $h_n^{(\alpha,\beta)}(x)$

leads to the following orthogonality relation for the polynomials $w_i^{(\alpha,\beta)}(t)$, which we naturally call the ***dual Hahn polynomials***:

$$\sum_{n=0}^{N-1} w_k^{(\alpha,\beta)}(t_n) w_l^{(\alpha,\beta)}(t_n)\tilde{\rho}_n = D_k^2 \delta_{kl},$$

where

$$\tilde{\rho}_n = \frac{\alpha+\beta+2n+1}{n!(N-n-1)!} \frac{\Gamma(\alpha+\beta+n+1)\Gamma(\beta+n+1)}{\Gamma(\alpha+\beta+n+N+1)\Gamma(\alpha+n+1)}$$

and

$$D_k^2 = \frac{\Gamma(\beta+k+1)}{k!(N-k-1)!\Gamma(N+\alpha-k)}$$

($\tilde{\rho}_n$ and D_k^2 are obtained by taking d_n^2 for the Hahn polynomials — see part 5, Table 3.)

Consequently the dual orthogonality relation for the Hahn polynomials $h_n^{(\alpha,\beta)}(x)$ leads to the dual Hahn polynomials, which are orthogonal on a nonuniform lattice. The theory of these polynomials will be discussed in the next section. We note that the dual orthogonality relation for the Kravchuk polynomials does not lead to a new system of orthogonal polynomials, i.e. the Kravchuk polynomials are self-dual.

5. Calculation of leading coefficients and squared norms. Tables of data. Let us obtain the fundamental constants for the Hahn, Meixner, Kravchuk, and Charlier polynomials. We first find the leading coefficients a_n and b_n in the expansions

$$y_n(x) = a_n x^n + b_n x^{n-1} + \dots .$$

Here will be sufficient to use (24) and (25).

The squared norm d_n^2 can be found from (31). Its calculation leads to the evaluation of the sums

$$S_n = \sum_{x_i=a}^{b-n-1} \rho_n(x_i).$$

The calculation of S_n is especially simple for the Meixner, Kravchuk, and Charlier polynomials. For these polynomials

$$\rho_n(x) = \rho(x+n)\prod_{k=1}^{n}\sigma(x+k) = \rho(x+n)\frac{\Gamma(1+x+n)}{\Gamma(1+x)}.$$

For the Meixner polynomials

$$\sum_i \rho_n(x_i) = \sum_{i=0}^{\infty} \frac{\mu^{i+n}\Gamma(\gamma+i+n)}{i!\Gamma(\gamma)}.$$

For $|\mu| < 1$ we have the Maclaurin expansion

$$(1-\mu)^{-(\gamma+n)} = \sum_{i=0}^{\infty} \frac{\Gamma(\gamma+i+n)}{\Gamma(\gamma+n)} \frac{\mu^i}{i!},$$

and so, for the polynomials $m_n^{(\gamma,\mu)}(x)$, we obtain

$$d_n^2 = \frac{n!(\gamma)_n}{\mu^n(1-\mu)^\gamma}. \tag{53a}$$

For the Kravchuk polynomials,

$$\sum_i \rho_n(x_i) = \sum_{i=0}^{N-n} \frac{N!p^{i+n}q^{N-i-n}}{i!\Gamma(N+1-i-n)}$$
$$= \frac{N!p^n}{(N-n)!} \sum_{i=0}^{N-n} C_{N-n}^i p^i q^{N-n-i} = \frac{N!p^n}{(N-n)!},$$

whence

$$d_n^2 = C_N^n (pq)^n. \tag{53b}$$

For the Charlier polynomials,

$$\sum_i \rho_n(x_i) = \sum_{i=0}^{\infty} \frac{e^{-\mu}\mu^{i+n}}{i!} = \mu^n,$$

whence

$$d_n^2 = n!/\mu^n. \tag{53c}$$

For the Hahn polynomials the squared norms can be found from (31a), since in this case the sum S_n reduces to a single term. We then find the following expressions for d_n^2:

$$d_n^2 = \begin{cases} \dfrac{\Gamma(\alpha+n+1)\Gamma(\beta+n+1)(\alpha+\beta+n+1)_N}{(\alpha+\beta+2n+1)n!(N-n-1)!} \\ \qquad (\text{for } h_n^{(\alpha,\beta)}(x,N)), \\ \dfrac{(\mu+\nu+N-n)_N}{(\mu+\nu+2N-2n-1)n!\Gamma(\mu+N-n)\Gamma(\nu+N-n)(N-n-1)!} \\ \qquad (\text{for } \tilde{h}_n^{(\mu,\nu)}(x)). \end{cases} \tag{53d}$$

Since the orthogonality property (27) of the classical orthogonal polynomials of a discrete variable is obtained from the general orthogonality property of orthogonal polynomials by replacing integration by summation, it follows that, with the corresponding scalar product (y_n, y_m) for the Hahn, Meixner, Kravchuk and Charlier polynomials, all the general properties of orthogonal polynomials are preserved. In particular, we have the *recursion relation*

$$xy_n(x) = \alpha_n y_{n+1}(x) + \beta_n y_n(x) + \gamma_n y_{n-1}(x), \tag{54}$$

whose coefficients can be found from the known values of a_n, b_n, and d_n^2 by the formulas

$$\alpha_n = \frac{a_n}{a_{n+1}}, \quad \beta_n = \frac{b_n}{a_n} - \frac{b_{n+1}}{a_{n+1}}, \quad \gamma_n = \frac{a_{n-1}}{a_n}\frac{d_n^2}{d_{n-1}^2}.$$

The basic information about the Hahn, Chebyshev, Meixner, Kravchuk, and Charlier polynomials is displayed in Tables 3, 4, and 5.

Table 3. Data for the Hahn polynomials $h_n^{(\alpha,\beta)}(x,N)$ and the Chebyshev polynomials $t_n(x)$

$y_n(x)$	$h_n^{(\alpha,\beta)}(x,N)$	$t_n(x)$
(a,b)	$(0,N)$	$(0,N)$
$\rho(x)$	$\dfrac{\Gamma(N+\alpha-x)\Gamma(\beta+1+x)}{\Gamma(x+1)\Gamma(N-x)} \quad (\alpha>-1,\beta>-1)$	1
$\sigma(x)$	$x(N+\alpha-x)$	$x(N-x)$
$\tau(x)$	$(\beta+1)(N-1)-(\alpha+\beta+2)x$	$N-1-2x$
λ_n	$n(\alpha+\beta+n+1)$	$n(n+1)$
B_n	$\dfrac{(-1)^n}{n!}$	$\dfrac{(-1)^n}{n!}$
$\rho_n(x)$	$\dfrac{\Gamma(N+\alpha-x)\Gamma(n+\beta+1+x)}{\Gamma(x+1)\Gamma(N-n-x)}$	$\dfrac{\Gamma(N-x)\Gamma(n+1+x)}{\Gamma(N-n-x)\Gamma(x+1)}$
a_n	$\dfrac{1}{n!}(\alpha+\beta+n+1)_n$	$\dfrac{1}{n!}(n+1)_n$
b_n	$-\dfrac{1}{(n-1)!}\left[(\beta+1)(N-1)+\dfrac{n-1}{2}(\alpha-\beta+2N-2)\right] \times(\alpha+\beta+n+1)_{n-1}$	$-\dfrac{N-1}{(n-1)!}(n)_n$
d_n^2	$\dfrac{\Gamma(\alpha+n+1)\Gamma(\beta+n+1)(\alpha+\beta+n+1)_N}{(\alpha+\beta+2n+1)n!(N-n-1)!}$	$\dfrac{(N+n)!}{(2n+1)(N-n-1)!}$
α_n	$\dfrac{(n+1)(\alpha+\beta+n+1)}{(\alpha+\beta+2n+1)(\alpha+\beta+2n+2)}$	$\dfrac{n+1}{2(2n+1)}$
β_n	$\dfrac{\alpha-\beta+2N-2}{4}+\dfrac{(\beta^2-\alpha^2)(\alpha+\beta+2N)}{4(\alpha+\beta+2n)(\alpha+\beta+2n+2)}$	$\dfrac{N-1}{2}$
γ_n	$\dfrac{(\alpha+n)(\beta+n)(\alpha+\beta+N+n)(N-n)}{(\alpha+\beta+2n)(\alpha+\beta+2n+1)}$	$\dfrac{n(N^2-n^2)}{2(2n+1)}$

Table 4. Data for the Hahn polynomials $\tilde{h}_n^{\mu,\nu}(x,N)$

$y_n(x)$	$\tilde{h}_n^{(\mu,\nu)}(x,N)$
(a,b)	$(0,N)$
$\rho(x)$	$\dfrac{1}{\Gamma(x+1)\Gamma(x+\mu+1)\Gamma(N+\nu-x)\Gamma(N-x)}$ $(\mu>-1,\ \nu>-1)$
$\sigma(x)$ $\tau(x)$ $\lambda_n(x)$	$x(x+\mu)$ $(N+\nu+1)(N-1)-(2N+\mu+\nu-2)x$ $n(2N+\mu+\nu-n-1)$
B_n	$\dfrac{1}{n!}$
$\rho_n(x)$	$\dfrac{1}{\Gamma(x+1)\Gamma(x+\mu+1)\Gamma(N+\nu-n-x)\Gamma(N-n-x)}$
a_n b_n d_n^2	$\dfrac{(-1)^n}{n!}(2N+\mu+\nu-2n)_n$ $\dfrac{(-1)^{n-1}}{(n-1)!}\Big[(N+\nu-1)(N-1) - \dfrac{n-1}{2}(2N+\nu-\mu-2)\Big](2N+\mu+\nu-2n+1)_{n-1}$ $\dfrac{(N+\mu+\nu-n)_N}{(2N+\mu+\nu-2n-1)n!\Gamma(N+\mu-n)\Gamma(N+\nu-n)(N-n-1)!}$
α_n β_n γ_n	$-\dfrac{(n+1)(2N+\mu+\nu-n-1)}{(2N+\mu+\nu-2n-1)(2N+\mu+\nu-2n-2)}$ $\dfrac{2(N-1)+\nu-\mu}{4}+\dfrac{(\mu^2-\nu^2)(2N+\mu+\nu)}{4(2N+\mu+\nu-2n)(2N+\mu+\nu-2n-2)}$ $-\dfrac{(N-n)(N-n+\mu)(N-n+\nu)(N-n+\mu+\nu)}{(2N+\mu+\nu-2n)(2N+\mu+\nu-2n-1)}$

Table 5. Data for the Meixner, Kravchuk and Charlier polynomials

$y_n(x)$	$m_n^{(\gamma,\mu)}(x)$	$k_n^{(p)}(x)$	$c_n^{(\mu)}(x)$
(a,b)	$(0,\infty)$	$(0,N+1)$	$(0,\infty)$
$\rho(x)$	$\dfrac{\mu^x\Gamma(\gamma+x)}{\Gamma(1+x)\Gamma(\gamma)}$ $(\gamma>0, 0<\mu<1)$	$\dfrac{N!p^x q^{N-x}}{\Gamma(1+x)\Gamma(N+1-x)}$ $(p>0, q>0, p+q=1)$	$\dfrac{e^{-\mu}\mu^x}{\Gamma(1+x)}$ $(\mu>0)$
$\sigma(x)$	x	x	x
$\tau(x)$	$\gamma\mu-x(1-\mu)$	$\dfrac{1}{q}(Np-x)$	$\mu-x$
λ_n	$n(1-\mu)$	$\dfrac{n}{q}$	n
B_n	$\dfrac{1}{\mu^n}$	$\dfrac{(-1)^n q^n}{n!}$	$\dfrac{1}{\mu^n}$
$\rho_n(x)$	$\dfrac{\mu^{x+n}\Gamma(n+\gamma+x)}{\Gamma(\gamma)\Gamma(x+1)}$	$\dfrac{N!p^{x+n}q^{N-n-x}}{\Gamma(x+1)\Gamma(N+1-n-x)}$	$\dfrac{e^{-\mu}\mu^{x+n}}{\Gamma(x+1)}$
a_n	$\left(\dfrac{\mu-1}{\mu}\right)^n$	$\dfrac{1}{n!}$	$\dfrac{1}{(-\mu)^n}$
b_n	$n\left(\gamma+\dfrac{n-1}{2}\dfrac{\mu+1}{\mu}\right)\left(\dfrac{\mu-1}{\mu}\right)^{n-1}$	$-\dfrac{Np+(n-1)(\frac{1}{2}-p)}{(n-1)!}$	$\dfrac{n(1+\frac{n-1}{2\mu})}{(-\mu)^{n-1}}$
d_n^2	$\dfrac{n!(\gamma)_n}{\mu^n(1-\mu)^\gamma}$	$\dfrac{N!}{n!(N-n)!}(pq)^n$	$\dfrac{n!}{\mu^n}$
α_n	$\dfrac{\mu}{\mu-1}$	$n+1$	$-\mu$
β_n	$\dfrac{n+\mu(n+\gamma)}{1-\mu}$	$n+p(N-2n)$	$n+\mu$
γ_n	$\dfrac{n(n-1+\gamma)}{\mu-1}$	$pq(N-n+1)$	$-n$

6. Connection with the Jacobi, Laguerre and Hermite polynomials. It is natural to expect that when $h \to 0$ the polynomial solutions of (2), properly normalized, will converge to the polynomial solutions of (1), i.e. to the Jacobi, Laguerre, and Hermite polynomials. The validity of this proposition is most easily established by induction by using the corresponding convergence of the recursion relations (54) for the respective polynomials. As an example, we carry out the limiting process for the Hahn and Jacobi polynomials.

To begin with, the linear change of variable $x = N(1+s)/2$ carries the interval $(0, N)$ of orthogonality for the Hahn polynomials to $(-1, 1)$. Then the difference equation (3) for the polynomials $h_n^{(\alpha,\beta)}(x) = u(s)$ takes the form

$$\begin{aligned}(1+s)(1-s+\alpha h)&\frac{u(s+h)-2u(s)+u(s-h)}{h^2}\\ &-[(\alpha+\beta+2)s+\alpha-\beta+(\beta+1)h]\frac{u(s+h)-u(s)}{h}\\ &+n(n+\alpha+\beta+1)u(s)=0,\end{aligned} \tag{55}$$

with $h = 2/N$.

As $h \to 0$, equation (55) goes over formally to the differential equation for the Jacobi polynomials $P_n^{(\alpha,\beta)}(s)$. Hence we expect the limit relation

$$\lim_{N\to\infty} C_n(N)h_n^{(\alpha,\beta)}\left[\frac{1}{2}N(1+s)\right] = P_n^{(\alpha,\beta)}(s). \tag{56}$$

where $C_n(N)$ is a normalizing factor.

To establish the validity of (56) and find $C_n(N)$ we compare the recursion relations for $P_n^{(\alpha,\beta)}(s)$ and $v_n(s,N) = C_n(N)h_n^{(\alpha,\beta)}[\frac{1}{2}N(1+s)]$:

$$\begin{aligned}sP_n^{(\alpha,\beta)}(s) =&\frac{2(n+1)(n+\alpha+\beta+1)}{(2n+\alpha+\beta+1)(2n+\alpha+\beta+2)}P_{n+1}^{(\alpha,\beta)}(s)\\ &+\frac{\beta^2-\alpha^2}{(2n+\alpha+\beta)(2n+\alpha+\beta+2)}P_n^{(\alpha,\beta)}(s)\\ &+\frac{2(n+\alpha)(n+\beta)}{(2n+\alpha+\beta)(2n+\alpha+\beta+1)}P_{n-1}^{(\alpha,\beta)}(s);\end{aligned}$$

$$\begin{aligned}sv_n =& \frac{2(n+1)(n+\alpha+\beta+1)}{(2n+\alpha+\beta+1)(2n+\alpha+\beta+2)}\frac{C_n}{NC_{n+1}}v_{n+1}\\ &+\frac{\beta^2-\alpha^2+(2/N)[n(n+\alpha+\beta+1)(\alpha-\beta-2)-(\beta+1)(\alpha+\beta)]}{(2n+\alpha+\beta)(2n+\alpha+\beta+2)}v_n\\ &+\frac{2(n+\alpha)(n+\beta)}{(2n+\alpha+\beta)(2n+\alpha+\beta+1)}\left(1-\frac{n}{N}\right)\left(1+\frac{n+\alpha+\beta}{N}\right)\frac{NC_n}{C_{n-1}}v_{n-1}.\end{aligned}$$

If we compare these recursion relations, it is clear that (56) will hold for all n if it is satisfied for $n = 0$ and $n = 1$, and if $C_n/C_{n+1} = N$. This yields $C = N^{-n}$.

Hence we obtain the following limit relation:

$$N^{-n} h_n^{(\alpha,\beta)} \left[\frac{1}{2} N(1+s)\right] = P_n^{(\alpha,\beta)}(s) + O(1/N). \tag{57}$$

In particular, for $t_n(x)$, the Chebyshev polynomials of a discrete variable, (57) takes the form

$$N^{-n} t_n \left[\frac{1}{2} N(1+s)\right] = P_n(s) + O(1/N), \tag{58}$$

where $P_n(s)$ are the Legendre polynomials.

By the same method we can obtain the following more precise asymptotic formula for the Hahn polynomials $h_n^{(\alpha,\beta)}(x, N)$:

$$\tilde{N}^{-1} h_n^{(\alpha,\beta)} \left[\frac{1}{2} \tilde{N}(1+s) - \frac{1}{2}(\beta+1), N\right] = P_n^{(\alpha,\beta)}(s) + O(\tilde{N}^{-2}), \tag{57a}$$

where $\tilde{N} = N + \frac{1}{2}(\alpha+\beta)$ $(N \to \infty)$.

In just the same way, if we put $y(x) = u(s)$, $x = sh$, $\mu = 1 - h$ in the equation for the Meixner polynomials we obtain a difference equation which for $h \to 0$ goes over to the differential equation

$$su'' + (\gamma - s)u' + nu = 0.$$

Polynomial solutions of this equation have the form $L_n^{\gamma-1}(s)$. Hence we expect a limit relation

$$\lim_{h \to 0} c_n m_n^{(\gamma, 1-h)}(s/h) = L_n^{(\gamma-1)}(s).$$

Equating coefficients for terms of higher degree, we obtain $c_n = 1/n!$. By the recursion relations for the Meixner and Laguerre polynomials we obtain

$$\frac{1}{n!} m_n^{(\alpha+1, 1-h)} \left(\frac{s}{h}\right) = L_n^{\alpha}(s) + O(h). \tag{59}$$

We now find the limit relation for the Kravchuk polynomials $k_n^{(p)}(x)$. Here it is convenient to appeal to a well known theorem on the binomial distribution from probability theory, namely that as $N \to \infty$ we have

$$\rho(x_i) = C_n^i p^i q^{n-i} \approx \frac{1}{\sqrt{2\pi N pq}} \exp\left\{-\frac{(i - Np)^2}{2Npq}\right\},$$

i.e. the weight function $\rho(x)$ for the Kravchuk polynomials, with $x = x_i = i$, tends, except for a normalizing factor, to the weight function for the Hermite polynomials,

$$\rho(s) = e^{-s^2} \text{ with } s = \frac{x - Np}{\sqrt{2Npq}}.$$

Corresponding to this, we put

$$x = Np + \sqrt{2Npq}s, \quad y(x) = u(s), \quad h = \frac{1}{\sqrt{2Npq}}$$

in the equation for the Kravchuk polynomials. Then this equation takes the form

$$\left(1 + \sqrt{\frac{2q}{Np}}s\right) \frac{u(s+h) - 2u(s) + u(s-h)}{h^2} - 2s\frac{u(s+h) - u(s)}{h} + 2nu(s) = 0.$$

As $N \to \infty$ this equation goes over formally to the differential equation

$$u'' - 2su' + 2nu = 0,$$

whose polynomial solutions are the Hermite polynomials $H_n(s)$. Repeating the arguments used for (57)–(59), we obtain

$$\lim_{N\to\infty} \left(\frac{2}{Npq}\right)^{n/2} n! k_n^{(p)}(Np + \sqrt{2Npq}s) = H_n(s). \tag{60}$$

7. Relation between generalized spherical harmonics and Kravchuk polynomials. Let us show that there is a relation between the generalized spherical harmonics and the Kravchuk polynomials, a relation that immediately implies the unitary property of the spherical harmonics. For this purpose we use formula (10.37) and the explicit expression of the generalized spherical harmonics in terms of the Jacobi polynomials. We have

$$\sum_{m'=-l}^{l} d^l_{mm'}(\beta) d^l_{m_1 m'}(\beta) = \delta_{mm_1}, \tag{61}$$

where

$$\begin{aligned} d^l_{mm'}(\beta) = {} & \frac{1}{2^m}\sqrt{\frac{(l+m)!(l-m)}{(l+m')!(l-m')!}} \\ & \times (1-\cos\beta)^{(m-m')/2}(1+\cos\beta)^{(m+m')/2}\, P_{l-m}^{(m-m',m+m')}(\cos\beta). \end{aligned} \tag{62}$$

To determine the nature of the dependence of $d^l_{mm'}(\beta)$ on m', with l, m and β fixed, we use the Rodrigues formula for the Jacobi polynomials (see §5). By applying Leibniz's rule for differentiating a product, it is easy to see that for fixed n and x the Jacobi polynomial $P_n^{(\alpha,\beta)}(x)$ is a polynomial in α and β and that the sum of the highest powers of α and β is n. In our case α and β depend linearly on m'. Consequently

$$d^l_{mm'}(\beta) = \frac{1}{\sqrt{(l+m')!(l-m')!}} \left(\frac{1-\cos\beta}{1+\cos\beta}\right)^{(l-m')/2} q_{l-m}(l-m'), \quad (63)$$

where $q_n(x)$ is a polynomial of degree n in x. Hence we can write (61) in the form

$$\sum_{x=0}^{2l} q_{l-m}(x) q_{l-m_1}(x) \omega(x) = \delta_{mm_1}, \quad (64)$$

where

$$\omega(x) = \frac{1}{x!(2l-x)!} \left(\frac{1-\cos\beta}{1+\cos\beta}\right)^x.$$

Formula (64) shows that the polynomials $q_n(x)$ are orthogonal on the discrete set $x = 0, 1, \ldots, 2l$ with weight function $\omega(x)$. It is easy to see that $\omega(x)$ coincides, up to a constant factor, with the weight function $\rho(x)$ with respect to which the Kravchuk polynomials $k_n^{(p)}(x, N)$ are orthogonal over the same set of points x, if we take

$$\frac{p}{1-p} = \frac{1-\cos\beta}{1+\cos\beta}, \quad N = 2l.$$

Therefore, by the uniqueness of the system of orthogonal polynomials with a given weight function, $q_n(x)$ coincides, up to a multiple independent of x, with the Kravchuk polynomials $k_n^{(p)}(x, N)$ with $p = \sin^2(\beta/2), N = 2l$. Thus (63) implies

$$d^l_{mm'}(\beta) = C\sqrt{\rho(x)}\, k_n^{(p)}(x, N) \quad (65)$$

with $x = l - m', n = l - m$. The normalizing constant $C = C(l, m, \beta)$ can be found, except for sign, from (61) with $m = m_1$:

$$C^2 d_n^2 = 1,$$

where d_n^2 is the squared norm of the Kravchuk polynomial. To determine the sign of C it is enough to find the sign of the coefficient of the term of highest degree m' on the left-hand and right-hand sides of (65). If we use the expression (62) for $d^l_{m,m'}(\beta)$, we obtain $C > 0$.

Finally, we have the following connection between generalized spherical harmonics and the Kravchuk polynomials [K2]:

$$d^l_{mm'}(\beta) = d_n^{-1}\sqrt{\rho(x)}k_n^{(p)}(x, N), \tag{66}$$

where

$$x = l - m', n = l - m, N = 2l, p = \sin^2(\beta/2),$$
$$\rho(x) = C^x_{2l}p^x(1-p)^{2l-x}.$$

8. Particular solutions for the difference equation of hypergeometric type. It was shown in §3 how we can obtain particular solutions for the equation of hypergeometric type (2.1) by generalizing the Rodrigues formula for classical orthogonal polynomials. There is a similar theorem for difference equations of hypergeometric type.

Theorem. *The difference equation of hypergeometric type*

$$\sigma(z)\Delta\nabla y + \tau(z)\Delta y + \lambda y = 0 \tag{67}$$

has particular solutions of the form

$$y = y_\nu(z) = \frac{C_\nu}{\rho(z)}\sum_{s=a}^{b-1}\frac{\rho_\nu(s)}{(s-z)_{\nu+1}} \tag{68}$$

if the condition

$$\left.\frac{\sigma(s)\rho_\nu(s)}{(s-z-1)_{\nu+2}}\right|_a^b = 0, \tag{69}$$

is satisfied, and has solutions of the form

$$y = y_\nu(z) = \frac{C_\nu}{\rho(z)}\int_C \frac{\rho_\nu(s)ds}{(s-z)_{\nu+1}} \tag{70}$$

if the condition

$$\int_C \frac{\sigma(s+1)\rho(s+1)}{(s-z)_{\nu+2}}ds = \int_C \frac{\sigma(s)\rho(s)}{(s-z-1)_{\nu+2}}ds \tag{71}$$

is satisfied. Here $(\omega)_\nu = \Gamma(\omega+\nu)/\Gamma(\omega)$, C is a contour in the complex s-plane, C_ν is a constant, the functions $\rho(z)$ and $\rho_\nu(z)$ are solutions of the equations

$$\Delta(\sigma\rho) = \tau\rho, \quad \Delta(\sigma\rho_\nu) = \tau_\nu\rho_\nu, \tag{72}$$

where

$$\tau_\nu = \tau_\nu(z) = \sigma(z+\nu) - \sigma(z) + \tau(z+\nu), \tag{73}$$

and ν is a root of the equation

$$\lambda + \nu\tau' + \frac{1}{2}\nu(\nu-1)\sigma'' = 0. \tag{74}$$

Proof. We multiply the both sides of the equation

$$\Delta[\sigma(s)\rho_\nu(s)] = \tau_\nu(s)\rho_\nu(s)$$

by $1/(s-z)_{\nu+2}$ and transform the left-hand side of the resulting equality, with a given z, using the formula

$$\Delta[f(s)g(s)] = f(s)\Delta g(s) + g(s+1)\Delta f(s).$$

Setting

$$g(s+1) = \frac{1}{(s-z)_{\nu+2}}, \quad f(s) = \sigma(s)\rho_\nu(s)$$

we obtain

$$\Delta g(s) = -\frac{(\nu+2)}{[s-(z+1)]_{\nu+3}}, \quad \Delta f(s) = \tau_\nu(s)\rho_\nu(s),$$

whence

$$\Delta\left[\frac{\sigma(s)\rho_\nu(s)}{(s-z-1)_{\nu+2}}\right] + (\nu+2)\frac{\sigma(s)\rho_\nu(s)}{[s-(z+1)]_{\nu+3}} = \frac{\tau_\nu(s)\rho_\nu(s)}{(s-z)_{\nu+2}}. \tag{75}$$

Let us consider the first case, i.e. when the solution is represented as the sum (68). Putting $s = a, a+1, \ldots, b-1$ in (75) and summing from $s=a$ to $s=b-1$, we obtain

$$\left.\frac{\sigma(s)\rho_\nu(s)}{(s-z-1)_{\nu+2}}\right|_a^b + (\nu+2)\sum_s \frac{\sigma(s)\rho_\nu(s)}{[s-(z+1)]_{\nu+3}} = \sum_s \frac{\tau_\nu(s)\rho_\nu(s)}{(s-z)_{\nu+2}}.$$

The result of substituting a and b is zero if (69) is satisfied. Therefore

$$(\nu+2)\sum_s \frac{\sigma(s)\rho_\nu(s)}{[s-(z+1)]_{\nu+3}} = \sum_s \frac{\tau_\nu(s)\rho_\nu(s)}{(s-z)_{\nu+2}}. \tag{76}$$

Using the expansions

$$\sigma(s)=\sigma(z+1)+(s-z-1)\Delta\sigma(z)+(s-z)(s-z-1)\sigma''/2,$$
$$\tau_\nu(s)=\tau_\nu(z)+(s-z)\tau'_\nu,$$

we can rewrite (76) in the form

$$\frac{1}{\nu+1}\sigma(z+1)\Delta\nabla u(z)+\frac{\nu+2}{\nu+1}\Delta\sigma(z)\nabla u(z)+\frac{\nu+2}{2}\sigma''u(z-1)$$
$$=\frac{1}{\nu+1}\tau_\nu(z)\nabla u(z)+\tau'_\nu u(z-1),$$

where

$$u(z)=\sum_s\frac{\rho_\nu(s)}{(s-z)_{\nu+1}}.$$

Taking account of (73) and (74), we obtain

$$\sigma(z+1)u(z+1)+[\lambda-\tau(z)-2\sigma(z)]u(z)+[\sigma(z-1)+\tau(z-1)]u(z-1)=0.$$

The relation obtained for the function $u(z)$ is equivalent to the difference equation (67), as can easily be verified if one replaces $u(z)$ by $(1/C_\nu)\rho(z)y(z)$ and then uses equation (72) for $\rho(z)$. The proof in the second case is similar except that we use integration instead of summation and (71) instead of (69). This completes the proof.

Remark. Let us discuss conditions (69) and (71). The first condition will be valid if

$$\left.\frac{\sigma(s)\rho_\nu(s)}{(s-z-1)_{\nu+2}}\right|_{s=a,b}=0. \tag{69a}$$

In some cases this condition is satisfied if $\sigma(a)=0$, $b=\infty$.

Condition (71) can be rewritten as

$$\int_{C'}\frac{\sigma(s')\rho_\nu(s')}{(s-z-1)_{\nu+2}}ds'=\int_C\frac{\sigma(s)\rho_\nu(s)}{(s-z-1)_{\nu+2}}ds, \tag{71a}$$

where C' is the contour obtained from C by the shift $s'=s+1$. The condition (71a) is satisfied if, for example, C and C' are closed and surround all the singularities of the integrand.

Examples. By using the theorem just proved, we construct some particular solutions of equation (67) corresponding to various choices of the form of the solution, the parameters a, b, ν, and the contour C.

1°. *The classical orthogonal polynomials of a discrete variable.* Let us consider equation (67) with $\lambda = \lambda_n = -n\tau' - \frac{1}{2}n(n-1)\sigma''$ $(n = 0, 1, \ldots)$. We choose the particular solution in the form (70):

$$y_n(z) = \frac{B_n n!}{2\pi i \rho(z)} \int_C \frac{\rho_n(s)}{(s-z)_{n+1}} ds, \tag{77}$$

where B_n is a constant, and C is a closed contour in the complex s-plane surrounding the points $s = z, z-1, \ldots, z-n$, i.e., the zeros of the denominator in the integrand, $(s-z)_{n+1} = (s-z)(s-z+1)\ldots(s-z+n)$. We shall also assume that the integrand has no other singularities inside C and that C' has the same properties as C.

Let us show that the solution (77) determines the classical orthogonal polynomials of a discrete variable. By induction, it can easily be verified that

$$\nabla_z^n \left(\frac{1}{s-z}\right) = \frac{n!}{(s-z)_{n+1}}.$$

Hence

$$y_n(z) = \frac{B_n}{2\pi i \rho(z)} \int_C \rho_n(s) \nabla_z^n \left(\frac{1}{s-z}\right) ds = \frac{B_n}{\rho(z)} \nabla_z^n \left[\frac{1}{2\pi i} \int_C \frac{\rho_n(s)}{s-z} ds\right].$$

Using the Cauchy integral formula

$$\rho_n(z) = \frac{1}{2\pi i} \int_C \frac{\rho_n(s)}{s-z} ds$$

we obtain *the Rodrigues formula for the classical orthogonal polynomials of a discrete variable*:

$$y_n(z) = \frac{B_n}{\rho(z)} \nabla^n [\rho_n(z)]. \tag{78}$$

Hence (77) is *the integral representation* for the classical orthogonal polynomials of a discrete variable.

2°. *The functions of the second kind of a discrete variable.* As the second linearly independent solution of equation (67) for $\lambda = \lambda_n$ we take a function of the form (68):

$$y = Q_n(z) = \frac{B_n n!}{\rho(z)} \sum_{s=a}^{b-1} \frac{\rho_n(s)}{(s-z)_{n+1}} \qquad (79)$$
$$(z \neq a, a+1, \ldots, b-1),$$

where B_n is the constant in the Rodrigues formula (78). The constants a and b are to be subject to the condition

$$\sigma(s)\rho(s)s^n|_{s=a,b} = 0 \quad (n = 0, 1, \ldots), \qquad (80)$$

which holds for the classical orthogonal polynomials of a discrete variable. It can be shown that in this case the conditions of the theorem are satisfied.

The function $Q_n(z)$ determined by (79) will be called *the function of the second kind of a discrete variable.* Let us find the connection between the functions $Q_n(z)$ and the polynomials $y_n(z)$. We have

$$\Delta_s \left[\frac{1}{(s-z)_n}\right] = -\frac{n}{(s-z)_{n+1}}.$$

Therefore it follows from (79) that

$$Q_n(z) = -\frac{B_n(n-1)!}{\rho(z)} \sum_{s=a}^{b-1} \rho_n(s)\Delta_s \left[\frac{1}{(s-z)_n}\right]. \qquad (81)$$

Using the formula for summation by parts, namely

$$\sum_{s=a}^{b-1} f(s)\Delta g(s) = f(s-1)g(s)\Big|_a^b - \sum_{s=a}^{b-1} g(s)\nabla f(s)$$

for $f(s) = \rho_n(s)$, $g(s) = 1/(s-z)_n$, we obtain from (81)

$$Q_n(z) = -\frac{B_n(n-1)!}{\rho(z)} \left\{ \frac{\rho_n(s-1)}{(s-z)_n}\Big|_a^b - \sum_{s=a}^{b-1} \frac{\nabla\rho_n(s)}{(s-z)_n} \right\}.$$

The result of substituting a and b is zero by (80). Therefore

$$Q_n(z) = \frac{B_n(n-1)!}{\rho(z)} \sum_{s=a}^{b-1} \frac{\nabla\rho_n(s)}{(s-z)_n}.$$

Similarly we find that

$$Q_n(z) = \frac{B_n(n-2)!}{\rho(z)} \sum_{s=a}^{b-1} \frac{\nabla^2 \rho_n(s)}{(s-z)_{n-1}} = \cdots = \frac{B_n}{\rho(z)} \sum_{s=a}^{b-1} \frac{\nabla^n \rho_n(s)}{s-z}.$$

From this we obtain, by using (78),

$$Q_n(z) = \frac{1}{\rho(z)} \sum_{s=a}^{b-1} \frac{y_n(s)\rho(s)}{s-z} \qquad (z \neq a, a+1, \ldots, b-1). \tag{82}$$

By setting $y_n(s) = [y_n(s) - y_n(z)] + y_n(z)$, we can rewrite formula (82) in the form

$$Q_n(z) = \frac{1}{B_0} y_n(z) Q_0(z) + \frac{1}{\rho(z)} r_n(z) \qquad (z \neq a, a+1, \ldots, b-1), \tag{83}$$

where

$$Q_0(z) = \frac{B_0}{\rho(z)} \sum_{s=a}^{b-1} \frac{\rho(s)}{s-z},$$

and

$$r_n(z) = \sum_{s=a}^{b-1} \frac{y_n(s) - y_n(z)}{s-z} \rho(s)$$

is a polynomial of degree $n-1$ in z. From (83) it follows that all the singularities of the second solution $Q_n(z)$ in the complex z plane are determined by the behavior of the functions $Q_0(z)$ and $1/\rho(z)$.

The values of the function $Q_n(z)$ were not determined on the set of points $z = x_k = a, a+1, \ldots, b-1$. At these points we put

$$\rho(x_k)Q_n(x_k) = \frac{1}{2}[\rho(x_k + i0)Q_n(x_k + i0) + \rho(x_k - i0)Q_n(x_k - i0)]$$
$$= \lim_{\epsilon \to 0} \frac{1}{2}[\rho(x_k + i\epsilon)Q_n(x_k + i\epsilon) + \rho(x_k - i\epsilon)Q_n(x_k - i\epsilon)],$$

which according to (82) yields

$$Q_n(x_k) = \frac{1}{\rho(x_k)} \sum_{s_i=a}^{b-1} \frac{y_n(s_i)\rho(s_i)}{s_i - x_k} \qquad (i \neq k).$$

§ 13 Classical orthogonal polynomials of a discrete variable on nonuniform lattices

1. The difference equation of hypergeometric type on a nonuniform lattice. In section 12 we considered the theory of solutions of the differential equation of hypergeometric type

$$\tilde{\sigma}(x)y'' + \tilde{\tau}(x)y' + \lambda y = 0, \tag{1}$$

generalized to the difference equation

$$\begin{aligned} &\tilde{\sigma}(x)\frac{1}{h}\left[\frac{y(x+h)-y(x)}{h} - \frac{y(x)-y(x-h)}{h}\right] + \\ &+ \frac{\tilde{\tau}(x)}{2}\left[\frac{y(x+h)-y(x)}{h} + \frac{y(x)-y(x-h)}{h}\right] + \lambda y(x) = 0, \end{aligned} \tag{2}$$

which approximates (1) on a lattice of constant mesh $\Delta x = h$. After the change of independent variable $x = x(s)$, we can obtain a further generalization to the case when (1) is replaced by a difference equation on a class of lattices with variable mesh $\Delta x = x(s+h) - x(s)$:

$$\begin{aligned} &\tilde{\sigma}[x(s)]\frac{1}{x(s+h/2)-x(s-h/2)}\left[\frac{y(s+h)-y(s)}{x(s+h)-x(s)} - \frac{y(s)-y(s-h)}{x(s)-x(s-h)}\right] \\ &+\frac{\tilde{\tau}[x(s)]}{2}\left[\frac{y(s+h)-y(s)}{x(s+h)-x(s)} + \frac{y(s)-y(s-h)}{x(s)-x(s-h)}\right] + \lambda y(s) = 0. \end{aligned} \tag{3}$$

Equation (3) approximates (1) to second order in h, as is easily seen by expanding $x(s \pm h), x(s \pm h/2)$ and $y(s \pm h)$ by Taylor's formula.

1°. Let us show that under certain requirements for the function $x(s)$ the difference equation (3) has a property similar to the fundamental property of the differential equation (1): *the difference derivative*

$$v_1(s) = \frac{y(s+h)-y(s)}{x(s+h)-x(s)},$$

which is approximately equal to the derivative dy/dx at the point $x(s+h/2)$, satisfies an equation of the form (3) *with $x(s)$ replaced by $x_1(s) = x(s+h/2)$, i.e. the equation*

$$\begin{aligned} &\frac{\tilde{\sigma}_1[x_1(s)]}{x_1(s+h/2)-x_1(s-h/2)}\left[\frac{v_1(s+h)-v_1(s)}{x_1(s+h)-x_1(s)} - \frac{v_1(s)-v_1(s-h)}{x_1(s)-x_1(s-h)}\right] \\ &+\frac{\tilde{\tau}_1[x_1(s)]}{2}\left[\frac{v_1(s+h)-v_1(s)}{x_1(s+h)-x_1(s)} + \frac{v_1(s)-v_1(s-h)}{x_1(s)-x_1(s-h)}\right] + \mu_1 v_1(s) = 0. \end{aligned} \tag{4}$$

Here $\tilde{\tau}_1(x_1)$ and $\tilde{\sigma}_1(x_1)$ are polynomials of respective degrees at most 1 and 2 in x_1; μ_1 is a constant.

For the proof of this statement it is convenient to replace s by hs in (3) and (4); as a result equations (3) and (4) become analogous equations with $h = 1$:

$$\tilde{\sigma}[x(s)]\frac{\Delta}{\Delta x(s-1/2)}\left[\frac{\nabla y(s)}{\nabla x(s)}\right] + \frac{\tilde{\tau}[x(s)]}{2}\left[\frac{\Delta y(s)}{\Delta x(s)} + \frac{\nabla y(s)}{\nabla x(s)}\right] + \lambda y(s) = 0, \tag{5}$$

$$\tilde{\sigma}_1[x_1(s)]\frac{\Delta}{\Delta x_1(s-1/2)}\left[\frac{\nabla v_1(s)}{\nabla x_1(s)}\right] + \frac{\tilde{\tau}_1[x_1(s)]}{2}\left[\frac{\Delta v_1(s)}{\Delta x_1(s)} + \frac{\nabla v_1(s)}{\nabla x_1(s)}\right] + \mu_1 v_1(s) = 0, \tag{6}$$

where

$$x_1(s) = x(s+1/2), \quad \Delta f(s) = f(s+1) - f(s), \quad \nabla f(s) = f(s) - f(s-1),$$
$$\frac{\Delta}{\Delta x(s-1/2)} f(s) = \frac{\Delta f(s)}{\Delta x(s-1/2)}, \quad v_1(s) = \frac{\Delta y(s)}{\Delta x(s)}.$$

Let us apply the operator $\Delta/\Delta x(s)$ to equation (5). It is convenient to use the relation

$$\Delta[f(s)g(s)] = \frac{g(s+1)+g(s)}{2}\Delta f(s) + \frac{f(s+1)+f(s)}{2}\Delta g(s). \tag{7}$$

which follows from (12.6). Since

$$\tilde{\sigma}[x(s)]\frac{\Delta}{\Delta x(s-1/2)}\left[\frac{\nabla y(s)}{\nabla x(s)}\right] = \tilde{\sigma}[x(s)]\frac{\nabla v_1(s)}{\nabla x_1(s)},$$

applying the operator $\Delta/\Delta x(s)$ to the first summand in (5) yields

$$\frac{\Delta}{\Delta x(s)}\left\{\tilde{\sigma}[x(s)]\frac{\nabla v_1(s)}{\nabla x_1(s)}\right\} = \frac{1}{2}\left[\frac{\Delta v_1(s)}{\Delta x_1(s)} + \frac{\nabla v_1(s)}{\nabla x_1(s)}\right]\frac{\Delta\tilde{\sigma}[x(s)]}{\Delta x(s)}$$
$$+\frac{1}{2}\left\{\tilde{\sigma}[x(s+1)] + \tilde{\sigma}[x(s)]\right\}\frac{\Delta}{\Delta x_1(s-1/2)}\left[\frac{\nabla v_1(s)}{\nabla x_1(s)}\right].$$

The expression obtained will have a form analogous to the left-hand side of (6) if we require that the functions

$$\frac{\Delta\tilde{\sigma}[x(s)]}{\Delta x(s)}, \quad \frac{1}{2}\left\{\tilde{\sigma}[x(s+1)] + \tilde{\sigma}[x(s)]\right\}$$

are, respectively, polynomials of degrees at most 1 and 2 in $x_1(s) = x(s+1/2)$. Since

$$\frac{\Delta}{\Delta x(s)} x^2(s) = x(s+1) + x(s),$$

these requirements are satisfied if the functions

$$1)\quad x(s+1) + x(s), \qquad 2)\quad x^2(s+1) + x^2(s)$$

are polynomials of degrees 1 and 2 in $x_1(s)$.

Let us show that if these two requirements for the function $x(s)$ are satisfied then applying the operator $\Delta/\Delta x(s)$ to equation (5) actually leads to an equation of the form (6). Applying the operator $\Delta/\Delta x(s)$ to the remaining terms of equation (5), we obtain

$$\frac{\Delta}{\Delta x(s)}[\lambda y(s)] = \lambda v_1(s),$$

$$\frac{\Delta}{\Delta x(s)}\left\{\tilde{\tau}[x(s)]\left[\frac{\Delta y(s)}{\Delta x(s)} + \frac{\nabla y(s)}{\nabla x(s)}\right]\right\} = \frac{\Delta}{\Delta x(s)}\{\tilde{\tau}[x(s)][v_1(s) + v_1(s-1)]\}$$
$$= \frac{1}{2}[v_1(s+1) + 2v_1(s) + v_1(s-1)]\frac{\Delta\tilde{\tau}[x(s)]}{\Delta x(s)}$$
$$+ \frac{\tilde{\tau}[x(s+1)] + \tilde{\tau}[x(s)]}{2}\,\frac{\Delta v_1(s) + \nabla v_1(s)}{\Delta x(s)}.$$

Now we express the functions

$$\frac{1}{2}[v_1(s+1) + 2v_1(s) + v_1(s-1)], \qquad \frac{\Delta v_1(s) + \nabla v_1(s)}{\Delta x(s)}$$

in terms of the difference derivatives

$$\frac{\Delta}{\Delta x_1(s-1/2)}\left[\frac{\nabla v_1(s)}{\nabla x_1(s)}\right], \quad \frac{1}{2}\left[\frac{\Delta v_1(s)}{\Delta x_1(s)} + \frac{\nabla v_1(s)}{\nabla x_1(s)}\right]$$

which appear in equation (6). For this purpose we use the easily verified relations:

$$\frac{1}{2}\left[\frac{\Delta v_1(s)}{\Delta x_1(s)} + \frac{\nabla v_1(s)}{\nabla x_1(s)}\right] \pm \frac{1}{2}\Delta\left[\frac{\nabla v_1(s)}{\nabla x_1(s)}\right] = \frac{\Delta v_1(s)}{\Delta x_1(s)} \quad \text{or} \quad \frac{\nabla v_1(s)}{\nabla x_1(s)}.$$

As a result we have

$$\Delta v_1(s) = \Delta x_1(s)\left\{\frac{1}{2}\left[\frac{\Delta v_1(s)}{\Delta x_1(s)}+\frac{\nabla v_1(s)}{\nabla x_1(s)}\right]+\frac{\Delta x(s)}{2}\frac{\Delta}{\Delta x_1(s-1/2)}\left[\frac{\nabla v_1(s)}{\nabla x_1(s)}\right]\right\},$$

$$\nabla v_1(s) = \nabla x_1(s)\left\{\frac{1}{2}\left[\frac{\Delta v_1(s)}{\Delta x_1(s)}+\frac{\nabla v_1(s)}{\nabla x_1(s)}\right]-\frac{\Delta x(s)}{2}\frac{\Delta}{\Delta x_1(s-1/2)}\left[\frac{\nabla v_1(s)}{\nabla x_1(s)}\right]\right\},$$

$$v_1(s+1)+2v_1(s)+v_1(s-1) = \Delta v_1(s)-\nabla v_1(s)+4v_1(s).$$

Thus we obtain the equation

$$\bar{\sigma}_1(s)\frac{\Delta}{\Delta x_1(s-1/2)}\left[\frac{\nabla v_1(s)}{\nabla x_1(s)}\right]+\frac{\bar{\tau}_1(s)}{2}\left[\frac{\Delta v_1(s)}{\Delta x_1(s)}+\frac{\nabla v_1(s)}{\nabla x_1(s)}\right]+\mu_1 v_1(s)=0, \tag{6a}$$

where

$$\bar{\sigma}_1(s) = \frac{\tilde{\sigma}[x(s+1)]+\tilde{\sigma}[x(s)]}{2}+\frac{1}{4}\frac{\Delta\tilde{\tau}[x(s)]}{\Delta x(s)}\frac{\Delta x_1(s)+\nabla x_1(s)}{2\Delta x(s)}[\Delta x(s)]^2 + \frac{\tilde{\tau}[x(s+1)]+\tilde{\tau}[x(s)]}{2}\frac{\Delta x_1(s)-\nabla x_1(s)}{4}, \tag{8}$$

$$\bar{\tau}_1(s) = \frac{\Delta\tilde{\sigma}[x(s)]}{\Delta x(s)}+\frac{\Delta\tilde{\tau}[x(s)]}{\Delta x(s)}\frac{\Delta x_1(s)-\nabla x_1(s)}{4} + \frac{\tilde{\tau}[x(s+1)]+\tilde{\tau}[x(s)]}{2}\frac{\Delta x_1(s)+\nabla x_1(s)}{2\Delta x(s)}, \tag{9}$$

$$\mu_1 = \lambda+\frac{\Delta\tilde{\tau}[x(s)]}{\Delta x(s)}. \tag{10}$$

Since

$$\frac{\Delta\tilde{\tau}[x(s)]}{\Delta x(s)} = \text{const},$$

and since $[\Delta x(s)]^2 = 2[x^2(s+1)+x^2(s)]-[x(s+1)+x(s)]^2$ is a polynomial of degree at most 2 in $x_1(s)$, and $\frac{1}{2}\{\tilde{\tau}[x(s+1)]+\tilde{\tau}[x(s)]\}$ is a polynomial of degree at most 1 in $x_1(s)$, equation (6a) for $v_1(s)$ will have the form (6) if

$$\frac{\Delta x_1(s)+\nabla x_1(s)}{2\Delta x(s)} = \text{const},$$

and $\frac{1}{4}[\Delta x_1(s)-\nabla x_1(s)]$ is a polynomial of degree at most 1 in $x_1(s)$.

According to the first requirement for $x(s)$ we have

$$\frac{x(s+1)+x(s)}{2}=\alpha x\left(s+\frac{1}{2}\right)+\beta \tag{11}$$

where α and β are constants. Hence

$$\begin{aligned}\frac{\Delta x_1(s)+\nabla x_1(s)}{2\Delta x(s)}&=\frac{\Delta[x(s+1/2)+x(s-1/2)]}{2\Delta x(s)}=\frac{\Delta[\alpha x(s)+\beta]}{\Delta x(s)}=\alpha,\\ \frac{\Delta x_1(s)-\nabla x_1(s)}{4}&=\frac{1}{2}\left[\frac{x_1(s+1)+x_1(s)}{2}+\frac{x_1(s)+x_1(s-1)}{2}\right]-x_1(s)\end{aligned} \tag{12}$$

$$\begin{aligned}&=\frac{1}{2}\left[\alpha x_1\left(s+\frac{1}{2}\right)+\beta+\alpha x_1\left(s-\frac{1}{2}\right)+\beta\right]-x_1(s)\\ &=\alpha[\alpha x_1(s)+\beta]+\beta-x_1(s).\end{aligned} \tag{13}$$

From this we see that $\tilde{\tau}_1(s)=\tilde{\tau}_1[x_1(s)]$ and $\tilde{\sigma}_1(s)=\tilde{\sigma}_1[x_1(s)]$, where $\tilde{\tau}_1(x_1)$ and $\tilde{\sigma}_1(x_1)$ are polynomials of respective degrees at most 1 and 2 in x_1.

Therefore equation (5) may be called *the difference equation of hypergeometric type* because it has a property, analogous to the property of the differential equation of hypergeometric type, of retaining its form after differentiation. Let us recall that for the difference equation this property is satisfied only on lattices $x(s)$ that satisfy certain requirements (the form of these lattices will be found in part 4).

Thus we come to an important theorem by means of which we can now construct a theory of classical orthogonal polynomials of a discrete variable on nonuniform lattices.

Theorem 1. *Let the lattice function $x(s)$ satisfy the following conditions: the functions $x(s+1)+x(s)$ and $x^2(s+1)+x^2(s)$ are, respectively, polynomials of degrees 1 and 2 in $x_1(s)=x(s+1/2)$. Then the difference derivative $v_1(s)=\Delta y(s)/\Delta x(s)$ of the solution $y(s)$ of equation (5) satisfies a difference equation of the form (6), where $\tilde{\sigma}_1(x_1)$ and $\tilde{\tau}_1(x_1)$ are polynomials of at most second and first degrees in $x_1(s)$, respectively, and $\mu_1=const$*.*

The converse is also valid: every solution of equation (6) with $\lambda\neq 0$ can be represented in the form $v_1(s)=\Delta y(s)/\Delta x(s)$, where $y(s)$ is a solution of equation (5). For the proof, in accordance with (5), we assume

$$y(s)=-\frac{1}{\lambda}\left\{\tilde{\sigma}[x(s)]\frac{\nabla v_1(s)}{\nabla x_1(s)}+\frac{\tilde{\tau}[x(s)]}{2}[v_1(s)+v_1(s-1)]\right\}. \tag{14}$$

* In part 4, below, we classify the forms of lattices $x(s)$ and show that in cases of practical interest the second condition for the function $x(s)$ is a consequence of the first.

We apply the operator $\Delta/\Delta x(s)$ to both sides of this equality and repeat the transformations considered above. We then use equation (6) for $v_1(s)$ with

$$\mu_1 = \lambda + \frac{\Delta\tilde{\tau}[x(s)]}{\Delta x(s)}, \quad \tilde{\sigma}_1[x_1(s)] = \bar{\sigma}_1(s), \quad \tilde{\tau}_1[x_1(s)] = \bar{\tau}_1(s)$$

($\bar{\sigma}_1(s)$ and $\bar{\tau}_1(s)$ are determined by formulas (8) and (9)). As a result, we obtain $v_1(s) = \Delta y(s)/\Delta x(s)$. Substitution of this relation into expression (14) yields equation (5) for the function $y(s)$, which was to be proved.

2°. We may show by induction that when the above requirements for $x(s)$ are satisfied, the functions $v_k(s)$ connected with the solution $y = y(s)$ of equation (5) by the relations

$$v_k(s) = \frac{\Delta v_{k-1}(s)}{\Delta x_{k-1}(s)}, \quad v_0(s) = y(s), \quad x_k(s) = x\left(s + \frac{k}{2}\right) \quad (k = 1, 2, \ldots)$$

satisfy the equations

$$\tilde{\sigma}_k[x_k(s)]\frac{\Delta}{\Delta x_k(s-\frac{1}{2})}\left[\frac{\nabla v_k(s)}{\nabla x_k(s)}\right] + \frac{\tilde{\tau}_k[x_k(s)]}{2}\left[\frac{\Delta v_k(s)}{\Delta x_k(s)} + \frac{\nabla v_k(s)}{\nabla x_k(s)}\right] + \mu_k v_k(s) = 0, \tag{16}$$

where $\tilde{\sigma}_k(x_k)$ and $\tilde{\tau}_k(x_k)$ are polynomials of at most second and first degrees in x_k, respectively; $\mu_k = \text{const}$; and

$$\begin{aligned}\tilde{\sigma}_k[x_k(s)] &= \frac{\tilde{\sigma}_{k-1}[x_{k-1}(s+1)] + \tilde{\sigma}_{k-1}[x_{k-1}(s)]}{2} \\ &+ \frac{1}{4}\frac{\Delta\tilde{\tau}_{k-1}[x_{k-1}(s)]}{\Delta x_{k-1}(s)}\frac{\Delta x_k(s) + \nabla x_k(s)}{2\Delta x_{k-1}(s)}[\Delta x_{k-1}(s)]^2 \\ &+ \frac{\tilde{\tau}_{k-1}[x_{k-1}(s+1)] + \tilde{\tau}_{k-1}[x_{k-1}(s)]}{2}\frac{\Delta x_k(s) - \nabla x_k(s)}{4}, \\ &\tilde{\sigma}_0(x_0) = \tilde{\sigma}(x);\end{aligned} \tag{17}$$

$$\begin{aligned}\tilde{\tau}_k[x_k(s)] &= \frac{\Delta\tilde{\sigma}_{k-1}[x_{k-1}(s)]}{\Delta x_{k-1}(s)} + \frac{\Delta\tilde{\tau}_{k-1}[x_{k-1}(s)]}{\Delta x_{k-1}(s)}\frac{\Delta x_k(s) - \nabla x_k(s)}{4} \\ &+ \frac{\tilde{\tau}_{k-1}[x_{k-1}(s+1)] + \tilde{\tau}_{k-1}[x_{k-1}(s)]}{2}\frac{\Delta x_k(s) + \nabla x_k(s)}{2\Delta x_{k-1}(s)}, \\ &\tilde{\tau}_0(x_0) = \tilde{\tau}(x);\end{aligned} \tag{18}$$

$$\mu_k = \mu_{k-1} + \frac{\Delta\tilde{\tau}_{k-1}[x_{k-1}(s)]}{\Delta x_{k-1}(s)}, \quad \mu_0 = \lambda. \tag{19}$$

The converse is also valid: every solution of equation (16) with $\mu_{k-1} \neq 0$ can be represented in the form

$$v_k(s) = \frac{\Delta v_{k-1}(s)}{\Delta x_{k-1}(s)},$$

where $v_{k-1}(s)$ is a solution of the equation obtained from (16) by replacing k by $k-1$.

3°. To study additional properties of solutions of (5) it is convenient to use the equation

$$\frac{1}{2}\left[\frac{\Delta y(s)}{\Delta x(s)}+\frac{\nabla y(s)}{\nabla x(s)}\right]=\frac{\Delta y(s)}{\Delta x(s)}-\frac{1}{2}\Delta\left[\frac{\nabla y(s)}{\nabla x(s)}\right],$$

and to rewrite equation (5) in the form

$$\sigma(s)\frac{\Delta}{\Delta x(s-1/2)}\left[\frac{\nabla y(s)}{\nabla x(s)}\right]+\tau(s)\frac{\Delta y(s)}{\Delta x(s)}+\lambda y(s)=0, \tag{20}$$

where

$$\sigma(s)=\tilde{\sigma}[x(s)]-\frac{1}{2}\tilde{\tau}[x(s)]\Delta x\left(s-\frac{1}{2}\right), \tag{21}$$

$$\tau(s)=\tilde{\tau}[x(s)]. \tag{22}$$

An analogous form for equation (16) is

$$\sigma_k(s)\frac{\Delta}{\Delta x_k(s-1/2)}\left[\frac{\nabla v_k(s)}{\nabla x_k(s)}\right]+\tau_k(s)\frac{\Delta v_k(s)}{\Delta x_k(s)}+\mu_k v_k(s)=0, \tag{23}$$

$$\sigma_k(s)=\tilde{\sigma}_k[x_k(s)]-\frac{1}{2}\tilde{\tau}_k[x_k(s)]\Delta x_k\left(s-\frac{1}{2}\right), \tag{24}$$

$$\tau_k(s)=\tilde{\tau}[x_k(s)]. \tag{25}$$

From (17) and (18) it follows that $\sigma_k(s)=\sigma_{k-1}(s)$, i.e.

$$\sigma_k(s)=\sigma(s). \tag{26}$$

In addition, from (17) and (18) we can also obtain the relation

$$\sigma(s)+\tau_k(s)\Delta x_k\left(s-\frac{1}{2}\right)=\sigma(s+1)+\tau_{k-1}(s+1)\Delta x_{k-1}\left(s+\frac{1}{2}\right), \tag{27}$$

which enables us, by using the functions $\sigma(s),\tau(s)$ and $x(s)$, to find $\tau_k(s)$ in the form

$$\tau_k(s)=\frac{\sigma(s+k)-\sigma(s)+\tau(s+k)\Delta x(s+k-1/2)}{\Delta x(s+(k-1)/2)}. \tag{28}$$

An expression for μ_k may be obtained directly from (19) as

$$\mu_k = \lambda + \sum_{m=0}^{k-1} \frac{\Delta \tau_m(s)}{\Delta x_m(s)}. \tag{19a}$$

2. The Rodrigues formula. 1°. Theorem 1 enables us to construct a family of polynomial solutions corresponding to certain values of λ in the same way as in the case of the differential equation. Proceeding from the conditions for $x(s)$ formulated above and using the relations

$$\frac{\Delta x^n(s)}{\Delta x(s)} = \frac{x(s+1)+x(s)}{2} \frac{\Delta x^{n-1}(s)}{\Delta x(s)} + \frac{x^{n-1}(s+1)+x^{n-1}(s)}{2}, \tag{29}$$

$$\frac{x^n(s+1)+x^n(s)}{2} = \frac{x(s+1)+x(s)}{2} \frac{x^{n-1}(s+1)+x^{n-1}(s)}{2} + \frac{[\Delta x(s)]^2}{4} \frac{\Delta x^{n-1}(s)}{\Delta x(s)} \tag{30}$$

as well as the obvious equality

$$\frac{[\Delta x(s)]^2}{4} = \frac{x^2(s+1)+x^2(s)}{2} - \left[\frac{x(s+1)+x(s)}{2}\right]^2, \tag{31}$$

it is easy to prove, by induction, general properties of the lattice function $x(s)$. Let $p_n(x)$ be an arbitrary polynomial of degree n; then

$$\frac{\Delta p_n[x(s)]}{\Delta x(s)} = q_{n-1}[x_1(s)], \tag{32}$$

$$\frac{p_n[x(s+1)]+p_n[x(s)]}{2} = r_n[x_1(s)], \tag{33}$$

where $q_{n-1}(x_1)$, $r_n(x_1)$ are polynomials of the appropriate degrees in x_1. It is obvious that equalities (32) and (33) remain valid under replacement of $x(s)$ and $x_1(s)$ by $x_k(s)$ and $x_{k+1}(s)$.

Note that relations (29) and (30) may be obtained from (7) by putting $f(s) = x^{n-1}(s)$, $g(s) = x(s)$ and $f(s) = e^{i\pi s}x^{n-1}(s)$, $g(s) = x(s)$, respectively.

To find a particular solution of equation (5) we observe that equation (16) with $k = n$, $\mu_n = 0$ has the particular solution $v_n(s) = \text{const}$. Let us show that when $k < n$ the functions $v_k(s)$ connected by the relations

$$v_{k+1}(s) = \frac{\Delta v_k(s)}{\Delta x_k(s)}$$

will be polynomials of degree $n-k$ in $x_k(s)$ if $v_n(s) = \text{const}$, provided that $\mu_k \neq 0$ for $k = 0, 1, \ldots, n-1$. We shall prove this by induction, assuming that the function $v_{k+1}(s)$ is a polynomial of degree $n-k-1$ in $x_{k+1}(s)$. From equation (16) we have

$$v_k(s) = -\frac{1}{\mu_k}\left\{\tilde{\sigma}_k[x_k(s)]\frac{\nabla v_{k+1}(s)}{\nabla x_{k+1}(s)} + \frac{\tilde{\tau}_k[x_k(s)]}{2}[v_{k+1}(s) + v_{k+1}(s-1)]\right\}.$$

By virtue of (32) and (33) the functions

$$\frac{\nabla v_{k+1}(s)}{\nabla x_{k+1}(s)} = \left.\frac{\Delta v_{k+1}(t)}{\Delta x_{k+1}(t)}\right|_{t=s-1}$$

and

$$\frac{1}{2}[v_{k+1}(s) + v_{k+1}(s-1)] = \left.\frac{1}{2}[v_{k+1}(t+1) + v_{k+1}(t)]\right|_{t=s-1}$$

are, respectively, polynomials of degrees $n-k-2$ and $n-k-1$ in $x_{k+2}(s-1)$. Since $x_{k+2}(s-1) = x_k(s)$, the function $v_k(s)$ is obviously a polynomial of degree $n-k$ in $x_k(s)$.

By applying this argument successively for $k = n-1, n-2$, etc. we find that there exists a solution $y(s) = v_0(s)$ of (5) which is a polynomial of degree n in $x(s)$ for those $\lambda = \lambda_n$ for which $\mu_n = 0$. We see from (19a) that

$$\lambda_n = -\sum_{m=0}^{n-1}\frac{\Delta\tau_m(s)}{\Delta x_m(s)} = -\sum_{m=0}^{n-1}\tilde{\tau}'_m. \tag{19b}$$

2°. To obtain an explicit expression for the polynomial $y(s) = y_n[x(s)]$ we rewrite equations (5) and (16) in the form of (20) and (23) by taking account of (26). Multiplying equations (20) and (23) by appropriate functions $\rho(s)$ and $\rho_k(s)$, and using (12.6), we reduce these equations to self-adjoint form:

$$\frac{\Delta}{\Delta x(s-1/2)}\left[\sigma(s)\rho(s)\frac{\nabla y(s)}{\nabla x(s)}\right] + \lambda\rho(s)y(s) = 0, \tag{34}$$

$$\frac{\Delta}{\Delta x_k(s-1/2)}\left[\sigma(s)\rho_k(s)\frac{\nabla v_k(s)}{\nabla x_k(s)}\right] + \mu_k\rho_k(s)v_k(s) = 0. \tag{35}$$

Here $\rho(s)$ and $\rho_k(s)$ satisfy the difference equations

$$\frac{\Delta}{\Delta x(s-1/2)}[\sigma(s)\rho(s)] = \tau(s)\rho(s), \tag{36}$$

$$\frac{\Delta}{\Delta x_k(s-1/2)}[\sigma(s)\rho_k(s)] = \tau_k(s)\rho_k(s). \tag{37}$$

By using (37) and (27) we can determine the connection between $\rho_k(s)$ and $\rho(s)$:

$$\frac{\sigma(s+1)\rho_k(s+1)}{\rho_k(s)} = \sigma(s) + \tau_k(s)\Delta x_k(s-1/2)$$
$$= \sigma(s+1) + \tau_{k-1}(s+1)\Delta x_{k-1}\left(s+\frac{1}{2}\right) = \frac{\sigma(s+2)\rho_{k-1}(s+2)}{\rho_{k-1}(s+1)}.$$

From this we obtain

$$\frac{\sigma(s+1)\rho_{k-1}(s+1)}{\rho_k(s)} = \frac{\sigma(s+2)\rho_{k-1}(s+2)}{\rho_k(s+1)} = C(s),$$

where $C(s)$ is an arbitrary function of period 1. By assuming $C(s) = 1$, we obtain

$$\rho_k(s) = \sigma(s+1)\rho_{k-1}(s+1). \tag{38}$$

Hence

$$\rho_k(s) = \rho(s+k)\prod_{i=1}^{k}\sigma(s+i). \tag{39}$$

By means of (38), equation (35) can be rewritten as a recurrent relation between the functions $v_k(s)$ and $v_{k+1}(s)$. In fact,

$$\rho_k(s)v_k(s) = -\frac{1}{\mu_k}\frac{\nabla}{\nabla x_{k+1}(s)}[\rho_{k+1}(s)v_{k+1}(s)]. \tag{40}$$

From this we obtain

$$\rho_k(s)v_k(s) = \frac{A_k}{A_n}\nabla_n^{(n-k)}[\rho_n(s)v_n(s)], \tag{41}$$

where

$$A_k = (-1)^k\prod_{i=0}^{k-1}\mu_i, \quad A_0 = 1, \tag{42}$$

$$\nabla_n^{(m)}[f(s)] = \nabla_{n-m+1}\dots\nabla_{n-1}\nabla_n[f(s)], \nabla_k = \frac{\nabla}{\nabla x_k(s)}. \tag{43}$$

Let us recall that, by definition,

$$v_k(s) = \frac{\Delta v_{k-1}(s)}{\Delta x_{k-1}(s)},$$

i.e.

$$v_k(s) = \Delta^{(k)}[y(s)], \tag{44}$$

where

$$\Delta^{(k)}[f(s)] = \Delta_{k-1}\Delta_{k-2}\dots\Delta_0[f(s)], \quad \Delta_k = \frac{\Delta}{\Delta x_k(s)}.$$

If $v_n(s) = C_n$, where C_n is a constant, then, as was shown above, $y(s)$ is a polynomial of degree n in $x(s)$, i.e. $y = y_n(s) \equiv \tilde{y}_n[x(s)]$. In this case, for the function

$$v_{kn}(s) = \Delta^{(k)}[y_n(s)]$$

in (41) we obtain

$$v_{kn}(s) = \frac{A_{kn}B_n}{\rho_k(s)}\nabla_n^{(n-k)}[\rho_n(s)], \tag{45}$$

where

$$A_{kn} = A_k(\lambda)\bigg|_{\lambda=\lambda_n} = (-1)^k \prod_{m=0}^{k-1} \mu_{mn}; \qquad \mu_{mn} = \mu_m(\lambda)\bigg|_{\lambda=\lambda_n} = \lambda_n - \lambda_m;$$

$A_{0n} = 1,\ B_n = C_n/A_{nn}$.

Here, in particular, from (45) with $k = 0$ we obtain an *analog of the Rodrigues formula* for the polynomial $\tilde{y}_n[x(s)] \equiv y_n(s)$:

$$y_n(s) = \frac{B_n}{\rho(s)}\nabla_n^{(n)}[\rho_n(s)] = \frac{B_n}{\rho(s)}\frac{\nabla}{\nabla x_1(s)}\cdots\frac{\nabla}{\nabla x_{n-1}(s)}\frac{\nabla}{\nabla x_n(s)}[\rho_n(s)]. \tag{46}$$

It can easily be seen that the arbitrariness in the choice of a periodic multiplier for the function $\rho(s)$ has no influence on the explicit form of the polynomial $\tilde{y}_n(x)$ obtained by the Rodrigues formula (46).

Thus, the polynomial solutions $y = \tilde{y}_n[x(s)]$ of equation (5) are uniquely determined by formula (46) up to the normalizing factor B_n. These solutions correspond to the values $\lambda = \lambda_n$ for which $\mu_n = 0$.

3. The orthogonality property. 1°. We now prove that the polynomial solutions of equations (5) have the orthogonality property. For this purpose we write the equations for $y_m(s)$ and $y_n(s)$ in the self-adjoint form (34):

$$\frac{\Delta}{\Delta x(s-1/2)}\left[\sigma(s)\rho(s)\frac{\nabla y_m(s)}{\nabla x(s)}\right] + \lambda_m\rho(s)y_m(s) = 0,$$

$$\frac{\Delta}{\Delta x(s-1/2)}\left[\sigma(s)\rho(s)\frac{\nabla y_n(s)}{\nabla x(s)}\right] + \lambda_n\rho(s)y_n(s) = 0.$$

Multiplying the first equation by $y_n(s)$ and the second by $y_m(s)$, and subtracting the second from the first, we obtain:

$$\begin{aligned}&(\lambda_m - \lambda_n)y_m(s)y_n(s)\rho(s)\Delta x(s-1/2)\\ &= \Delta\left\{\sigma(s)\rho(s)\left[y_m(s)\frac{\nabla y_n(s)}{\nabla x(s)} - y_n(s)\frac{\nabla y_m(s)}{\nabla x(s)}\right]\right\}.\end{aligned} \tag{47}$$

If we now put $s = s_i$ and $s_{i+1} = s_i + 1$, where $i = 0, 1, \ldots$, and sum over the indices i for which $a \le s_i \le b-1$, we obtain

$$\begin{aligned}&(\lambda_m - \lambda_n)\sum_{s_i=a}^{b-1} y_m(s_i)y_n(s_i)\rho(s_i)\Delta x\left(s_i - \frac{1}{2}\right)\\ &= \sigma(s)\rho(s)\left[y_m(s)\frac{\nabla y_n(s)}{\nabla x(s)} - y_n(s)\frac{\nabla y_m(s)}{\nabla x(s)}\right]\Bigg|_a^b.\end{aligned} \tag{48}$$

Since

$$y_m(s) = \frac{y_m(s) + y_m(s-1)}{2} + \frac{1}{2}\nabla y_m(s),$$
$$y_n(s) = \frac{y_n(s) + y_n(s-1)}{2} + \frac{1}{2}\nabla y_n(s),$$

and the functions $y_m(s)$ and $y_n(s)$ are polynomials in $x(s)$, by virtue of (32) and (33) the expression

$$\begin{aligned}&y_m(s)\frac{\nabla y_n(s)}{\nabla x(s)} - y_n(s)\frac{\nabla y_m(s)}{\nabla x(s)}\\ &= \frac{y_m(s) + y_m(s-1)}{2}\frac{\nabla y_n(s)}{\nabla x(s)} - \frac{y_n(s) + y_n(s-1)}{2}\frac{\nabla y_m(s)}{\nabla x(s)}\end{aligned}$$

is a polynomial in $x(s-1/2)$. Hence under the boundary conditions

$$\sigma(s)\rho(s)x^k\left(s - \frac{1}{2}\right)\Bigg|_{s=a,b} = 0 \quad (k = 0, 1, \ldots) \tag{49}$$

the right-hand side of (48) is zero. As a result, we obtain

$$\sum_{s_i=a}^{b-1} y_m(s_i)y_n(s_i)\rho(s_i)\Delta x\left(s_i - \frac{1}{2}\right) = \delta_{mn}d_n^2. \tag{50}$$

The polynomial solutions of equation (5) satisfying the orthogonality relations (50) under the additional conditions

$$\rho(s_i)\Delta x\left(s_i - \frac{1}{2}\right) > 0 \qquad (a \le s_i \le b-1), \tag{50a}$$

will be called *classical orthogonal polynomials of a discrete variable on non-uniform lattices.*

Since, by the Rodrigues formula (46),

$$\tilde{y}_1(x) = B_1\tilde{\tau}[x(s)] \equiv B_1\tau(s),$$

for classical orthogonal polynomials, the function $\tau(s)$ is a polynomial of the first degree in $x = x(s)$ with a nonzero coefficient for $x(s)$.

If a and b are finite, the boundary conditions (49) can be presented in the simpler form

$$\sigma(a)\rho(a) = 0, \qquad \sigma(b)\rho(b) = 0, \tag{49a}$$

because $x(s - 1/2)$ is bounded. If we take $\rho(s_i) \neq 0$ for $a \leq s_i \leq b-1$, the boundary condition at $s = a$ is satisfied, provided that

$$\sigma(a) = 0. \tag{49b}$$

On the other hand, by virtue of the equality

$$\sigma(s+1)\rho(s+1) = \rho(s)[\sigma(s) + \tau(s)\Delta x(s - 1/2)]$$

the boundary condition $\sigma(b)\rho(b) = 0$ is satisfied if

$$\sigma(s) + \tau(s)\Delta x\left(s - \frac{1}{2}\right)\bigg|_{s=b-1} = 0. \tag{49c}$$

2°. Proceeding similarly for equation (35) with $k = 1$ we can show that for the functions $v_{1n}(s)$ the orthogonality relation

$$\sum_{s_i=a_1}^{b_1-1} v_{1m}(s_i)v_{1n}(s_i)\rho_1(s_i)\Delta x_1\left(s_i - \frac{1}{2}\right) = \delta_{mn}d_{1n}^2 \tag{51}$$

is valid if the boundary conditions

$$\sigma(s)\rho_1(s)x_1^k\left(s - \frac{1}{2}\right)\bigg|_{s=a_1,b_1} = 0 \quad (k = 0, 1, \ldots) \tag{52}$$

are satisfied. For finite a and b, because of the relations

$$\sigma(a) = 0, \qquad \sigma(b)\rho(b) = 0,$$
$$\sigma(s)\rho_1(s) = \sigma(s)\sigma(s+1)\rho(s+1)$$

the boundary conditions (52) are satisfied for $a_1 = a$ and $b_1 = b - 1$. In a similar way, by induction, we find that the polynomials $v_{kn}(s)$ satisfy the orthogonality relations

$$\sum_{s_i=a}^{b-k-1} v_{km}(s_i)v_{kn}(s_i)\rho_k(s_i)\Delta x_k\left(s_i - \frac{1}{2}\right) = \delta_{mn}d_{kn}^2, \tag{53}$$

where d_{kn}^2 is the squared norm of the polynomial $v_{kn}(s)$.

We shall also assume that the polynomials $v_{kn}(s)$, which satisfy the orthogonality relation (53), also satisfy the conditions

$$\rho_k(s_i)\Delta x_k(s_i - 1/2) > 0 \qquad (a \le s_i \le b - k - 1), \tag{53a}$$

similar to (50a), which were considered above for $k = 0$.

4. Classification of lattices. Let us determine the possible forms of the functions $x(s)$ for which equation (5) is a difference equation of hypergeometric type. In accordance with the conditions of Theorem 1 in part 1, the function $x(s)$ may be determined by solving equation (11) under the additional condition that the function $x^2(s+1) + x^2(s)$ is a quadratic polynomial in $x_1(s) = x(s + 1/2)$. Here it is convenient to use the fact that the form of equation (11) is preserved under the linear transformations

$$x(s) \to Ax(s) + b, \quad s \to \pm s + s_0,$$

where A, B, and s_0 are constants.

When $\alpha \neq 1$, the transformation

$$x(s) \to x(s) + \frac{\beta}{1-\alpha}$$

carries (11) to the form

$$x(s+1) + x(s) = 2\alpha x(s + 1/2). \tag{54}$$

The general solution of this equation is

$$x(s) = c_1 q_1^{2s} + c_2 q_2^{2s},$$

where q_1 and q_2 are the roots of the equation

$$q^2 - 2\alpha q + 1 = 0, \tag{55}$$

and c_1 and c_2 are arbitrary functions of period 1/2. Let us show that the additional condition on $x(s)$ will be satisfied if c_1 and c_2 are constants. Since

$$x^2(s+1) + x^2(s) = [x(s+1) + x(s)]^2 - 2x(s)x(s+1),$$

it is sufficient to show that the product $x(s)x(s+1)$ is a polynomial in $x_1(s)$. By virtue of (55), $q_1 q_2 = 1$. Hence

$$\begin{aligned} x(s)x(s+1) &= (c_1 q_1^{2s} + c_2 q_2^{2s})(c_1 q_1^{2s+2} + c_2 q_2^{2s+2}) \\ &= (c_1 q_1^{2s+1} + c_2 q_2^{2s+1})^2 + c_1 c_2 (q_1^{2s} q_2^{2s+2} + q_2^{2s} q_1^{2s+2} - 2q_1^{2s+1} q_2^{2s+1}) \\ &= [x(s+1/2)]^2 + c_1 c_2 (q_1 - q_2)^2, \end{aligned}$$

i.e. $x(s)x(s+1)$ is a quadratic polynomial in $x(s+1/2)$. When $\alpha = 1$, the general solution of (11) has the form

$$x(s) = as^2 + bs + c,$$

where $a = 4\beta$, b and c are arbitrary functions of period 1/2. It is easy to verify that the additional condition on $x(s)$ is satisfied if a, b and c are constants.

By using the linear transformations $x(s) \to Ax(s) + B$, $s \to s + s_0$, indicated above, where A, B, and s_0 are constants, we can reduce the expressions for the functions $x(s)$ to simpler forms.

1) Let

$$x(s) = C_1 q_1^{2s} + C_2 q_2^{2s},$$

where q_1, q_2 are the roots of the equation

$$q^2 - 2\alpha q + 1 = 0,$$

and C_1 and C_2 are constants. If $\alpha > 1$, we have $q_1 = e^{\omega}$, $q_2 = e^{-\omega}$, with $\omega > 0$. If $C_1 \cdot C_2 > 0$ the function $x(s)$ can be represented in the form $x(s) = \cosh 2\omega s$ by using the transformation $s \to s + s_0$, $x(s) \to Ax(s)$, provided that the constants s_0 and A are chosen to satisfy the conditions

$$C_1 e^{2\omega s_0} = C_2 e^{-2\omega s_0} = A/2.$$

If, however, $C_1 \cdot C_2 < 0$, the function $x(s)$ can be represented, in a similar way, in the form $x(s) = \sinh 2\omega s$. Now suppose that $C_2 = 0$. Then if $s_0 = 0$ and $A = C$, the function $x(s)$ can be represented in the form $x(s) = e^{2\omega s}$. If $C_1 = 0$, by making the transformation $x(s) \to C_2 x(s)$, $s \to -s$, we can again represent $x(s)$ in the form $x(s) = e^{2\omega s}$.

If $\alpha < 1$, then $q_1 = e^{i\omega}$, $q_2 = e^{-i\omega}$, $C_2 = \bar{C}_1 = |C_1| e^{i\delta}$ (the bar denotes the complex conjugate), and $x(s)$ has the form $x(s) = |C_1| \cos(2\omega s - \delta)$. The transformation $x(s) \to |C_1| x(s)$, $s \to s + s_0$, $s_0 = \delta/(2\omega)$ carries $x(s)$ to the form $x(s) = \cos 2\omega s$.

2) Now let $\alpha = 1$. Then $x(s) = as^2+bs+c$, where a, b, and c are constants. If $a = 0$, the transformation $x(s) \to Ax(s) + B$, with $A = b$, $B = C$, carries $x(s)$ to the form $x(s) = s$. On the other hand, if $a \neq 0$, the transformation $x(s) \to Ax(s)$, $s \to s + s_0$, with $A = b + 2as_0 = a$, $B = as_0^2 + bs_0 + c$, will carry $x(s)$ to the form $x(s) = s(s+1)$.

In the last case, $x(s)$ was expressed in the form $x(s) = s(s+1)$, rather than $x(s) = s^2$, since the polynomials in a discrete variable on the lattice $x(s) = s(s+1)$ are connected in a simple way with the Racah coefficients, which are extensively used in atomic physics. In this way we arrive at the following canonical forms of the functions $x(s)$:

I. $x(s) = s \quad (\alpha = 1, \beta = 0);$ (56)

II. $x(s) = s(s+1) \quad (\alpha = 1, \beta = 1/4);$ (56a)

III. $x(s) = e^{2\omega s} \quad (\alpha > 1, \alpha = \cosh\omega, \beta = 0);$ (57)

IV. $x(s) = \sinh 2\omega s \quad (\alpha > 1, \alpha = \cosh\omega, \beta = 0);$ (57a)

V. $x(s) = \cosh 2\omega s \quad (\alpha > 1, \alpha = \cosh\omega, \beta = 0);$ (57b)

VI. $x(s) = \cos 2\omega s \quad (0 < \alpha < 1, \alpha = \cos\omega, \beta = 0).$ (58)

The case $\alpha \leq 0$ is usually not of interest. The form of the lattices for $x(s)$ is chosen so that the function $x(s)$ will be real at real s.

5. Classification of polynomial systems on linear and quadratic lattices. Let us consider the basic systems of classical orthogonal polynomials on nonuniform lattices $x(s) = s$ and $x(s) = s(s+1)$. In order to find explicit expressions for the weight functions $\rho(s)$ for which the polynomials (46) are orthogonal, we rewrite (36) in the form

$$\frac{\rho(s+1)}{\rho(s)} = \frac{\sigma(s) + \tau(s)\Delta x(s-1/2)}{\sigma(s+1)}. \tag{59}$$

So that a one-to-one correspondence will exist between $x = x(s)$ and s we shall assume that $x(s)$ is monotonic on the interval $a \leq s \leq b$.

1°. *The lattice* $x(s) = s$. The case of a linear lattice $x(s) = s$ was discussed in detail in §12. Depending on the degrees of the polynomials $\sigma(s)$ and $\sigma(s) + \tau(s)\Delta x(s-1/2) = \sigma(s) + \tau(s)$, by solving (59) we obtain the Hahn polynomials $h_n^{(\alpha,\beta)}(x, N)$ and $\tilde{h}_n^{(\mu,\nu)}(x, N)$, the Meixner polynomial $m_n^{(\gamma,\mu)}(x)$, the Kravchuk polynomial $k_n^{(p)}(x, N)$, and the Charlier polynomial $c_n^{(\mu)}(x)$, the basic data for which are given in Table 3, 4 and 5.

2°. *The lattice* $x(s) = s(s+1)$. For $x(s) = s(s+1)$ $(s > -1/2)$ equation (59) can be transformed into a more convenient form. Under the transformation $s \to -s-1$ we have

$$x(s) = x(-s-1), \quad \Delta x(s-1/2) = -\Delta x(t-1/2)|_{t=-s-1};$$

then according to (21) and (22) we obtain

$$\sigma(s)+\tau(s)\Delta x(s-1/2)=\sigma(-s-1), \tag{60}$$

$$\tilde{\sigma}[x(s)]=\frac{1}{2}[\sigma(s)+\sigma(-s-1)], \qquad \tilde{\tau}[x(s)]=\frac{\sigma(-s-1)-\sigma(s)}{\Delta x(s-1/2)}. \tag{61}$$

The equation for $\rho(s)$ has the form

$$\frac{\rho(s+1)}{\rho(s)}=\frac{\sigma(-s-1)}{\sigma(s+1)}. \tag{62}$$

By virtue of (21) and (22), $\sigma(s)$ is a polynomial of the fourth or third degree in s in this case. Let

$$\sigma(s)=A\prod_{j=1}^{4}(s-s_j). \tag{63}$$

Then (62) for $\rho(s)$ has the form

$$\frac{\rho(s+1)}{\rho(s)}=\frac{\prod_{j=1}^{4}(s+1+s_j)}{\prod_{j=1}^{4}(s+1-s_j)}. \tag{64}$$

Since $\sigma(a)=0$, $\sigma(-b)=0$ according to (49b), (49c), and (60), we may take $s_1=a$ and $s_2=-b$.

a) Let

$$\sigma(s)=(s-a)(s+b)(s-c)(d-s) \tag{65}$$

(in (63) we put $A=-1$, $s_3=c$, $s_4=d$). Then

$$\rho(s)=\frac{\Gamma(s+a+1)\Gamma(s+c+1)\Gamma(s+d+1)\Gamma(d-s)}{\Gamma(s-a+1)\Gamma(s-c+1)\Gamma(s+b+1)\Gamma(b-s)}. \tag{66}$$

Since $\Delta x(s-1/2)=2s+1>0$ for $s>-1/2$, condition (50a) will be satisfied when

$$-1/2<a<b<1+d, \quad |c|<1+a.$$

b) Let

$$\sigma(s) = (s-a)(s+b)(s-c)(s+d)$$
$$(A=1,\ s_3=c,\ s_4=-d). \tag{67}$$

Then

$$\rho(s) = \frac{\Gamma(s+a+1)\Gamma(s+c+1)}{\Gamma(s-a+1)\Gamma(s-c+1)\Gamma(s+b+1)\Gamma(b-s)\Gamma(s+d+1)\Gamma(d-s)}$$
$$(-1/2 < a < b < 1+d, \quad |c| < 1+a). \tag{68}$$

We denote by $u_n^{(c,d)}(x)$ and $\tilde{u}_n^{(c,d)}(x)$, respectively, the polynomials $\tilde{y}_n(x)$, with $B_n = (-1)^n/n!$ and $B_n = 1/n!$, corresponding to the weight functions in (66) and (68). We call these the Racah polynomials, because they are connected, by a simple relation, with the Racah coefficients which are widely used in atomic physics.

c) For $x(s) = s(s+1)$ it may occur that $\sigma(s)$ is a cubic polynomial, i.e.

$$\sigma(s) = (s-a)(s+b)(s-c). \tag{69}$$

Then

$$\frac{\rho(s+1)}{\rho(s)} = \frac{(s+1+a)(b-s-1)(s+1+c)}{(s+1-a)(s+1+b)(s+1-c)}, \tag{70}$$

whence

$$\rho(s) = \frac{\Gamma(s+a+1)\Gamma(s+c+1)}{\Gamma(s-a+1)\Gamma(s+b+1)\Gamma(b-s)\Gamma(s-c+1)} \tag{71}$$
$$(-1/2 < a < b, \quad |c| < 1+a).$$

We denote by $w_n^{(c)}(x)$ the orthogonal polynomials with $B_n = (-1)^n/n!$. Comparing the corresponding orthogonality relations and the coefficients of the leading terms of the polynomials $w_n^{(c)}(x) \equiv w_n^{(c)}(x, a, b)$ with those of the dual Hahn polynomials $w_n^{(\alpha,\beta)}(x)$ (see §12), we see that they coincide if

$$a = (\alpha+\beta)/2, \quad b = a+N, \quad c = (\beta-\alpha)/2,$$

i.e. the Hahn polynomials and the $w_n^{(c)}(x, a, b)$ are connected by*

$$h_n^{(\alpha,\beta)}(i)$$
$$= (-1)^{i+n} \frac{i!\,(N-i-1)!\,\Gamma(\beta+n+1)}{n!\,(N-n-1)!\,\Gamma(\beta+i+1)} w_i^{(\beta-\alpha)/2}\left(t_n, \frac{\alpha+\beta}{2}, \frac{\alpha+\beta}{2}+N\right) \tag{72}$$
$$\left(t_n = s_n(s_n+1), \quad s_n = \frac{\alpha+\beta}{2}+n\right).$$

3°. We obtained the difference equation (3) from the differential equation (1) for the classical orthogonal polynomials. Consequently it is natural to expect that the polynomial solutions of (3) and the weight functions will, in the limit $h \to 0$, become (with appropriate normalization) the polynomial solutions of (1) and the corresponding weight functions.

Let us consider this limiting process for the Racah polynomials. Setting $h \to 0$ in (3) corresponds to $N = b - a \to \infty$ for the Racah polynomials. It is easy to show that the weight function $\rho(s)$ for the Racah polynomials $u_n^{(c,d)}[x(s)]$ becomes, in the limit $N \to \infty$, the weight function $(1-t)^\alpha(1+t)^\beta$ for the Jacobi polynomials $P_n^{(\alpha,\beta)}(t)$, where

$$t = 2\frac{x(s)-x(a)}{x(b)-x(a)} - 1, \quad \alpha = d - b, \beta = a + c.$$

For the proof it is sufficient to use the relation

$$\frac{\Gamma(z+\gamma)}{\Gamma(z+\delta)} \to z^{\gamma-\delta} \text{ as } z \to \infty.$$

In fact, for a fixed $t \in (-1, 1)$ and $N \to \infty$ we have

$$b = N + a \approx N,$$
$$1 + t = 2\frac{(s+1/2)^2 - (a+1/2)^2}{(b+1/2)^2 - (a+1/2)^2} \approx \frac{2}{N^2}s^2,$$
$$1 - t \approx \frac{2}{N^2}(b^2 - s^2),$$

* The Racah polynomials and dual Hahn polynomials, with a different normalization, were originally introduced in [A5] by means of the apparatus of generalized hypergeometric functions.

$$\frac{\Gamma(s+a+1)\,\Gamma(s+c+1)}{\Gamma(s-a+1)\,\Gamma(s-c+1)} \approx s^{2(a+c)} = (s^2)^\beta \approx \left[\frac{N^2}{2}(1+t)\right]^\beta,$$
$$\frac{\Gamma(s+d+1)\,\Gamma(d-s)}{\Gamma(s+b+1)\,\Gamma(b-s)} \approx (s+b)^{d-b}(b-s)^{d-b}$$
$$= (b^2-s^2)^\alpha \approx \left[\frac{N^2}{2}(1-t)\right]^\alpha.$$

Consequently

$$\lim_{N\to\infty}\left(\frac{2}{N^2}\right)^{\alpha+\beta}\rho(s) = (1-t)^\alpha(1+t)^\beta. \tag{73}$$

A similar limit relation must connect the Racah polynomials $u_n^{(c,d)}[x(s)]$ and the Jacobi polynomials $P_n^{(\alpha,\beta)}(t)$:

$$\lim_{N\to\infty} C_n(N)u_n^{(\beta-a,b+\alpha)}[x(s)] = P_n^{(\alpha,\beta)}(t). \tag{74}$$

The constants $C_n(N)$ are easily determined by equating the coefficients of the leading terms on the two sides of (74):

$$C_n(N) = N^{-2n}.$$

Because of the limit relation (74) we shall now refer to the Racah polynomials $u_n^{(c,d)}(x)$ as $u_n^{(\alpha,\beta)}(x)$, taking $\alpha = d-b$, $\beta = a+c$.

6. Construction of q-analogs of polynomials that are orthogonal on linear and quadratic lattices. We shall consider polynomials that are orthogonal on the lattices $x(s) = s$ (Hahn, Meixner, Kravchuk and Charlier polynomials) and $x(s) = s(s+1)$ (Racah and dual Hahn polynomials). In constructing a theory of classical orthogonal polynomials of a discrete variable, we use difference equations of hypergeometric type which retain this form after difference differentiation. It is possible to introduce difference equations of this kind only for certain types of lattices $x(s)$. Beyond the linear and quadratic lattices, the requirements are satisfied (as we showed in part 5) by the lattice functions

$$x(s) = \begin{cases} q^s = e^{2\omega s} & (q = e^{2\omega}); \\ \frac{1}{2}(q^s - q^{-s}) = \sinh 2\omega s & (q = e^{2\omega}); \\ \frac{1}{2}(q^s + q^{-s}) = \cosh 2\omega s & (q = e^{2\omega}); \\ \frac{1}{2}(q^s + q^{-s}) = \cos 2\omega s & (q = e^{2i\omega}). \end{cases}$$

When $q \to 1$ $(\omega \to 0)$ we have

$$e^{2\omega s} \approx 1 + 2\omega s, \qquad \sinh 2\omega s \approx 2\omega s;$$
$$\cosh 2\omega s \approx 1 + 2\omega^2 s^2, \qquad \cos 2\omega s \approx 1 - 2\omega^2 s^2,$$

i.e. the lattices become either linear or quadratic in s. The polynomials whose limits as $q \to 1$ are polynomials which are orthogonal on linear or quadratic lattices $x(s) = s$ or $x(s) = s(s+1)$ are called q-analogs of the corresponding polynomials. We shall discuss methods for constructing weight functions for the q-analogs of the Hahn, Meixner, Kravchuk and Charlier polynomials on the lattices $x(s) = e^{2\omega s}$ and $x(s) = \sinh 2\omega s$, as well as for the q-analogs of the Racah and dual Hahn polynomials on the lattices $x(s) = \cosh 2\omega s$ and $x(s) = \cos 2\omega s$. In constructing the weight functions $\rho(s)$, we shall start from equation (59).

1°. Construction of q-analogs of the Hahn, Meixner, Kravchuk and Charlier polynomials on the lattices $x(s) = \exp(2\omega s)$ and $x(s) = \sinh 2\omega s$.

a) The lattice $x(s) = q^s$ $(q = e^{2\omega})$. Replacing s by $s - a$ does not change the form of (5) with $x(s) = q^s$, and consequently, when considering the orthogonality property, we may take $a = 0$, as in the case $x(s) = s$. Since

$$\sigma(s) = \tilde{\sigma}[x(s)] - \frac{1}{2}\tilde{\tau}[x(s)]\Delta x\left(s - \frac{1}{2}\right), \qquad \Delta x\left(s - \frac{1}{2}\right) = \frac{q-1}{\sqrt{q}}x(s)$$

where $\tilde{\sigma}(x)$ and $\tilde{\tau}(x)$ are polynomials of at most the second and first degrees, respectively, the functions $\sigma(s)$ and $\sigma(s) + \tau(s)\Delta x(s - 1/2)$ are polynomials of degree 2 at most in $x = x(s)$, and the right-hand side of (59) is a rational function in $x(s)$. We are now going to decompose the functions $\sigma(s+1)$ and $\sigma(s) + \tau(s)\Delta x(s - 1/2)$ into simple factors, linear in $x = q^s$, and use the property of the solutions of the equation

$$\frac{\rho(s+1)}{\rho(s)} = F(s) \qquad \text{with } F(s) = \frac{\prod\limits_i f_i(s)}{\prod\limits_i g_i(s)},$$

which was considered in §12, part 4. Then the solutions of (59) for the lattice $x = q^s$ can always be determined if we know particular solutions of the equations

$$\frac{\rho(s+1)}{\rho(s)} = \begin{cases} \dfrac{q^{s+\gamma} - 1}{q - 1}, & \dfrac{q^{\gamma - s} - 1}{q - 1}; \\[2ex] \dfrac{q^{s+\gamma} + 1}{q + 1}, & \dfrac{q^{\gamma - s} + 1}{q + 1}; \\[2ex] \alpha^s, & \beta, \end{cases} \tag{75}$$

where α, β and γ are constants. These solutions are

$$\rho(s) = \begin{cases} \Gamma_q(s+\gamma), & \dfrac{1}{\Gamma_q(\gamma - s + 1)}; \\ \Pi_q(s+\gamma), & \dfrac{1}{\prod_q(\gamma - s + 1)}; \\ \alpha^{s(s-1)/2}, & \beta^s. \end{cases} \tag{76}$$

The function $\Gamma_q(s)$ is called the q-gamma function; it is a generalization of Euler's gamma-function $\Gamma(s)$ [J2]. It is defined by

$$\Gamma_q(s) = \begin{cases} (1-q)^{1-s} \dfrac{\prod_{k=0}^{\infty}(1-q^{k+1})}{\prod_{k=0}^{\infty}(1-q^{s+k})}, & |q| < 1; \\ q^{(s-1)(s-2)/2}\Gamma_{1/q}(s), & |q| > 1. \end{cases} \tag{77}$$

For the function $\Gamma_q(s)$ we have the relations

$$\Gamma_q(s+1) = \frac{q^s - 1}{q-1}\Gamma_q(s), \tag{78}$$

$$\lim_{q\to 1}\Gamma_q(s) = \Gamma(s). \tag{79}$$

The functions $\Pi_q(s)$ and $\Gamma_q(s)$ are connected by the relation

$$\Pi_q(s) = \frac{\Gamma_{q^2}(s)}{\Gamma_q(s)}. \tag{80}$$

From (78) and (79) we have

$$\Pi_q(s+1) = \frac{q^s + 1}{q+1}\Pi_q(s), \tag{81}$$

$$\lim_{q\to 1}\Pi_q(s) = 1. \tag{82}$$

We shall also use, instead of $\Gamma_q(s)$ and $\Pi_q(s)$, the related functions $\tilde{\Gamma}_q(s)$ and $\tilde{\Pi}_q(s)$ satisfying the equations

$$\frac{\tilde{\Gamma}_q(s+1)}{\tilde{\Gamma}_q(s)} = \psi_q(s), \quad \psi_q(s) = \frac{q^{s/2} - q^{-s/2}}{q^{1/2} - q^{-1/2}} = \frac{\sinh \omega s}{\sinh \omega}; \tag{78a}$$

$$\frac{\tilde{\Pi}_q(s+1)}{\tilde{\Pi}_q(s)} = \phi_q(s), \quad \phi_q(s) = \frac{q^{s/2} + q^{-s/2}}{q^{1/2} + q^{-1/2}} = \frac{\cosh \omega s}{\cosh \omega}. \tag{81}$$

The functions $\tilde{\Gamma}_q(s)$ and $\tilde{\Pi}_q(s)$ are connected with the functions $\Gamma_q(s)$ and $\Pi_q(s)$ by the relations

$$\left.\begin{aligned}\tilde{\Gamma}_q(s) &= q^{-(s-1)(s-2)/4}\Gamma_q(s),\\ \tilde{\Pi}_q(s) &= \frac{\tilde{\Gamma}_{q^2}(s)}{\tilde{\Gamma}_q(s)}.\end{aligned}\right\} \tag{83}$$

These functions have more symmetry than $\Gamma_q(s)$ and $\Pi_q(s)$. For example, from the relation $\psi_q(-s) = -\psi_q(s)$ it follows that

$$\left.\frac{\tilde{\Gamma}_q(t+1)}{\tilde{\Gamma}_q(t)}\right|_{t=-s} = -\frac{\tilde{\Gamma}_q(s+1)}{\tilde{\Gamma}_q(s)},$$

which corresponds to a similar relation for Euler's gamma-function:

$$\left.\frac{\Gamma(t+1)}{\Gamma(t)}\right|_{t=-s} = -\frac{\Gamma(s+1)}{\Gamma(s)}.$$

For the function $\Gamma_q(s)$ the analogous equality has a more complicated form:

$$\left.\frac{\Gamma_q(t+1)}{\Gamma_q(t)}\right|_{t=-s} = -q^{-s}\frac{\Gamma_q(s+1)}{\Gamma_q(s)}.$$

Furthermore, the relation

$$\Gamma_q(s) = q^{(s-1)(s-2)/2}\Gamma_{1/q}(s)$$

becomes the more symmetric relation:

$$\tilde{\Gamma}_q(s) = \tilde{\Gamma}_{1/q}(s).$$

This relation allows a natural generalization of the definition of $\tilde{\Gamma}_q(s)$ and $\tilde{\Pi}_q(s)$ to the case when $q = e^{2i\omega} (\omega > 0)$, i.e. to the case $|q| = 1$:

$$\left.\begin{aligned}\tilde{\Gamma}_q(s) &= \lim_{q'\to q}\tilde{\Gamma}_{q'}(s), \quad (|q'| \neq 1)\\ \tilde{\Pi}_q(s) &= \lim_{q'\to q}\tilde{\Pi}_{q'(s)}, \quad (|q'| \neq 1)\end{aligned}\right\} \tag{83a}$$

Note that for the function $\tilde{\Gamma}_q(s)$ we have

$$\lim_{q'\to q, |q'|<1}\tilde{\Gamma}_{q'}(s) = \lim_{q'\to q, |q'|>1}\tilde{\Gamma}_{q'}(s).$$

We shall also use the following asymptotic properties of $\tilde{\Gamma}_q(s)$ when $s \to +\infty$, $0 < q < 1$:

$$\tilde{\Gamma}_q(s+1) \approx q^{-s(s-1)/4}(1-q)^{-s}e^{-C_q},$$
$$\frac{\tilde{\Gamma}_q(s+a)}{\tilde{\Gamma}_q(s)} \approx q^{-a(a+2s-3)/4}(1-q)^{-a}.$$

Here $C_q = -\sum\limits_{k=0}^{\infty} \ln(1-q^{k+1})$. For the function $\tilde{\Pi}_q(s)$ when $s \to +\infty$, and $0 < q < 1$, by using the preceding properties and (83) we obtain

$$\tilde{\Pi}_q(s+1) \approx q^{-s(s-1)/4}(1+q)^{-s}\exp[-(C_{q^2} - C_q)],$$
$$\frac{\tilde{\Pi}_q(s+a)}{\tilde{\Pi}_q(s)} \approx q^{-a(a+2s-3)/4}(1+q)^{-a}.$$

The analogous relations for $q > 1$ can easily be obtained by using the connections between the functions $\tilde{\Gamma}_{1/q}(s)$ and $\tilde{\Pi}_{1/q}(s)$, and between the functions $\tilde{\Gamma}_q(s)$ and $\tilde{\Pi}_q(s)$:

$$\tilde{\Gamma}_{1/q}(s) = \tilde{\Gamma}_q(s), \quad \tilde{\Pi}_{1/q}(s) = \tilde{\Pi}_q(s).$$

By means of the simple relation

$$\lim_{q \to 1} \psi_q(s) = s \tag{84}$$

we can construct systems of orthogonal polynomials which when $q \to 1$ become the polynomials on the lattice $x(s) = s$ considered above. In accordance with (84), we shall choose the form of the functions $\sigma(s)$ and $\sigma(s) + \tau(s)\Delta x(s-1/2)$ so that the weight functions $\rho(s)$ and the polynomials $\tilde{y}_n[x(s)]$ (with a specific choice of the constant B_n in the Rodrigues formula) become, when $q \to 1$, the weight functions and polynomials on the lattice $x(s) = s$. The polynomials on the lattice $x(s) = q^s$ corresponding to the Hahn polynomials $h_n^{(\alpha,\beta)}(s)$ and $\tilde{h}_n^{(\mu,\nu)}(s)$, the Meixner polynomials $m_n^{(\gamma,\mu)}(s)$, the Kravchuk polynomials $k_n^{(p)}(s)$, and the Charlier polynomials $c_n^{(\mu)}(s)$ will be denoted by $h_n^{(\alpha,\beta)}(x,q)$ and $\tilde{h}_n^{(\mu,\nu)}(x,q)$; $m_n^{(\gamma,\mu)}(x,q)$; $k_n^{(p)}(x,q)$; and $c_n^{(\mu)}(x,q)$. In order to guess the form of the functions $\sigma(s)$ and $\sigma(s) + \tau(s)\Delta x(s-1/2)$ in (59) for these polynomials, we replace the factors of the form $\pm(s+\gamma)$ in the corresponding expressions on the lattice $x(s) = s$ by the function $q^{(s+\gamma)/2}\psi_q[\pm(s+\gamma)]$, which is a polynomial of first degree in $x(s) = q^s$. Under this substitution the functions $\sigma(s)$ and $\sigma(s) + \tau(s)\Delta x(s-1/2)$ will be polynomials of at most second degree in q^s.

In addition, in the process of choosing the form of the polynomials $\sigma(s)$ and $\sigma(s)+\tau(s)\Delta x(s-1/2)$ it should be kept in mind that their constant terms must obviously coincide, since the product

$$\tau(s)\Delta x(s-1/2)=\tilde{\tau}[x(s)][(q^{1/2}-q^{-1/2})x(s)]$$

is a polynomial of second degree in $x=x(s)$ with constant term zero. In order to satisfy this condition in the case when the functions $\sigma(s)$ and $\sigma(s)+\tau(s)$ on the lattice $x(s)=s$ are polynomials of at most first degree, we multiply the assumed expressions for $\sigma(s+1)$ and $\sigma(s)+\tau(s)\Delta x(s-1/2)$ in (59) by q^s (the introduction of this multiplier does not change equation (59) for $\rho(s)$). But if $\sigma(s)$ is a polynomial of second degree on the lattice $x(s)=s$, under the preceding transformations the constant terms in the polynomials $\sigma(s)$ and $\sigma(s)+\tau(s)\Delta x(s-1/2)$ coincide, as can easily be verified.

In order to express the function $\rho(s)$ in terms of functions of the form $\tilde{\Gamma}_q[\pm(s-\gamma)]$ and $\tilde{\Pi}_q[\pm(s-\gamma)]$, we shall use, instead of (75) and (76), particular solutions of the equations

$$\frac{\rho(s+1)}{\rho(s)}=\begin{cases}\psi_q(s+\gamma), & \psi_q(\gamma-s);\\ \phi_q(s+\gamma), & \phi_q(\gamma-s);\\ \alpha^s, & \beta\end{cases} \tag{75a}$$

which have the form

$$\rho(s)=\begin{cases}\tilde{\Gamma}_q(s+\gamma), & 1/\tilde{\Gamma}_q(\gamma-s+1);\\ \tilde{\Pi}_q(s+\gamma), & 1/\tilde{\Pi}_q(\gamma-s+1);\\ \alpha^{s(s-1)/2}, & \beta^s.\end{cases} \tag{76a}$$

Let us consider some specific cases of applying these methods.

1) *The Hahn polynomials* $h_n^{(\alpha,\beta)}(x,q)$ *and* $\tilde{h}_n^{(\mu,\nu)}(x,q)$.

For the Hahn polynomials $h_n^{(\alpha,\beta)}(s)$ we have

$$\sigma(s)=s(N+\alpha-s),\quad \sigma(s)+\tau(s)=(s+\beta+1)(N-1-s).$$

Therefore on the lattice $x(s)=q^s$ we take

$$\sigma(s)=q^{s/2}\psi_q(s)q^{(s-N-\alpha)/2}\psi_q(N+\alpha-s),$$

$$\sigma(s)+\tau(s)\Delta x\left(s-\frac{1}{2}\right)=q^{(s+\beta+1)/2}\psi_q(s+\beta+1)q^{(s-N+1)/2}\psi_q(N-1-s).$$

Equation (59) for $\rho(s)$ takes the form

$$\frac{\rho(s+1)}{\rho(s)}=q^{(\alpha+\beta)/2}\frac{\psi_q(s+\beta+1)\psi_q(N-1-s)}{\psi_q(s+1)\psi_q(N+\alpha-1-s)}.$$

By using (75a) and (76a) we obtain

$$\rho(s) = q^{(\alpha+\beta)s/2}\frac{\tilde{\Gamma}_q(s+\beta+1)\tilde{\Gamma}_q(N+\alpha-s)}{\tilde{\Gamma}_q(s+1)\tilde{\Gamma}(N-s)}. \tag{85}$$

Since, when $q \to 1$, the function $\rho(s)$ takes in the limit the form of the weight function for the Hahn polynomials $h_n^{(\alpha,\beta)}(s)$, it can be seen from the Rodrigues formula that the limiting relation

$$\lim_{q\to 1} h_n^{(\alpha,\beta)}(x(s),q) = h_n^{(\alpha,\beta)}(s)$$

will be satisfied if in the Rodrigues formula for $h_n^{(\alpha,\beta)}(x(s),q)$ we take $B_n = (1-q)^n/n!$.

In a similar way, for the Hahn polynomials $\tilde{h}_n^{(\mu,\nu)}(x(s),q)$ we obtain:

$$\sigma(s) = q^{s/2}\psi_q(s)q^{(s+\mu)/2}\psi_q(s+\mu),$$

$$\sigma(s)+\tau(s)\Delta x\left(s-\frac{1}{2}\right)$$
$$= q^{(s-N-\nu+1)/2}\psi_q(N+\nu-1-s)q^{(s-N+1)/2}\psi_q(N-1-s),$$

$$\rho(s) = \frac{q^{-(N+(\mu+\nu)/2)s}}{\tilde{\Gamma}_q(N+\nu-s)\tilde{\Gamma}_q(N-s)\tilde{\Gamma}_q(s+1)\tilde{\Gamma}_q(s+\mu+1)}; \tag{86}$$

$$\lim_{q\to 1} \tilde{h}_n^{(\mu,\nu)}(x(s),q) = \tilde{h}_n^{(\mu,\nu)}(s), \quad B_n = (q-1)^n/n!.$$

2) *The Meixner, Kravchuk and Charlier q-polynomials. For the Meixner polynomials* we have

$$\sigma(s) = s, \quad \sigma(s)+\tau(s) = \mu(s+\gamma) \quad (0<\mu<1, \gamma>0).$$

Hence, for the polynomials $m_n^{(\gamma,\mu)}(x,q)$, by assuming

$$\sigma(s) = q^{s/2}\psi_q(s)q^{s-1},$$

$$\sigma(s)+\tau(s)\Delta x\left(s-\frac{1}{2}\right) = \mu q^{(s+\gamma)/2}\psi_q(s+\gamma)q^s,$$

we obtain

$$\frac{\rho(s+1)}{\rho(s)} = \mu q^{(\gamma-1)/2}\frac{\psi_q(s+\gamma)}{\psi_q(s+1)}.$$

A solution of this equation has the form

$$\rho(s) = C\mu^s q^{s(\gamma-1)/2} \frac{\tilde{\Gamma}_q(s+\gamma)}{\tilde{\Gamma}_q(s+1)}, \tag{87}$$

where C is a constant. In order for the function $\rho(s)$, when $q \to 1$, to take the form of the weight function for the polynomials $m_n^{(\gamma,\mu)}(s)$, we suppose that $C = 1/\tilde{\Gamma}_q(\gamma)$. The limit relation

$$\lim_{q\to1} m_n^{(\gamma,\mu)}(x(s), q) = m_n^{(\gamma,\mu)}(s)$$

is satisfied for $B_n = ((q-1)/\mu)^n$.

3) *For the Kravchuk polynomials*

$$\sigma(s) = s, \quad \sigma(s) + \tau(s) = \frac{p}{1-p}(N-s).$$

Therefore, for the polynomials $k_n^{(p)}(x, q)$ we take

$$\sigma(s) = q^{s/2}\psi_q(s)q^{s-1},$$
$$\sigma(s) + \tau(s)\Delta x\left(s - \frac{1}{2}\right) = \frac{p}{1-p}q^{(s-N)/2}\psi_q(N-s)q^s,$$

from which

$$\frac{\rho(s+1)}{\rho(s)} = \frac{p}{1-p}q^{-(N+1)/2}\frac{\psi_q(N-s)}{\psi_q(s+1)},$$

and hence

$$\rho(s) = C\left(\frac{p}{1-p}\right)^s \frac{q^{-s(N+1)/2}}{\tilde{\Gamma}_q(s+1)\tilde{\Gamma}_q(N-s+1)} \tag{88}$$

where C is a constant. The limit relations will be satisfied if

$$C = (1-p)^N \tilde{\Gamma}_q(N+1), \quad B_n = \frac{(1-p)^n(1-q)^n}{n!}.$$

Similarly, *for the polynomials* $c_n^{(\mu)}(x, q)$ we obtain

$$\sigma(s) = q^{s/2}\psi_q(s)q^{s-1},$$
$$\sigma(s) + \tau(s)\Delta x(s - 1/2) = \mu q^s,$$
$$\rho(s) = e^{-\mu}\frac{\mu^s}{\tilde{\Gamma}_q(s+1)}q^{-s(s+1)/4}, \tag{89}$$
$$\lim_{q\to1} c_n^{(\mu)}(x(s), q) = c_n^{(\mu)}(s), \quad B_n = \left(\frac{q-1}{\mu}\right)^n.$$

All the q-polynomials considered on the lattice $x(s) = q^s$ are orthogonal (in s) for the same values of a and b as for the corresponding polynomials on the lattice $x(s) = s$ (under the additional restriction $q < 1$ for $m_n^{(\gamma,\mu)}(x, q)$, which follows directly from the asymptotic behavior of the function $\rho(s)$ when $s \to +\infty$).

In addition to the polynomials that we have considered on the lattice $x(s) = q^s$ we can also construct polynomial systems for which there are no analogs on the lattice $x(s) = s$. For example, for

$$\sigma(s) = q^{s/2}\psi_q(s)q^{(s-N-\alpha)/2}\psi_q(N+\alpha-s),$$

$$\sigma(s) + \tau(s)\Delta x\left(s-\frac{1}{2}\right) = \frac{q^{1/2}}{1-q}q^{(s-N+1)/2}\psi_q(N-1-s)$$

$$(\alpha > 0,\ 0 < q < 1)$$

we obtain

$$\rho(s) = \frac{q^{s(2\alpha+1-s)/4}}{(1-q)^s}\frac{\tilde{\Gamma}_q(N+\alpha-s)}{\tilde{\Gamma}_q(s+1)\tilde{\Gamma}_q(N-s)}. \tag{90}$$

The corresponding polynomials are orthogonal for $a = 0, b = N$. In this case these polynomials have no analogs on the lattice $x(s) = s$, since, when $q \to 1$, the function $\sigma(s)$ has a limit; however, $\sigma(s) + \tau(s)\Delta x(s - 1/2) \to \infty$, $\rho(s) \to \infty$.

4) *The lattice* $x(s) = \sinh 2\omega s$. Let $q = e^{2\omega}$; then

$$x(s) = \frac{1}{2}(q^s - q^{-s}), \quad \Delta x\left(s-\frac{1}{2}\right) = \frac{1}{2}(q^{1/2} - q^{-1/2})(q^s + q^{-s}).$$

It is also convenient to use the symmetry property of the lattice $x(s)$:

$$x(s) = x(t), \quad \Delta x(s-1/2) = -\Delta x(t-1/2),$$

where $t = -s - i\pi/\ln q$.

In accordance with (21) and (22)

$$\sigma(s) + \tau(s)\Delta x\left(s-\frac{1}{2}\right) = \sigma\left(-s-\frac{i\pi}{\ln q}\right), \tag{91}$$

from which

$$\tilde{\sigma}[x(s)] = \frac{1}{2}\left[\sigma(s) + \sigma\left(-s-\frac{i\pi}{\ln q}\right)\right], \tag{92}$$

$$\tilde{\tau}[x(s)] = \frac{\sigma(-s-i\pi/\ln q) - \sigma(s)}{\Delta x(s-1/2)}. \tag{93}$$

We note that by virtue of (21)

$$\sigma(s) = q^{-2s} p_4(q^s), \tag{94}$$

where $p_4(q^s)$ is a polynomial of fourth degree in q^s. By using (92) and (93) it can be easily verified by means of (94) that the functions $\tilde{\tau}[x(s)]$ and $\tilde{\sigma}[x(s)]$ are polynomials of at most first and second degrees in $x(s) = (q^s - q^{-s})/2$, respectively, for an arbitrary form of $p_4(q^s)$. Let us construct the systems of orthogonal polynomials which, when $q \to 1$, i.e. $\omega \to 0$, take the form of the Hahn, Meixner, Kravchuk and Charlier polynomials on the lattice $x(s) = s$. To do this we use the limit relations

$$\lim_{q\to 1} \psi_q(s) = s, \quad \lim_{q\to 1} \phi_q(s) = 1, \tag{95}$$

where the functions $\psi_q(s)$ and $\phi_q(s)$ are determined by (78a) and (81a). In order to guess the form of the functions $\sigma(s)$ and $\sigma(s) + \tau(s)\Delta x(s - 1/2)$ for these polynomials we shall observe the following rule: if on the lattice $x(s) = s$ the expression for $\sigma(s)$ has the factor $s - \gamma$ and the expression for $\sigma(s) + \tau(s)$ has the factor $s - \gamma_1$, then on the lattice $x(s) = \sinh 2\omega s$ we shall take the factor $\psi_q(s - \gamma)\phi_q(s + \gamma_1)$ in $\sigma(s)$. Then, in accordance with (91), the factor $\phi_q(s + \gamma)\psi_q(s - \gamma)$ will appear in $\sigma(s) + \tau(s)\Delta x(s - 1/2)$. This is because under the transformation $s \to -s - i\pi/\ln q$ we have

$$\psi_q(s - \gamma) \to -i\frac{q^{1/2} + q^{-1/2}}{q^{1/2} - q^{-1/2}} \phi_q(s + \gamma),$$

$$\phi_q(s + \gamma_1) \to i\frac{q^{1/2} - q^{-1/2}}{q^{1/2} + q^{-1/2}} \psi_q(s - \gamma_1),$$

$$\psi_q(s - \gamma)\phi_q(s + \gamma_1) \to \phi_q(s + \gamma)\psi_q(s - \gamma_1).$$

In addition, by virtue of (95)

$$\lim_{q\to 1} \psi_q(s - \gamma)\phi_q(s + \gamma_1) = s - \gamma,$$

$$\lim_{q\to 1} \phi_q(s + \gamma)\psi_q(s - \gamma_1) = s - \gamma_1.$$

In our further discussion, when the cases cannot be reduced to the one considered, we shall choose the form of the factors in $\sigma(s)$ specifically in each case.

5) *Analogs of the Hahn polynomials* $h_n^{(\alpha,\beta)}(s-a)$ *and* $\tilde{h}_n^{(\mu,\nu)}(s-a)$ *on the lattice* $x(s)=\sinh 2\omega s$ $(a\le s\le b-1)$. For the Hahn polynomials $h_n^{(\alpha,\beta)}(s-a)$ we have

$$\sigma(s)=(s-a)(b+\alpha-s),$$
$$\sigma(s)+\tau(s)=(s-a+\beta+1)(b-1-s).$$

Consequently on the lattice $x(s)=\sinh 2\omega s$ we suppose that

$$\sigma(s)=\psi_q(s-a)\phi_q(s+a-\beta-1)\psi_q(b+\alpha-s)\phi_q(b-1+s),$$
$$\sigma(s)+\tau(s)\Delta x(s-1/2)=\phi_q(s+a)\psi_q(s-a+\beta+1)\phi_q(b+\alpha+s)\psi_q(b-1-s).$$

It is evident that the expression for $\sigma(s)$ has the form (94), and that the functions $\tilde{\sigma}(x)$ and $\tilde{\tau}(x)$ will be polynomials of at most second and first degrees in x, respectively.

Equation (59) takes the form

$$\frac{\rho(s+1)}{\rho(s)}=\frac{\phi_q(s+a)\psi_q(s-a+\beta+1)\phi_q(b+\alpha+s)\psi_q(b-1-s)}{\psi_q(s+1-a)\phi_q(s+a-\beta)\psi_q(b+\alpha-1-s)\phi_q(s+b)}.$$

By using (75a) and (76a), we obtain

$$\rho(s)=\frac{\tilde{\Pi}_q(s+a)\tilde{\Gamma}_q(s-a+\beta+1)\tilde{\Pi}_q(b+\alpha+s)\tilde{\Gamma}_q(b+\alpha-s)}{\tilde{\Gamma}_q(s+1-a)\tilde{\Gamma}_q(b-s)\tilde{\Pi}_q(s+a-\beta)\tilde{\Pi}_q(s+b)}. \tag{96}$$

By the same argument, for analogs of the polynomials $\tilde{h}_n^{(\mu,\nu)}(s-a)$ we obtain

$$\sigma(s)=\psi_q(s-a)\phi_q(s+b+\nu-1)\psi_q(s-a+\mu)\phi_q(s+b-1),$$
$$\sigma(s)+\tau(s)\Delta x\left(s-\frac{1}{2}\right)$$
$$=\phi_q(s+a)\psi_q(b+\nu-1-s)\phi_q(s+a-\mu)\psi_q(b-1-s), \tag{97}$$
$$\rho(s)$$
$$=\frac{\tilde{\Pi}_q(s+a)\tilde{\Pi}_q(s+a-\mu)}{\tilde{\Gamma}_q(s+1-a)\tilde{\Gamma}_q(b-s)\tilde{\Gamma}_q(b+\nu-s)\tilde{\Gamma}_q(s-a+\mu+1)\tilde{\Pi}_q(s+b)\tilde{\Pi}_q(s+b+\nu)}.$$

2) *Analogs of the Meixner polynomials on the lattice* $x(s) = \sinh 2\omega s$, with $s \geq a$. For the Meixner polynomials $m_n^{(\gamma,\mu)}(s-a)$ we have

$$\sigma(s) = s - a, \quad \sigma(s) + \tau(s) = \mu(s + \gamma - a) \quad (0 < \mu < 1).$$

The factors $s - a$ and $s + \gamma - a$ in the expression for $\sigma(s)$ on the lattice $x(s) = \sinh 2\omega s$ correspond to the factor $\psi_q(s-a)\phi_q(s-\gamma+a)$. Furthermore, in the expression for $\sigma(s)$ we have to choose a factor which is a polynomial of first degree in q^s and such that, in the limit $q \to 1$, it will be equal to unity and, under the transformation $s \to -s - i\pi/\ln q$, to μ. The factor can be taken in the form

$$\frac{(1+\mu) + (1-\mu)q^{s+a}}{q^{1/2} + q^{-1/2}}.$$

It corresponds to the factor

$$\frac{(1+\mu) - (1-\mu)q^{-(s-a)}}{q^{1/2} + q^{-1/2}}$$

in the expression for $\sigma(s) + \tau(s)\Delta x(s - 1/2)$. As a result we obtain

$$\sigma(s) = \psi_q(s-a)\phi_q(s-\gamma+a)\frac{(1+\mu) + (1-\mu)q^{s+a}}{q^{1/2} + q^{-1/2}},$$

$$\sigma(s) + \tau(s)\Delta x\left(s - \frac{1}{2}\right) = \phi_q(s+a)\psi_q(s+\gamma-a)\frac{(1+\mu) - (1-\mu)q^{-(s-a)}}{q^{1/2} + q^{-1/2}}.$$

By assuming

$$\frac{1-\mu}{1+\mu} = q^{-\delta} \quad (\delta = \delta(q)),$$

we obtain an equation for $\rho(s)$ in the form

$$\frac{\rho(s+1)}{\rho(s)} = \frac{q-1}{q+1}q^{-(s+1/2)}\frac{\phi_q(s+a)\psi_q(s+\gamma-a)\psi_q(s+\delta-a)}{\psi_q(s+1-a)\phi_q(s-\gamma+a+1)\phi_q(s+a+1-\delta)}.$$

Using (75a) and (76a) yields

$$\rho(s) = \left(\frac{q-1}{q+1}\right)^s q^{-s^2/2}\frac{\tilde{\Pi}_q(s+a)\tilde{\Gamma}_q(s+\gamma-a)\tilde{\Gamma}_q(s+\delta-a)}{\tilde{\Gamma}_q(s+1-a)\tilde{\Pi}_q(s-\gamma+a+1)\tilde{\Pi}_q(s+a+1-\delta)}. \tag{98}$$

3) *Analogs of the Kravchuk polynomials on the lattice* $x(s) = \sinh 2\omega s$ $(a \leq s \leq b-1,\ b-a = N+1)$. For the Kravchuk polynomials $k_n^{(p)}(s-a)$ we have

$$\sigma(s) = s - a, \quad \sigma(s) + \tau(s) = \mu(b-1-s), \quad \mu = \frac{p}{1-p}.$$

By using the same argument as for the analogs of the Meixner polynomials, we can choose the functions $\sigma(s)$ and $\sigma(s)+\tau(s)\Delta x(s-1/2)$ in the form

$$\sigma(s) = \psi_q(s-a)\phi_q(s+b-1)F_1(s,\mu,q),$$
$$\sigma(s)+\tau(s)\Delta x(s-1/2) = \phi_q(s+a)\psi_q(b-1-s)F_2(s,\mu,q);$$

$$F_1(s,\mu,q) = \begin{cases} \dfrac{(1-\mu)+(1+\mu)q^{s+b-1}}{q^{1/2}+q^{-1/2}} & (0<\mu<1, \text{ i.e. } 0<p<\frac{1}{2}),\\ q^s & (\mu=1, \text{ i.e. } p=\frac{1}{2})\\ \dfrac{(\mu+1)q^{s-a}-(\mu-1)}{q^{1/2}+q^{-1/2}} & (\mu>1, \text{ i.e. } \frac{1}{2}<p<1);\end{cases}$$

$$F_2(s,\mu,q) = \begin{cases} \dfrac{(1+\mu)q^{b-1-s}-(1-\mu)}{q^{1/2}+q^{-1/2}} & (0<\mu<1),\\ q^{-s} & (\mu=1),\\ \dfrac{(\mu+1)q^{-(s+a)}+(\mu-1)}{q^{1/2}+q^{-1/2}} & (\mu>1).\end{cases}$$

By solving equation (59) we obtain

a)
$$\rho(s)=\left(\frac{q-1}{q+1}\right)^s \frac{q^{-s^2/2}\tilde{\Pi}_q(s+a)}{\tilde{\Gamma}_q(b-s)\tilde{\Gamma}_q(b-s+\delta)\tilde{\Gamma}_q(s-a+1)\tilde{\Pi}_q(s+b)\tilde{\Pi}_q(s+b+\delta)};$$
$$\left(0<\mu<1, \quad q^\delta = \frac{1+\mu}{1-\mu} = \frac{1}{1-2p}\right) \tag{99a}$$

b)
$$\rho(s) = \frac{q^{-s^2}\tilde{\Pi}_q(s+a)}{\tilde{\Gamma}_q(b-s)\tilde{\Gamma}_q(s-a+1)\tilde{\Pi}_q(s+b)}$$
$$(\mu=1) \tag{99b}$$

c)
$$\rho(s) = \left(\frac{q+1}{q-1}\right)^s \frac{q^{-s^2/2}\tilde{\Pi}_q(s+a)\tilde{\Pi}(s+a-\delta)}{\tilde{\Gamma}_q(b-s)\tilde{\Gamma}_q(s-a+1)\tilde{\Pi}_q(s+b)\tilde{\Gamma}_q(s-a+\delta+1)}$$
$$\left(\mu>1, \quad q^\delta = \frac{\mu+1}{\mu-1} = \frac{1}{2p-1}\right). \tag{99c}$$

4) *Analogs of the Charlier polynomials on the lattice* $x(s) = \sinh 2\omega s$ $(s \geq a)$. For the Charlier polynomials $c_n^{(\mu)}(s-a)$ we have

$$\sigma(s) = s - a, \quad \sigma(s) + \tau(s) = \mu.$$

We shall choose the functions $\sigma(s)$ and $\sigma(s) + \tau(s)\Delta x(s - 1/2)$ in the form

$$\sigma(s) = q^{(s-a)/2}\psi_q(s-a)F_1(s,\mu,q),$$
$$\sigma(s) + \tau(s)\Delta x\left(s - \frac{1}{2}\right) = q^{-(s+a)/2}\phi_q(s+a)F_2(s,\mu,q),$$
$$F_1(s,\mu,q) = \frac{q^s + q^{-s} + 2 + 2\mu(q-1)q^{-s}}{(q^{1/2}+q^{-1/2})^2},$$
$$F_2(s,\mu,q) = -\frac{q^{1/2}+q^{-1/2}}{q^{1/2}-q^{-1/2}}F_1\left(-s - \frac{i\pi}{\ln q}, \mu, q\right)$$
$$= \frac{q+1}{q-1}\frac{q^s + q^{-s} - 2 + 2\mu(q-1)q^s}{(q^{1/2}+q^{-1/2})^2}.$$

In order to rewrite (59) in a simpler form, we transform the expression for $F_1(s,\mu,q)$ into

$$F_1(s,\mu,q) = \frac{q^{s/2} + (1 + i\sqrt{2\mu(q-1)})q^{-s/2}}{q^{1/2}+q^{-1/2}}\frac{q^{s/2} + (1 - i\sqrt{2\mu(q-1)})q^{-s/2}}{q^{1/2}+q^{-1/2}}$$
$$= q^{(\alpha+i\beta)/2}\phi_q(s-\alpha-i\beta)q^{(\alpha-i\beta)/2}\phi_q(s-\alpha+i\beta) = q^{\alpha}|\phi_q(s-\alpha+i\beta)|^2,$$

where

$$1 \pm i\sqrt{2\mu(q-1)} = q^{\alpha \pm i\beta}.$$

By using the connection between the functions $F_2(s,\mu,q)$ and $F_1(s,\mu,q)$ we obtain

$$F_2(s,\mu,q) = \frac{q-1}{q+1}q^{\alpha}|\psi_q(s+\alpha+i\beta)|^2.$$

Solving equation (59) yields

$$\rho(s) = \left(\frac{q-1}{q+1}\right)^s q^{-s^2/2}\frac{\tilde{\Pi}_q(s+a)|\tilde{\Gamma}_q(s+\alpha+i\beta)|^2}{\tilde{\Gamma}_q(s-a+1)|\tilde{\Pi}_q(s+1-\alpha+i\beta)|^2}. \tag{100}$$

In order for the polynomials corresponding to these forms of $\rho(s)$ to become the analogous polynomials on the lattice $x(s) = s$ when $q \to 1$, it is enough to use the same values of the constants B_n as on the lattice $x(s) = q^s$ in the Rodrigues formula.

2°. Construction of q-analogs for the Racah and dual Hahn polynomials on the lattices $x(s) = \cosh 2\omega s$ and $x(s) = \cos 2\omega s$.

a) *The lattice* $x(s) = \cosh 2\omega s$. We suppose that $q = e^{2\omega}$; then

$$x(s) = \frac{1}{2}(q^s + q^{-s}), \quad \Delta x\left(s - \frac{1}{2}\right) = \frac{1}{2}(q^{1/2} - q^{-1/2})(q^s - q^{-s}).$$

Since

$$x(-s) = x(s), \quad \Delta x\left(t - \frac{1}{2}\right)\bigg|_{t=-s} = -\Delta x\left(s - \frac{1}{2}\right),$$

then according to (21) and (22), we have

$$\sigma(s) + \tau(s)\Delta x\left(s - \frac{1}{2}\right) = \sigma(-s).$$

From this,

$$\tilde{\sigma}[x(s)] = \frac{\sigma(s) + \sigma(-s)}{2}, \tag{101}$$

$$\tilde{\tau}[x(s)] = \frac{\sigma(-s) - \sigma(s)}{\Delta x(s - 1/2)}. \tag{102}$$

By virtue of (21)

$$\sigma(s) = q^{-2s} p_4(q^s), \tag{103}$$

where $p_4(q^s)$ is a polynomial of fourth degree in q^s. By using (101) and (102), we can easily verify by means of (103) that the functions $\tilde{\tau}[x(s)]$ and $\tilde{\sigma}[x(s)]$ are polynomials of at most first and second degrees, respectively, in $x(s) = (q^s + q^{-s})/2$, for an arbitrary form of the polynomial $p_4(q^s)$.

Since

$$2\frac{x(s+1/2) - 1}{(q-1)^2} - \frac{1}{4} \to s(s+1) \text{ as } q \to 1,$$

then on the lattice $x(s) = \cosh 2\omega s$ we may obtain analogs of the polynomials on the square lattice $x(s) = s(s+1)$ if, starting from the expressions for $\sigma(s)$ and $\sigma(s) + \tau(s)\Delta x(s - 1/2)$ on the lattice $s(s+1)$, we replace s by $s - 1/2$, a by $a - 1/2$, and b by $b - 1/2$.

1) *Analogs of the Racah polynomials* $u_n^{(\alpha,\beta)}(x)$ *on the lattice with* $x(s) =$ $\cosh 2\omega s$. For the Racah polynomials under the transformation $s \to s - 1/2$ we have

$$\sigma(s) = (s - a)(s + b - 1)(s + a - \beta - 1)(b + \alpha - s).$$

For their analogs we choose the functions $\sigma(s)$ and $\sigma(s)+\tau(s)\Delta x(s-1/2)$ in the form:

$$\sigma(s)=\psi_q(s-a)\psi_q(s+b-1)\psi_q(s+a-\beta-1)\psi_q(b+\alpha-s),$$
$$\sigma(s)+\tau(s)\Delta x\left(s-\frac{1}{2}\right)=\sigma(-s)$$
$$=\psi_q(s+a)\psi_q(b-s-1)\psi_q(s-a+\beta+1)\psi_q(b+\alpha+s).$$

Solving equation (59) yields

$$\rho(s)=\frac{\tilde{\Gamma}_q(s+a)\tilde{\Gamma}_q(s-a+\beta+1)\tilde{\Gamma}_q(s+b+\alpha)\tilde{\Gamma}_q(b+\alpha-s)}{\tilde{\Gamma}_q(s-a+1)\tilde{\Gamma}_q(b-s)\tilde{\Gamma}_q(s+b)\tilde{\Gamma}_q(s+a-\beta)}. \tag{104}$$

2) *Analogs of the dual Hahn polynomials on the lattice* $x(s)=\cosh 2\omega s$. For the dual Hahn polynomials we obtain, after the transformation $s\to s-1/2$,

$$\sigma(s)=(s-a)(s+b-1)(s-c-1/2).$$

For their analogs we choose the functions $\sigma(s)$ and $\sigma(s)+\tau(s)\Delta x(s-1/2)$ in the form

$$\sigma(s)=\psi_q(s-a)\psi_q(s+b-1)\psi_q(s-c-1/2)\phi_q(s-a),$$
$$\sigma(s)+\tau(s)\Delta x(s-1/2)=\psi_q(s+a)\psi_q(b-s-1)\psi_q(s+c+1/2)\phi_q(s+a).$$

Solving equation (59) yields

$$\rho(s)=\frac{\tilde{\Gamma}_q(s+a)\tilde{\Pi}_q(s+a)\tilde{\Gamma}_q(s+c+1/2)}{\tilde{\Gamma}_q(s-a+1)\tilde{\Pi}_q(s-a+1)\tilde{\Gamma}_q(s+b)\tilde{\Gamma}_q(b-s)\tilde{\Gamma}_q(s-c+1/2)}. \tag{105}$$

Since

$$\frac{\Delta x(s-1/2)}{(q-1)^2}\to s \text{ as } q\to 1,$$

which coincides with $\frac{1}{2}\Delta x(s-1/2)$ on the lattice

$$x(s)=[t(t+1)-1/4]\big|_{t=s-1/2},$$

in the both cases we should take

$$B_n=\frac{(-1)^n}{2^n n!}(q-1)^{2n}.$$

b) *The lattice* $x(s) = \cos 2\omega s$. In order to obtain expressions for the weight function with which the analogs of the Racah polynomials and the dual Hahn polynomials are orthogonal on the lattice $x(s) = \cos 2\omega s$, it is natural to replace, in the formulas for the lattice $x(s) = \cosh 2\omega s$, the parameter ω by $i\omega$,

$$\psi_q(s) = \frac{\sinh \omega s}{\sinh \omega} \quad (q = e^{2\omega})$$

by

$$\psi_q(s) = \frac{\sin \omega s}{\sin \omega} \quad (q = e^{2i\omega}),$$

$$\phi_q(s) = \frac{\cosh \omega s}{\cosh \omega}$$

by

$$\phi_q(s) = \frac{\cos \omega s}{\cos \omega},$$

and then to use the definitions of the functions $\tilde{\Gamma}_q(s), \tilde{\Pi}_q(s)$ for $q = e^{2i\omega}$, i.e. for $|q| = 1$:

$$\tilde{\Gamma}_q(s) = \lim_{q' \to q} \tilde{\Gamma}_{q'}(s), \quad \tilde{\Pi}_q(s) = \lim_{q' \to q} \tilde{\Pi}_{q'}(s) \qquad (|q'| \neq 1).$$

1) *Analogs of the Racah polynomials* $u_n^{(\alpha,\beta)}(x)$ *on the lattice* $x(s) = \cos 2\omega s$.

$$\sigma(s) = \psi_q(s-a)\psi_q(s+b-1)\psi_q(s+a-\beta-1)\psi_q(b+\alpha-s),$$
$$\sigma(s) + \tau(s)\Delta x\left(s - \frac{1}{2}\right) = \sigma(-s)$$
$$= \psi_q(s+a)\psi_q(b-s-1)\psi_q(s-a+\beta+1)\psi_q(b+\alpha+s),$$

$$\rho(s) = \frac{\tilde{\Gamma}_q(s+a)\tilde{\Gamma}_q(s-a+\beta+1)\tilde{\Gamma}(s+b+\alpha)\tilde{\Gamma}(b+\alpha-s)}{\tilde{\Gamma}_q(s-a+1)\tilde{\Gamma}_q(b-s)\tilde{\Gamma}_q(s+b)\tilde{\Gamma}(s+a-\beta)}. \tag{106}$$

2) *Analogs of the dual Hahn polynomials on the lattice* $x(s) = \cos 2\omega s$.

$$\sigma(s) = \psi_q(s-a)\psi_q(s+b-1)\psi_q(s-c-1/2)\phi_q(s-a),$$
$$\sigma(s) + \tau(s)\Delta x\left(s - \frac{1}{2}\right)$$
$$= \psi_q(s+a)\psi_q(b-s-1)\psi_q\left(s+c+\frac{1}{2}\right)\phi_q(s+a),$$

$$\rho(s) = \frac{\tilde{\Gamma}_q(s+a)\tilde{\Gamma}_q(s+c+1/2)\tilde{\Pi}_q(s+a)}{\tilde{\Gamma}_q(s-a+1)\tilde{\Gamma}_q(b-s)\tilde{\Gamma}_q(s+b)\tilde{\Pi}_q(s-a+1)\tilde{\Gamma}_q(s-c+1/2)}. \quad (107)$$

In the both cases we should set $B_n = 2^n\omega^{2n}/n!$.

Remark. For the lattice $x(s) = \cos 2\omega s$, condition (50a) may fail to be satisfied with certain values of a, b and ω. If, for example, $\rho(s)\Delta x(s-1/2) < 0$ with $a \le s \le b-1$, then we must replace $\rho(s_i)\Delta x(s_i - 1/2)$ by the absolute value $|\rho(s_i)\Delta x(s_i - 1/2)|$ in the orthogonality condition (50), so that the squared norms d_n^2 of the polynomials will be positive.

3°. Tables of basic data for q-analogs. In conclusion we give tables (pp. 180–186) which contain the basic data about the polynomials which are analogs of the Hahn, Meixner, Kravchuk, Charlier, Racah and dual Hahn polynomials. We use the following notations: first we give the lattice number (I–VI) and then the conditional notation of the polynomial analog. The notations H_1, H_2, H_d correspond to the Hahn polynomials $h_n^{(\alpha,\beta)}(x)$, $\tilde{h}_n^{(\mu,\nu)}(x)$ and the dual Hahn polynomials $w_n^{(c)}(x)$. The notations M, K, C, R_1, R_2 correspond to the Meixner, Kravchuk, Charlier and Racah polynomials: $m_n^{(\gamma,\mu)}(x)$, $k_n^{(p)}(x)$, $c_n^{(\mu)}(x)$, $u_n^{(\alpha,\beta)}(x)$ and $\tilde{u}_n^{(c,d)}(x)$. For example, the systems of polynomials defined by the weight functions (85)–(89) will be denoted by III-H_1, III-H_2, III-M, III-K, III-C.

7. Calculation of the leading coefficients and squared norms. Tables of data.

We obtain the basic data for the classical orthogonal polynomials of a discrete variable on nonuniform lattices, supposing that the functions $\sigma(s)$, $\tau(s)$ and $\rho(s)$ are given for each form of the lattices.

1°. For *calculating the coefficients* a_n, b_n in the expansion

$$\tilde{y}_n(x) = a_n x^n + b_n x^{n-1} + \dots$$

we use the Rodrigues formula (45) with $k = n-1$. In accordance with (38) and (37) we have

$$\begin{aligned} v_{n-1,n}(s) &= \frac{A_{n-1,n}B_n}{\rho_{n-1}(s)} \frac{\nabla}{\nabla x_n(s)}[\rho_n(s)] \\ &= \frac{A_{n-1,n}B_n}{\rho_{n-1}(s)} \frac{\nabla}{\nabla x_n(s)}[\sigma(s+1)\rho_{n-1}(s+1)] \\ &= \frac{A_{n-1,n}B_n}{\rho_{n-1}(s)} \frac{\Delta}{\Delta x_{n-1}(s-1/2)}[\sigma(s)\rho_{n-1}(s)] = A_{n-1,n}B_n\tilde{\tau}_{n-1}[x_{n-1}(s)]. \end{aligned}$$

By formula (28) the first-degree polynomial $\tilde{\tau}_{n-1}[x_{n-1}(s)]$ can be expressed in terms of $\tau(s)$ and $\sigma(s)$. On the other hand, by virtue of (44) we have

$$v_{n-1,n}(s) = \Delta^{(n-1)}\tilde{y}_n[x(s)] = a_n\Delta^{(n-1)}[x^n(s)] + b_n\Delta^{(n-1)}[x^{n-1}(s)].$$

The operator $\Delta^{(k)}$ carries every polynomial of degree n in $x(s)$ to a polynomial of degree $n-k$ in $x_k(s)$. Consequently

$$\Delta^{(n-1)}[x^n(s)] = \alpha_n[x_{n-1}(s) + \beta_n] \quad (n = 2, 3, \ldots). \tag{108}$$

Hence, equating coefficients for different powers of $x_{n-1}(s)$ in the following equality

$$a_n\Delta^{(n-1)}[x^n(s)] + b_n\Delta^{(n-1)}[x^{n-1}(s)] = A_{n-1,n}B_n\tilde{\tau}_{n-1}[x_{n-1}(s)],$$

which yields

$$a_n\alpha_n[x_{n-1}(s) + \beta_n] + b_n\alpha_{n-1} = A_{n-1,n}B_n\tilde{\tau}_{n-1}(x_{n-1}),$$

we can find a_n and b_n in the form

$$a_n = \frac{A_{n-1,n}B_n}{\alpha_n}\tilde{\tau}'_{n-1}, \qquad \frac{b_n}{a_n} = \frac{\alpha_n}{\alpha_{n-1}}\left(\frac{\tilde{\tau}_{n-1}(0)}{\tilde{\tau}'_{n-1}} - \beta_n\right). \tag{109}$$

Let us determine the coefficients α_n and β_n. We have

$$\begin{aligned}
\alpha_{n+1}[x_n(s) + \beta_{n+1}] &= \Delta^{(n)}[x^{n+1}(s)] \\
&= \frac{\Delta}{\Delta x_{n-1}(s)}\frac{\Delta}{\Delta x_{n-2}(s)}\cdots\frac{\Delta}{\Delta x_1(s)}\left\{\frac{\Delta}{\Delta x(s)}[x^{n+1}(s)]\right\} \\
&= \frac{\Delta}{\Delta x_{n-2}(t)}\cdots\frac{\Delta}{\Delta x_1(t)}\frac{\Delta}{\Delta x(t)}[f(t)] = \Delta^{(n-1)}f(t),
\end{aligned}$$

where

$$t = s + \frac{1}{2}, f(t) = \frac{\Delta}{\Delta x(s)}[x^{n+1}(s)] = C_{n+1}x^n(t) + D_{n+1}x^{n-1}(t) + \cdots \tag{110}$$

(C_n and D_n evidently depend on the form of the lattice). From this it follows that

$$\begin{aligned}
\alpha_{n+1}[x_{n-1}(t) + \beta_{n+1}] &= C_{n+1}\alpha_n[x_{n-1}(t) + \beta_n] \\
&+ D_{n+1}\frac{\Delta}{\Delta x_{n-2}(t)}\left\{\alpha_{n-1}[x_{n-2}(t) + \beta_{n-1}]\right\},
\end{aligned}$$

Table 6. Lattice I, $x(s) = s$. Basic Data for the Hahn, Meixner, Kravchuk and Charlier polynomials

$No.$	$\tilde{y}_n(x)$	(a,b)	$\rho(s)$	$\sigma(s)$	$\sigma(s)+\tau(s)\Delta x(s-1/2)$	B_n
$\mathrm{I}-H_1$	$h_n^{(\alpha,\beta)}(s)$	$(0,N)$	$\dfrac{\Gamma(n+\alpha-s)\Gamma(\beta+1+s)}{\Gamma(s+1)\Gamma(N-s)}$ $(\alpha>-1,\beta>-1)$	$s(N+\alpha-s)$	$(s+\beta+1)(N-1-s)$	$\dfrac{(-1)^n}{n!}$
$\mathrm{I}-H_2$	$\tilde{h}_n^{(\mu,\nu)}(s)$	$(0,N)$	$\dfrac{1}{\Gamma(s+1)\Gamma(s+\mu+1)\Gamma(N+\nu-s)\Gamma(N-s)}$ $(\mu>-1,\nu>-1)$	$s(s+\mu)$	$(N+\nu-1-s)(N-1-s)$	$\dfrac{1}{n!}$
$\mathrm{I}-M$	$m_n^{(\gamma,\mu)}(s)$	$(0,+\infty)$	$\dfrac{\mu^s\Gamma(s+\gamma)}{\Gamma(s+1)\Gamma(\gamma)}$ $(\gamma>0,0<\mu<1)$	s	$\mu(\gamma+s)$	$\dfrac{1}{\mu^n}$
$\mathrm{I}-K$	$k_n^{(p)}(s)$	$(0,N+1)$	$\dfrac{N!p^s(1-p)^{N-s}}{\Gamma(s+1)\Gamma(N+1-s)}$ $(0<p<1)$	s	$\dfrac{p}{1-p}(N-s)$	$\dfrac{(-1)^n(1-p)^n}{n!}$
$\mathrm{I}-C$	$c_n^{(\mu)}(s)$	$(0,+\infty)$	$\dfrac{e^{-\mu}\mu^s}{\Gamma(s+1)}$ $(\mu>0)$	s	μ	$\dfrac{1}{\mu^n}$

Table 7. Lattice II, $x(s)=s(s+1)$. Basic Data for the Racah and dual Hahn polynomials

$No.$	$\tilde{y}_n(x)$	(a,b)	$\rho(s)$	$\sigma(s)$	$\sigma(s)+\tau(s)\Delta x(s-1/2)$	B_n
$\mathrm{II}-R_1$	$u_n^{(\alpha,\beta)}(x)$	(a,b)	$\frac{\Gamma(s+a+1)\Gamma(s-a+\beta+1)\Gamma(b+\alpha-s)\Gamma(b+\alpha+s+1)}{\Gamma(s+a-\beta+1)\Gamma(s-a+1)\Gamma(b-s)\Gamma(b+s+1)}$ $(-1/2<a<b,\ \alpha>-1,\ -1<\beta<2a+1,\ b=a+N)$	$(s-a)(s+b)$ $\times(s+a-\beta)(b+\alpha-s)$	$(s+a+1)(s+1-a+\beta)$ $\times(b-s-1)(b+\alpha+s+1)$	$\frac{(-1)^n}{n!}$
$\mathrm{II}-R_2$	$\tilde{u}_n^{(c,d)}(x)$	(a,b)	$\frac{\Gamma(s+a+1)\Gamma(s+c+1)}{\Gamma(s-a+1)\Gamma(s-c+1)\Gamma(s+d+1)\Gamma(s+b+1)\Gamma(b-s)\Gamma(d-s)}$ $(-1/2<a<b<1+d,\ \lvert c\rvert<1+a,\ b=a+N)$	$(s-a)(s+b)$ $\times(s-c)(s+d)$	$(s+a+1)(s+c+1)$ $\times(b-s-1)(d-s-1)$	$\frac{1}{n!}$
$\mathrm{II}-H_d$	$w_n^{(c)}(x)$	(a,b)	$\frac{\Gamma(s+a+1)\Gamma(s+c+1)}{\Gamma(s-a+1)\Gamma(s-c+1)\Gamma(b-s)\Gamma(b+s+1)}$ $(-1/2<a<b,\ \lvert c\rvert<a+1,\ b=a+N)$	$(s-a)(s+b)(s-c)$	$(s+a+1)(s+c+1)(b-s-1)$	$\frac{(-1)^n}{n!}$

Table 8. Lattice III, $x(s) = q^s$, $q = e^{2\omega}$. Basic Data for the q-analogs of the Hahn, Meixner, Kravchuk and Charlier polynomials

No.	$\tilde{y}_n(x)$	(a,b)	$\rho(s)$
III − H_1	$h_n^{(\alpha,\beta)}(x,q)$	$(0,N)$	$q^{(\alpha+\beta)s/2}\dfrac{\tilde{\Gamma}_q(s+\beta+1)\tilde{\Gamma}_q(N+\alpha-s)}{\tilde{\Gamma}_q(s+1)\tilde{\Gamma}_q(N-s)}$ $(\alpha > -1,\ \beta > -1)$
III − H_2	$\tilde{h}_n^{(\mu,\nu)}(x,q)$	$(0,N)$	$\dfrac{q^{-(N+(\mu+\nu)/2)s}}{\tilde{\Gamma}_q(N+\nu-s)\tilde{\Gamma}_q(N-s)\tilde{\Gamma}_q(s+1)\tilde{\Gamma}_q(s+\mu+1)}$ $(\mu > -1,\ \nu > -1)$
III − M	$m_n^{(\gamma,\mu)}(x,q)$	$(0,+\infty)$	$q^{(\gamma-1)s/2}\dfrac{\mu^s\tilde{\Gamma}_q(s+\gamma)}{\tilde{\Gamma}_q(s+1)\tilde{\Gamma}_q(\gamma)}$ $(0 < q < 1,\ 0 < \mu < 1)$
III − K	$k_n^{(p)}(x,q)$	$(0,N+1)$	$\dfrac{\tilde{\Gamma}_q(N+1)p^s(1-p)^{N-s}q^{-(N+1)s/2}}{\tilde{\Gamma}_q(s+1)\tilde{\Gamma}_q(N-s+1)}$ $(0 < p < 1)$
III − C	$c_n^{(\mu)}(x,q)$	$(0,+\infty)$	$q^{-s(s+1)/4}\dfrac{e^{-\mu}\mu^s}{\tilde{\Gamma}_q(s+1)}$ $\begin{pmatrix} 0 < q < 1, & \mu < 1/[(1-q)\sqrt{q}]; \\ q > 1, & \mu > 0 \end{pmatrix}$

Table 8. (cont).

$\sigma(s)$	$\sigma(s)+\tau(s)\Delta x(s-1/2)$	B_n
$q^{s/2}\psi_q(s)q^{(s-N-\alpha)/2}$ $\times\psi_q(N+\alpha-s)$	$q^{(s+\beta+1)/2}\psi_q(s+\beta+1)$ $\times q^{(s-N+1)/2}\psi_q(N-1-s)$	$\dfrac{(1-q)^n}{n!}$
$q^{s/2}\psi_q(s)$ $\times q^{(s+\mu)/2}\psi_q(s+\mu)$	$q^{(s-N-\nu+1)/2}$ $\times\psi_q(N+\nu-1-s)$ $\times q^{(s-N+1)/2}\psi_q(N-1-s)$	$\dfrac{(q-1)^n}{n!}$
$q^{s/2}\psi_q(s)q^{s-1}$	$\mu q^{(s+\gamma)/2}\psi_q(s+\gamma)q^s$	$\left(\dfrac{q-1}{\mu}\right)^n$
$q^{s/2}\psi_q(s)q^{s-1}$	$\dfrac{p}{1-p}q^{(s-N)/2}\psi_q(N-s)q^s$	$\dfrac{(1-p)^n(1-q)^n}{n!}$
$q^{s/2}\psi_q(s)q^{s-1}$	μq^s	$\left(\dfrac{q-1}{\mu}\right)^n$

Table 9. Lattice IV, $x(s) = \sinh(2\omega s) = (q^s - q^{-s})/2, q = e^{2\omega}$. Basic Data for the q-analogs of the Hahn, Meixner, Kravchuk and Charlier polynomials

$No.$	(a, b)	$\rho(s)$
$IV - H_1$	(a, b)	$\dfrac{\tilde{\Pi}_q(s+a)\tilde{\Gamma}_q(s-a+\beta+1)\tilde{\Pi}_q(b+\alpha+s)\tilde{\Gamma}_q(b+\alpha-s)}{\tilde{\Gamma}_q(s-a+1)\tilde{\Gamma}_q(b-s)\tilde{\Pi}_q(s+a-\beta)\tilde{\Pi}_q(s+b)}$ $(\alpha > -1,\ \beta > -1,\ b = a + N)$
$IV - H_2$	(a, b)	$\dfrac{\tilde{\Pi}_q(s+a)\tilde{\Pi}_q(s+a-\mu)}{\tilde{\Gamma}_q(s+1-a)\tilde{\Gamma}_q(b-s)\tilde{\Gamma}_q(b+\nu-s)\tilde{\Gamma}_q(s-a+\mu+1)\tilde{\Pi}_q(s+b)\tilde{\Pi}_q(s+b+\nu)}$ $(\mu > -1,\ \nu > -1,\ b = a + N)$
$IV - M$	$(a, +\infty)$	$\left(\dfrac{q-1}{q+1}\right)^s q^{-s^2/2} \dfrac{\tilde{\Pi}_q(s+a)\tilde{\Gamma}_q(s+\gamma-a)\tilde{\Gamma}_q(s+\delta-a)}{\tilde{\Gamma}_q(s+1-a)\tilde{\Pi}_q(s+a+1-\gamma)\tilde{\Pi}_q(s+a+1-\delta)}$ $(0 < \mu < 1,\ \gamma > 0,\ q^{-\delta} = (1-\mu)/(1+\mu))$
$IV - K$	(a, b)	$\dfrac{\tilde{\Pi}_q(s+a)F_1(s,q)}{\tilde{\Gamma}_q(s-a+1)\tilde{\Gamma}_q(b-s+1)\tilde{\Pi}_q(s+b+1)}$ $(b = a + N + 1)$, $F_1(s,q) = \begin{cases} \left(\dfrac{q-1}{q+1}\right)^s \dfrac{q^{-s^2/2}}{\tilde{\Gamma}_q(b-s+\delta+1)\tilde{\Pi}_q(s+b+\delta+1)} \\ \left(0 < p < 1/2, q^\delta = \dfrac{1}{1-2p}\right); \\ q^{-s^2}(p = 1/2); \\ \left(\dfrac{q+1}{q-1}\right)^s \dfrac{q^{-s^2/2}\tilde{\Pi}_q(s+a-\delta)}{\tilde{\Gamma}_q(s-a+\delta+1)} \\ \left(1/2 < p < 1, q^\delta = \dfrac{1}{2p-1}\right) \end{cases}$
$IV - C$	$(a, +\infty)$	$\left(\dfrac{q-1}{q+1}\right)^s q^{-s^2/2} \dfrac{\tilde{\Pi}_q(s+a)\lvert\tilde{\Gamma}_q(s+\alpha+i\beta)\rvert^2}{\tilde{\Gamma}_q(s-a+1)\lvert\tilde{\Pi}_q(s+1-\alpha+i\beta)\rvert^2}$ $\left(q^{\alpha+i\beta} = 1 + i\sqrt{2\mu(q-1)}\right)$

Table 9. (cont).

$\sigma(s)$	$\sigma(s)+\tau(s)\Delta x(s-1/2)$	B_n
$\psi_q(s-a)\phi_q(s+a-\beta-1)$ $\times\psi_q(b+\alpha-s)\phi_q(b-1+s)$	$\phi_q(s+a)\psi_q(s-a+\beta+1)$ $\times\phi_q(b+\alpha+s)\psi_q(b-1-s)$	$\dfrac{(1-q)^n}{n!}$
$\psi_q(s-a)\phi_q(s+b+\nu-1)$ $\times\psi_q(s-a+\mu)\phi_q(b-1+s)$	$\phi_q(s+a)\psi_q(b+\nu-1-s)$ $\times\phi_q(s+a-\mu)\psi_q(b-1-s)$	$\dfrac{(q-1)^n}{n!}$
$\psi_q(s-a)\phi_q(s-\gamma+a)$ $\dfrac{(1+\mu)+(1-\mu)q^{s+a}}{q^{1/2}+q^{-1/2}}$	$\phi_q(s+a)\psi_q(s+\gamma-a)$ $\dfrac{(1+\mu)-(1-\mu)q^{-(s-a)}}{q^{1/2}+q^{-1/2}}$	$\dfrac{(q-1)^n}{\mu^n}$
$\psi_q(s-a)\phi_q(s+b)F_2(s,q)$ $F_2(s,q)=\begin{cases}\dfrac{(1-\mu)+(1+\mu)q^{s+b}}{q^{1/2}+q^{-1/2}} \\ \left(0<p<\dfrac{1}{2},\mu=\dfrac{p}{1-p}\right); \\ q^s\,(p=1/2); \\ \dfrac{(\mu+1)q^{(s-a)}-(\mu-1)}{q^{1/2}+q^{-1/2}} \\ \left(\dfrac{1}{2}<p<1,\mu=\dfrac{p}{1-p}\right)\end{cases}$	$\phi_q(s+a)\psi_q(b-s)F_3(s,q)$ $F_3(s,q)=\begin{cases}\dfrac{(1+\mu)q^{b-s}-(1-\mu)}{q^{1/2}+q^{-1/2}} \\ \left(0<p<\dfrac{1}{2},\mu=\dfrac{p}{1-p}\right); \\ q^{-s}\,(p=1/2); \\ \dfrac{(\mu+1)q^{-(s+a)}+(\mu-1)}{q^{1/2}+q^{-1/2}} \\ \left(\dfrac{1}{2}<p<1,\mu=\dfrac{p}{1-p}\right)\end{cases}$	$\dfrac{(1-p)^n(1-q)^n}{n!}$
$q^{\frac{s-a}{2}}\psi_q(s-a)$ $\times\dfrac{q^s+q^{-s}+2+2\mu(q-1)q^{-s}}{(q^{1/2}+q^{-1/2})^2}$	$q^{-\left(\frac{s+a}{2}\right)}\phi_q(s+a)\left(\dfrac{q+1}{q-1}\right)$ $\times\dfrac{q^s+q^{-s}-2+2\mu(q-1)q^s}{(q^{1/2}+q^{-1/2})^2}$	$\left(\dfrac{q-1}{\mu}\right)^n$

Table 10. Lattice V, $x(s) = \cosh(2\omega s)$, $q = e^{2\omega}$ and lattice VI, $x(s) = \cos 2\omega s$, $q = e^{2i\omega}$. Basic Data for the q-analogs of the Racah and dual Hahn polynomials

$No.$	(a,b)	$\rho(s)$	$\sigma(s)$	$\sigma(s)+\tau(s)\Delta x(s-1/2)$	B_n
$V - R_1$	(a,b)	$\dfrac{\tilde{\Gamma}_q(s+a)\tilde{\Gamma}_q(s-a+\beta+1)\tilde{\Gamma}_q(s+b+\alpha)\tilde{\Gamma}_q(b+\alpha-s)}{\tilde{\Gamma}_q(s-a+1)\tilde{\Gamma}_q(b-s)\tilde{\Gamma}_q(s+b)\tilde{\Gamma}_q(s+a-\beta)}$ $(\alpha > -1,\ \beta > -1,\ a > 0,\ b = a+N)$	$\psi_q(s-a)\psi_q(s+b-1)$ $\times\psi_q(s+a-\beta-1)\psi_q(b+\alpha-s)$	$\psi_q(s+a)\psi_q(b-s-1)$ $\times\psi_q(s-a+\beta+1)\psi_q(b+\alpha+s)$	$\dfrac{(-1)^n}{2^n n!}$ $\times(q-1)^{2n}$
$V - H_d$	(a,b)	$\dfrac{\tilde{\Gamma}_q(s+a)\tilde{\Pi}_q(s+a)\tilde{\Gamma}_q(s+c+1/2)}{\tilde{\Gamma}_q(s-a+1)\tilde{\Pi}_q(s-a+1)\tilde{\Gamma}_q(s+b)\tilde{\Gamma}_q(s-c+1/2)}$ $(a > 0,\ \lvert c\rvert < a+1/2,\ b = a+N)$	$\psi_q(s-a)\psi_q(s+b_1)$ $\times\psi_q(s-c-1/2)\phi_q(s-a)$	$\psi_q(s+a)\psi_q(b-s-1)$ $\times\psi_q(s+c+1/2)\phi_q(s+a)$	$\dfrac{(-1)^n}{2^n n!}$ $\times(q-1)^{2n}$
$VI - R_1$	(a,b)	$\dfrac{\tilde{\Gamma}_q(s+a)\tilde{\Gamma}_q(s-a+\beta+1)\tilde{\Gamma}_q(s+b+\alpha)\tilde{\Gamma}_q(b+\alpha-s)}{\tilde{\Gamma}_q(s-a+1)\tilde{\Gamma}_q(b-s)\tilde{\Gamma}_q(s+b)\tilde{\Gamma}_q(s+a-\beta)}$ $(\alpha > -1,\ -1 < \beta < 2a,\ b = a+N,\ a > 0)$	$\psi_q(s-a)\psi_q(s+b-1)$ $\times\psi_q(s+a-\beta-1)\psi_q(b+\alpha-s)$	$\psi_q(s+a)\psi_q(b-s-1)$ $\times\psi_q(s-a+\beta+1)\psi_q(b+\alpha+s)$	$\dfrac{2^n\omega^{2n}}{n!}$
$VI - H_d$	(a,b)	$\dfrac{\tilde{\Gamma}_q(s+a)\tilde{\Pi}_q(s+a)\tilde{\Gamma}_q(s+c+1/2)}{\tilde{\Gamma}_q(s-a+1)\tilde{\Pi}_q(s-a+1)\tilde{\Gamma}_q(s+b)\tilde{\Gamma}_q(b-s)\tilde{\Gamma}_q(s-c+1/2)}$ $(a > 0,\ \lvert c\rvert < a+1/2,\ b = a+N)$	$\psi_q(s-a)\psi_q(s+b-1)$ $\times\psi_q(s-c-1/2)\phi_q(s-a)$	$\psi_q(s+a)\psi_q(b-s-1)$ $\times\psi_q(s+c+1/2)\phi_q(s+a)$	$\dfrac{2^n\omega^{2n}}{n!}$

which for the given C_n and D_n yields a system of equations for α_n and β_n:

$$\begin{aligned} \alpha_{n+1} &= C_{n+1}\alpha_n, \\ \alpha_{n+1}\beta_{n+1} &= C_{n+1}\alpha_n\beta_n + D_{n+1}\alpha_{n-1}. \end{aligned} \tag{111}$$

System (111) may be represented in the form of separate equations for each value of interest:

$$\alpha_{n+1} = C_{n+1}\alpha_n, \tag{111a}$$

$$\beta_{n+1} = \beta_n + \frac{D_{n+1}}{C_n C_{n+1}}. \tag{111b}$$

It is not difficult to solve equations (111a) and (111b).

In order to find the coefficients C_n and D_n it is convenient to use the relations (29) and (30).

a) In the case of the quadratic lattice $x(s) = s(s+1)$, i.e. lattice II, relations (29) and (30) have the form

$$\frac{\Delta x^n(s)}{\Delta x(s)} = \left[x\left(s+\frac{1}{2}\right)+\frac{1}{4}\right]\frac{\Delta x^{n-1}(s)}{\Delta x(s)} + \frac{x^{n-1}(s+1)+x^{n-1}(s)}{2}, \tag{29a}$$

$$\frac{x^n(s+1)+x^n(s)}{2} = \left[x\left(s+\frac{1}{2}\right)+\frac{1}{4}\right]\left\{\frac{x^{n-1}(s+1)+x^{n-1}(s)}{2} + \frac{\Delta x^{n-1}(s)}{\Delta x(s)}\right\}. \tag{30a}$$

By supposing in accordance with (110) that

$$\frac{\Delta x^n(s)}{\Delta x(s)} = C_n x^{n-1}\left(s+\frac{1}{2}\right) + D_n x^{n-2}\left(s+\frac{1}{2}\right) + \cdots,$$

$$\frac{x^n(s+1)+x^n(s)}{2} = A_n x^n\left(s+\frac{1}{2}\right) + B_n x^{n-1}\left(s+\frac{1}{2}\right) + \cdots,$$

and equating the coefficients of the same powers of x in the relations (29a) and (30a), we obtain

$$\begin{aligned} C_n &= C_{n-1} + A_{n-1}, \\ D_n &= D_{n-1} + \frac{1}{4}C_{n-1} + B_{n-1}, \\ A_n &= A_{n-1}, \\ B_n &= \frac{1}{4}A_{n-1} + B_{n-1} + C_{n-1} \end{aligned}$$

$$(A_1 = 1, B_1 = 1/4, C_1 = 1, D_1 = 0).$$

From this it is easy to find that

$$A_n = 1,\ C_n = n,\ B_n = \frac{n(2n-1)}{4},\ D_n = \frac{n(n-1)(2n-1)}{12}.$$

By means of (111a) and (111b) we obtain

$$\frac{\alpha_{n+1}}{\alpha_n} = n+1, \qquad \beta_{n+1} - \beta_n = \frac{2n+1}{12} \quad (\alpha_1 = 1, \beta_1 = 0),$$

whence

$$\alpha_n = n!, \qquad \beta_n = \frac{n^2-1}{12}.$$

b) In the remaining cases, i.e. for lattices III–VI, we have $x(s) = Aq^s + Bq^{-s}$; A and B are constants;

$$x(s+1) + x(s) = 2\alpha x(s+1/2) \qquad (\alpha \neq 0);$$
$$x(s)x(s+1) = (Aq^s + Bq^{-s})(Aq^{s+1} + Bq^{-(s+1)})$$
$$= (Aq^{s+1/2} + Bq^{-(s+1/2)})^2 + \text{ const } = x^2\left(s+\frac{1}{2}\right) + \text{ const },$$
$$[\Delta x(s)]^2 = [x(s+1) + x(s)]^2 - 4x(s+1)x(s)$$
$$= \left[2\alpha x\left(s+\frac{1}{2}\right)\right]^2 - 4x^2\left(s+\frac{1}{2}\right) + \text{ const }.$$

As a result the relations (29) and (30) take the form

$$\frac{\Delta x^n(s)}{\Delta x(s)} = \alpha x\left(s+\frac{1}{2}\right)\frac{\Delta x^{n-1}(s)}{\Delta x(s)} + \frac{x^{n-1}(s+1) + x^{n-1}(s)}{2}, \tag{29b}$$
$$\frac{x^n(s+1) + x^n(s)}{2} = \alpha x\left(s+\frac{1}{2}\right)\frac{x^{n-1}(s+1) + x^{n-1}(s)}{2}$$
$$+ \left[(\alpha^2-1)x^2\left(s+\frac{1}{2}\right) + \text{ const }\right]\frac{\Delta x^{n-1}(s)}{\Delta x(s)}. \tag{30b}$$

From these relations we can obtain, by induction, that

$$\frac{\Delta x^n(s)}{\Delta x(s)} = C_n x^{n-1}\left(s+\frac{1}{2}\right) + E_n x^{n-3}\left(s+\frac{1}{2}\right) + \cdots \tag{112}$$
$$\frac{x^n(s+1) + x^n(s)}{2} = A_n x^n\left(s+\frac{1}{2}\right) + F_n x^{n-2}\left(s+\frac{1}{2}\right) + \cdots \tag{113}$$

where A_n, C_n, E_n and F_n are constants. Substituting the expansions (112) and (113) into (29b) and (30b) yields a linear homogeneous system of first-order difference equations with constant coefficients for A_n and C_n:

$$\begin{aligned} C_n &= \alpha C_{n-1} + A_{n-1}, \\ A_n &= \alpha A_{n-1} + (\alpha^2 - 1)C_{n-1}, \end{aligned}$$

or

$$\begin{pmatrix} C_n \\ A_n \end{pmatrix} = \begin{pmatrix} \alpha & 1 \\ \alpha^2 - 1 & \alpha \end{pmatrix} \begin{pmatrix} C_{n-1} \\ A_{n-1} \end{pmatrix}. \tag{114}$$

As usual in solving a system of the form $X_n = AX_{n-1}$, where X_n is a column vector and A is a matrix, we seek for particular solutions in the form $X_n = \lambda^n X_0$. After substituting $X^n = \lambda^n X_0$ into the equation we obtain $AX_0 = \lambda X_0$. From this it is seen that λ is an eigenvalue of the secular equation $\det(A - \lambda E) = 0$, where E is a unit matrix and X_0 is an eigenvector.

In our case

$$\det(A - \lambda E) = \begin{vmatrix} \alpha - \lambda & 1 \\ \alpha^2 - 1 & \alpha - \lambda \end{vmatrix} = \lambda^2 - 2\alpha\lambda + 1 = 0, \tag{115}$$

whence $\lambda_{1,2} = \alpha \pm \sqrt{\alpha^2 - 1}$, while $\lambda_1\lambda_2 = 1$, and $\lambda_1 + \lambda_2 = 2\alpha$. We note that equation (115) coincides with (55), i.e. $\lambda_1 = q_1$, $\lambda_2 = q_2$.

The general solution of system (114) is a linear combination of particular solutions. Since $C_1 = 1$ and $A_1 = \alpha$, the solution of the difference equations for C_n and A_n has the form

$$\begin{aligned} C_n &= \frac{q_1^n - q_2^n}{q_1 - q_2} = \psi_q(n) = \frac{\sinh n\omega}{\sinh \omega}, \\ A_n &= \frac{1}{2}(q_1^n + q_2^n) = \alpha\phi_q(n) = \cosh n\omega, \end{aligned}$$

where

$$q_1 = e^{\omega} = q^{1/2}, \qquad q_2 = e^{-\omega} = q^{-1/2}, \qquad \alpha = \cosh \omega.$$

In this case we have $D_n = 0$ in (110), whence by means of (111a) and (111b) we obtain

$$\frac{\alpha_{n+1}}{\alpha_n} = \psi_q(n+1), \qquad \beta_{n+1} = \beta_n$$
$$(\alpha_1 = 1,\ \beta_1 = 0).$$

From this, $\alpha_n = C(q)\tilde{\Gamma}_q(n+1)$, $\beta_n = 0$. Since $\tilde{\Gamma}_q(2) = 1$, $\alpha_1 = 1$, we have $C(q) = 1$ i.e., $\alpha_n = \tilde{\Gamma}_q(n+1)$.

2°. To determine the *squared norm* $d_n^2 \equiv d_{0n}^2$ in (53) we first need to know the connection between d_{kn}^2 and $d_{k+1,n}^2$. By multiplying the both sides of equation (35), where we put $v_k(s) = v_{kn}(s)$, $s = s_i$, $\lambda = \lambda_n$, by the product $v_{kn}(s_i)\Delta x(s_i - 1/2)$ and using the formula for summation by parts, we obtain

$$\sum_{s_i=a}^{b-k-1} f(s_i)\Delta g(s_i) = f(s_i)g(s_i)\Big|_a^{b-k} - \sum_{s_i=a}^{b-k-1} g(s_{i+1})\Delta f(s_i),$$

and hence

$$\begin{aligned}\mu_{kn}d_{kn}^2 &= -\sum_i v_{kn}(s_i)\Delta\left[\sigma(s_i)\rho_k(s_i)\frac{\nabla v_{kn}(s_i)}{\nabla x_k(s_i)}\right]\\ &= -v_{kn}(s_i)\sigma(s_i)\rho_k(s_i)\frac{\nabla v_{kn}(s_i)}{\nabla x_k(s_i)}\Big|_a^{b-k}\\ &\quad + \sum_i \sigma(s_i+1)\rho_k(s_i+1)\left[\frac{\Delta v_{kn}(s_i)}{\Delta x_k(s_i)}\right]^2 \Delta x_k(s_i)\\ &= \sum_i v_{k+1,n}^2(s_i)\rho_{k+1}(s_i)\Delta x_{k+1}\left(s_i - \frac{1}{2}\right) = d_{k+1,n}^2.\end{aligned}$$

The terms evaluated at the limits are zero by virtue of the boundary conditions (53a) imposed on the function $\rho_k(s)$. From this we successively obtain

$$\begin{aligned} d_n^2 \equiv d_{0n}^2 &= \frac{1}{\mu_{0n}}d_{1n}^2 = \frac{1}{\mu_{0n}}\frac{1}{\mu_{1n}}d_{2n}^2 = \cdots = \frac{d_{nn}^2}{\prod_{k=0}^{n-1}\mu_{kn}}\\ &= \frac{v_{nn}^2(x)}{\prod_{k=0}^{n-1}\mu_{kn}}S_n = (-1)^n A_{nn}B_n^2 S_n, \end{aligned} \tag{116}$$

where

$$S_n = \sum_{s_i=a}^{b-n-1} \rho_n(s_i)\Delta x_n\left(s_i - \frac{1}{2}\right).$$

If a and b are finite, the squared norm is calculated very simply. In this case we have, in fact, $b - a = N$, where N is a positive integer. For $n = N - 1$ the sum S_n contains only one summand:

$$S_{N-1} = \rho_{N-1}(a)\Delta x_{N-1}(a - 1/2). \tag{117}$$

To determine S_n, when $n < N-1$, it is sufficient to know how to calculate the ratio S_{n-1}/S_n. For this purpose we transform the expression for S_n, using the connection between $\rho_n(s)$ and $\rho_{n-1}(s)$:

$$\rho_n(s) = \sigma(s+1)\rho_{n-1}(s+1) = \rho_{n-1}(s)[\sigma(s) + \tau_{n-1}(s)\Delta x_{n-1}(s-1/2)].$$

From this, on the one hand

$$\begin{aligned} S_n &= \sum_i \sigma(s_i+1)\rho_{n-1}(s_i+1)\Delta x_n\left(s_i - \frac{1}{2}\right) \\ &= \sum_i \sigma(s_i)\rho_{n-1}(s_i)\Delta x_n\left(s_i - \frac{3}{2}\right). \end{aligned}$$

On the other hand,

$$S_n = \sum_i \rho_{n-1}(s_i)\left[\sigma(s_i) + \tau_{n-1}(s_i)\Delta x_{n-1}\left(s_i - \frac{1}{2}\right)\right]\Delta x_n\left(s_i - \frac{1}{2}\right).$$

We take half the sum of these expressions and by appealing to (26), (24) and (25) we obtain

$$\begin{aligned} S_n = \frac{1}{2}\sum_i &\rho_{n-1}(s_i)\Delta x_{n-1}\left(s_i - \frac{1}{2}\right) \\ &\times\left\{\tilde{\sigma}_{n-1}[x_{n-1}(s_i)]\frac{\Delta x_n(s_i - 1/2) + \Delta x_n(s_i - 3/2)}{\Delta x_{n-1}(s_i - 1/2)}\right. \\ &\left. + \tilde{\tau}_{n-1}[x_{n-1}(s_i)]\,\frac{\Delta x_n(s_i - 1/2) - \Delta x_n(s_i - 3/2)}{2}\right\}. \end{aligned}$$

Using the relations (12) and (13), we obtain an expression for S_n in the form

$$S_n = \sum_i \rho_{n-1}(s_i)\Delta x_{n-1}\left(s_i - \frac{1}{2}\right) Q_n[x_{n-1}(s_i)],$$

where

$$\begin{aligned} Q_n[x_{n-1}(s)] &= \alpha\tilde{\sigma}_{n-1}[x_{n-1}(s)] \\ &\quad + \tilde{\tau}_{n-1}[x_{n-1}(s)][(\alpha^2 - 1)x_{n-1}(s) + (\alpha+1)\beta] \end{aligned} \tag{118}$$

is a polynomial of at most second degree in $x_{n-1}(s)$. We decompose the polynomial $Q_n(x_{n-1})$ into powers of the first-degree polynomial $\tilde{\tau}_{n-1}(x_{n-1})$:

$$Q_n(x_{n-1}) = A_n\tilde{\tau}^2_{n-1}(x_{n-1}) + B_n\tilde{\tau}_{n-1}(x_{n-1}) + C_n. \tag{119}$$

Then $S_n = S_n^{(1)} + C_n S_{n-1}$, where

$$\begin{aligned} S_n^{(1)} &= \sum_i \{A_n \tilde{\tau}_{n-1}[x_{n-1}(s_i)] + B_n\} \tilde{\tau}_{n-1}[x_{n-1}(s_i)] \rho_{n-1}(s_i) \Delta x_{n-1}\left(s_i - \tfrac{1}{2}\right) \\ &= \sum_i \{A_n \tilde{\tau}_{n-1}[x_{n-1}(s_i)] + B_n\} \Delta[\sigma(s_i)\rho_{n-1}(s_i)]. \end{aligned}$$

By using summation by parts, since the terms evaluated at the limits are zero and

$$\frac{\Delta \tilde{\tau}_{n-1}[x_{n-1}(s)]}{\Delta x_{n-1}(s)} = \tilde{\tau}'_{n-1} = \text{ const },$$

we obtain

$$S_n^{(1)} = -A_n \tilde{\tau}'_{n-1} \sum_i \rho_n(s_i) \Delta x_n \left(s_i - \frac{1}{2}\right) = -A_n \tilde{\tau}'_{n-1} S_n.$$

Thus $S_n = -A_n \tilde{\tau}'_{n-1} S_n + C_n S_{n-1}$, whence

$$\frac{S_{n-1}}{S_n} = \frac{1 + A_n \tilde{\tau}'_{n-1}}{C_n}.$$

To calculate the constant A_n it is sufficient to compare the coefficients of $x_{n-1}^2(s)$ in equations (118) and (119):

$$A_n = \frac{1}{2(\tilde{\tau}'_{n-1})^2} [\alpha \tilde{\sigma}''_{n-1} + 2(\alpha^2 - 1)\tilde{\tau}'_{n-1}].$$

For calculating C_n we set $x_{n-1} = x_{n-1}^*$ in (118) and (119), where x_{n-1}^* is a root of the equation

$$\tilde{\tau}_{n-1}(x_{n-1}) = 0;$$

this yields $C_n = \alpha \tilde{\sigma}_{n-1}(x_{n-1}^*)$. As a result, we obtain

$$\frac{S_{n-1}}{S_n} = \frac{\alpha + \dfrac{\tilde{\sigma}''_{n-1}}{2\tilde{\tau}'_{n-1}}}{\tilde{\sigma}_{n-1}(x_{n-1}^*)}. \tag{120}$$

3°. By using the values of a_n, b_n, d_n^2 we can construct *the recursion relation*

$$x y_n(x) = \alpha_n y_{n+1}(x) + \beta_n y_n(x) + \gamma_n y_{n-1}(x), \tag{121}$$

where

$$\alpha_n = \frac{a_n}{a_{n+1}}, \qquad \beta_n = \frac{b_n}{a_n} - \frac{b_{n+1}}{a_{n+1}}, \qquad \gamma_n = \frac{a_{n-1}}{a_n}\frac{d_n^2}{d_{n-1}^2}.$$

4°. We obtain *the basic data* for the classical orthogonal polynomials of a discrete variable on nonuniform lattices for the examples *of the Racah polynomials* $u_n^{(\alpha,\beta)}(x)$ and *the dual Hahn polynomials* $w_n^{(c)}(x)$.

For the Racah polynomials, $\sigma(s) = (s-a)(s+b)(s+a-\beta)(b+\alpha-s)$. In accordance with (61)

$$\begin{aligned}\tau(s) = \tilde{\tau}[x(s)] &= \frac{\sigma(-s-1)-\sigma(s)}{2s+1} \\ &= (\alpha+1)(a-\beta)a + (\beta+1)(b-1)(b+\alpha+1)s(s+1).\end{aligned}$$

Similarly, by using formulas (28) and (45) for $\tilde{\tau}_{n-1}(x)$ and $A_{n-1,n}$ we can calculate the constants a_n and b_n from (109).

To determine the squared norm d_n^2 we first need to find the value $x_{n-1}^* = -\tilde{\tau}_{n-1}(0)/\tilde{\tau}'_{n-1}$ for which $\tilde{\tau}_{n-1}(x_{n-1}) = 0$, and then use (116), (117), and (120). Similarly we can calculate the leading coefficients and squared norms of the dual Hahn polynomials. The results for the Racah and dual Hahn polynomials are presented in Tables 11 and 12 (pp. 195–196), where we also give the coefficients of the recursion relations (121).

Finally, we discuss the difference derivatives of the Racah and dual Hahn polynomials. Using (45), we obtain the following formulas for the Racah polynomials $u_n^{(\alpha,\beta)}(x) \equiv u_n^{(\alpha,\beta)}(x,a,b)$ and the dual Hahn polynomials $w_n^{(c)}(x) \equiv w_n^{(c)}(x,a,b)$:

$$\frac{\Delta u_n^{(\alpha,\beta)}[x(s),a,b]}{\Delta x(s)} = (\alpha+\beta+n+1)u_{n-1}^{(\alpha+1,\beta+1)}\left[x\left(s+\frac{1}{2}\right), a+\frac{1}{2}, b-\frac{1}{2}\right],$$

$$\frac{\Delta w_n^{(c)}[x(s),a,b]}{\Delta x(s)} = w_{n-1}^{(c-1/2)}\left[x\left(s+\frac{1}{2}\right), a+\frac{1}{2}, b-\frac{1}{2}\right].$$

8. Asymptotic properties. The classical orthogonal polynomials of a discrete variable, which are solutions of the difference equation (3), tend, as we have seen, to the polynomial solutions of (1) as $h \to 0$. This property was established in Part 5 for the Racah polynomials (see (74)).

It turns out that the Racah and Jacobi polynomials, $u_n^{(\alpha,\beta)}(x)$ and $P_n^{(\alpha,\beta)}(t)$, are connected by an asymptotic formula of higher precision than

(74). Let us choose a function connecting the arguments of these polynomials in the form

$$x = s(s+1) = -\frac{1}{4} + \left(a - \frac{\beta}{2}\right)^2 \frac{1-t}{2} + \left(b + \frac{\alpha}{2}\right)^2 \frac{1+t}{2} \tag{122}$$

and consider the polynomials $p_n(t) = c_n u_n^{(\alpha,\beta)}(x)$, where $c_n = \tilde{N}^{-2n}, \tilde{N}^2 = (b+\alpha/2)^2 - (a-\beta/2)^2$. It can be shown that when $b - a = N \to \infty$ the relation

$$p_n(t) = P_n^{(\alpha,\beta)}(t) + O(1/\tilde{N}^2) \tag{123}$$

is valid. We use the identity

$$\begin{aligned} &\frac{1}{4}[a^2 + b^2 + (a-\beta)^2 + (b+\alpha)^2 - 2] - \frac{1}{8}(\alpha+\beta)(\alpha+\beta+2) \\ &= -\frac{1}{4} + \frac{1}{2}\left[\left(a - \frac{\beta}{2}\right)^2 + \left(b + \frac{\alpha}{2}\right)^2\right] - \frac{1}{4}(\alpha+1)(\beta+1). \end{aligned} \tag{124}$$

Then recurrence relations for the polynomials $P_n^{(\alpha,\beta)}(t)$ and $p_n(t)$ may be presented in the form

$$tP_n^{(\alpha,\beta)}(t) = \alpha_n P_{n+1}^{(\alpha,\beta)}(t) + \beta_n P_n^{(\alpha,\beta)}(t) + \gamma_n P_{n-1}^{(\alpha,\beta)}(t), \tag{125}$$

$$\begin{aligned} tp_n(t) &= \alpha_n p_{n+1}(t) + \left[\beta_n - \frac{(\alpha+1)(\beta+1) + 2n(\alpha+\beta+n+1)}{2\tilde{N}^2}\right] p_n(t) \\ &+ \gamma_n \left[1 - \left(\frac{n + (\alpha+\beta)/2}{b + a + (\alpha-\beta)/2}\right)^2\right] \left[1 - \left(\frac{n + (\alpha+\beta)/2}{b - a + (\alpha+\beta)/2}\right)^2\right] p_{n-1}. \end{aligned} \tag{126}$$

The coefficients α_n, β_n and γ_n are given in Table 2.

When $b - a = N \to \infty$ the coefficients in the relation (126) are different from those in (125) by values of the order $O(\tilde{N}^{-2})$. Since

$$p_0(t) = P_0^{(\alpha,\beta)}(t), \quad p_1(t) = P_1^{(\alpha,\beta)}(t) + O(\tilde{N}^{-2}),$$

we obtain by induction the desired asymptotic formula

$$u_n^{(\alpha,\beta)}(x,a,b) = \tilde{N}^{2n}[P_n^{(\alpha,\beta)}(t) + O(\tilde{N}^{-2})], \qquad b - a = N \to \infty,$$

where

$$x = -\frac{1}{4} + \left(a - \frac{\beta}{2}\right)^2 \frac{1-t}{2} + \left(b + \frac{\alpha}{2}\right)^2 \frac{1+t}{2},$$

$$\tilde{N}^2 = \left(b + \frac{\alpha}{2}\right)^2 - \left(a - \frac{\beta}{2}\right)^2.$$

Table 11. Data for the Racah polynomials $u_n^{\alpha,\beta}(x)$

$\tilde{y}_n(x)$	$u_n^{(\alpha,\beta)}[x(s)], \qquad x(s)=s(s+1)$
(a,b) $\rho(s)$	(a,b) $\dfrac{\Gamma(a+s+1)\Gamma(s-a+\beta+1)\Gamma(b+\alpha-s)\Gamma(b+\alpha+s+1)}{\Gamma(a-\beta+s+1)\Gamma(s-a+1)\Gamma(b-s)\Gamma(b+s+1)}$ $(-1/2<a\le b-1,\ \alpha>-1,\ -1<\beta<2a+1)$
$\sigma(s)$ $\tau(s)$ λ_n	$(s-a)(s+b)(s+a-\beta)(b+\alpha-s)$ $(\alpha+1)a(a-\beta)+(\beta+1)b(b+\alpha)-(\alpha+1)(\beta+1)-(\alpha+\beta+2)x(s)$ $n(\alpha+\beta+n+1)$
B_n $\rho_n(s)$	$\dfrac{(-1)^n}{n!}$ $\dfrac{\Gamma(a+s+n+1)\Gamma(s-a+\beta+n+1)\Gamma(b+\alpha-s)\Gamma(b+\alpha+s+n+1)}{\Gamma(a-\beta+s+1)\Gamma(s-a+1)\Gamma(b-s-n)\Gamma(b+s+1)}$
a_n	$\dfrac{1}{n!}(\alpha+\beta+n+1)_n$
b_n	$\dfrac{(\alpha+\beta+n+1)_{n-1}}{(n-1)!}[-ab(b-a+\alpha+\beta+n)+(a-\beta)(b+\alpha)(b-a-n)$ $-(b-a)(b-a+\alpha+\beta)n+\frac{1}{3}(2n^2+1)(\alpha+\beta)+\frac{1}{3}n(n^2+2)]$
d_n^2	$\dfrac{\Gamma(\alpha+n+1)\Gamma(\beta+n+1)\Gamma(b-a+\alpha+\beta+n+1)\Gamma(a+b+\alpha+n+1)}{(\alpha+\beta+2n+1)n!\Gamma(\alpha+\beta+n+1)(b-a-n-1)!\Gamma(a+b-\beta-n)}$
α_n β_n γ_n	$\dfrac{(n+1)(\alpha+\beta+n+1)}{(\alpha+\beta+2n+1)(\alpha+\beta+2n+2)}$ $\frac{1}{4}[a^2+b^2+(a-\beta)^2+(b+\alpha)^2-2]-\frac{1}{8}(\alpha+\beta+2n)$ $\times(\alpha+\beta+2n+2)+\dfrac{(\beta^2-\alpha^2)[(b+\alpha/2)^2-(a-\beta/2)^2]}{2(\alpha+\beta+2n)(\alpha+\beta+2n+2)}$ $\dfrac{(\alpha+n)(\beta+n)}{(\alpha+\beta+2n)(\alpha+\beta+2n+1)}\left[\left(a+b+\frac{\alpha-\beta}{2}\right)^2-\left(n+\frac{\alpha+\beta}{2}\right)^2\right]$ $\times\left[\left(b-a+\frac{\alpha+\beta}{2}\right)^2-\left(n+\frac{\alpha+\beta}{2}\right)^2\right]$

Table 12. Data for the dual Hahn polynomials $w_n^{(c)}(x)$

$\tilde{y}_n(x)$	$w_n^{(c)}[x(s)], \qquad x(s) = s(s+1)$
(a, b)	(a, b)
$\rho(s)$	$\dfrac{\Gamma(a+s+1)\Gamma(c+s+1)}{\Gamma(s-a+1)\Gamma(b-s)\Gamma(b+s+1)\Gamma(s-c+1)}$ $(-1/2 < a \leq b-1,\ \lvert c\rvert < a+1)$
$\sigma(s)$	$(s-a)(s+b)(s-c)$
$\tau(s)$	$ab - ac + bc - a + b - c - 1 - x(s)$
λ_n	n
B_n	$\dfrac{(-1)^n}{n!}$
$\rho_n(s)$	$\dfrac{\Gamma(a+s+n+1)\Gamma(c+s+n+1)}{\Gamma(s-a+1)\Gamma(b-s-n)\Gamma(b+s+1)\Gamma(s-c+1)}$
a_n	$\dfrac{1}{n!}$
b_n	$-\dfrac{1}{(n-1)!}[ab - ac + bc - 1/3 + (b-a-c)n - 2n^2/3]$
d_n^2	$\dfrac{\Gamma(a+c+n+1)}{n!(b-a-n-1)!\Gamma(b-c-n)}$
α_n	$n+1$
β_n	$ab - ac + bc + (b-a-c-1)(2n+1) - 2n^2$
γ_n	$(a+c+n)(b-a-n)(b-c-n)$

Under the same conditions, by means of the asymptotic representation

$$\frac{\Gamma(s+a+1)}{\Gamma(s-a)} = s^{2a+1}[1+O(s^{-2})], \quad s \to \infty,$$

we obtain formulas for $\rho(s)$ and the squared norms of $u_n^{(\alpha,\beta)}(x)$:

$$\rho(s) = \left(\frac{1}{2}\tilde{N}^2\right)^{\alpha+\beta}(1-t)^{\alpha}(1+t)^{\beta}[1+O(\tilde{N}^{-2})],$$

$$d_n^2 = (\tilde{N}^2)^{\alpha+\beta+2n+1}\frac{\Gamma(\alpha+n+1)\Gamma(\beta+n+1)}{(\alpha+\beta+2n+1)n!\,\Gamma(\alpha+\beta+n+1)}[1+O(\tilde{N}^{-2})].$$

In a similar way we can obtain an asymptotic formula for the dual Hahn polynomials as $b \to \infty$:

$$(-1)^n b^{-n} w_n^{\alpha-a}(x) = L_n^{\alpha}(t) + O(1/b), \tag{128}$$

where $x = a(\alpha - a) + (b-1)t$.

9. Construction of some classes of nonuniform lattices by means of the Darboux-Christoffel formula. We have considered a class of lattices for which it is possible to construct a rather simple theory of orthogonal polynomials of a discrete variable by using a generalized theory of the classical orthogonal polynomials.

We now consider another method of constructing lattices for orthogonal polynomials of a discrete variable by using the Darboux-Christoffel formula. Let $\{p_n(x)\}$ be any system of orthogonal polynomials for which orthogonality is defined either by the integral of the product of the polynomials and a weight function $\rho(x)$, or by the corresponding sums. For such polynomials we already have the Darboux-Christoffel formula

$$\sum_{n=0}^{N-1}\frac{p_n(x)p_n(y)}{d_n^2} = \frac{a_{N-1}}{a_N}\frac{1}{d_{N-1}^2}\frac{p_N(x)p_{N-1}(y) - p_{N-1}(x)p_N(y)}{x-y}. \tag{129}$$

Here d_n^2 are the squared norms and a_n are the coefficients of the leading terms of the polynomials.

Let us show that by using this formula we can easily determine a lattice $\{x_i\}$ and weights $\bar{\rho}(x_i)$ for which the polynomials $p_n(x)$ are orthogonal in the following sense:

$$\sum_{i=0}^{N-1} p_m(x_i)p_n(x_i)\bar{\rho}(x_i) = d_n^2\delta_{mn}. \tag{130}$$

In fact, let $\{x_i\}$ be the zeros of $p_N(x)$, i.e. $p_N(x_i) = 0$. Then if we take $x = x_i$ and $y = x_j$ in (129), we obtain

$$\sum_{n=0}^{N-1} \frac{p_n(x_i)p_n(x_j)}{d_n^2} = D_i^2 \delta_{ij}, \tag{131}$$

where

$$D_i^2 = \sum_{n=0}^{N-1} \frac{p_n^2(x_i)}{d_n^2} = \frac{a_{N-1}}{a_N d_{N-1}^2} p_N'(x_i) p_{N-1}(x_i).$$

It is convenient to write (131) in the form

$$\sum_{n=0}^{N-1} c_{ni} c_{nj} = \delta_{ij}, \tag{132}$$

with

$$c_{ni} = \frac{p_n(x_i)}{d_n D_i}.$$

It follows from (132) that the matrix C with elements c_{ni} ($n, i = 0, 1, \ldots, N-1$) is unitary, and hence there is another orthogonality relation for C:

$$\sum_{i=0}^{N-1} c_{mi} c_{ni} = \delta_{mn} \qquad (m, n = 0, 1, \ldots, N-1). \tag{133}$$

It is evident that (133) is equivalent to the orthogonality relation (130) for the polynomials $p_n(x)$ if $\bar{\rho}(x_i) = 1/D_i^2$.

We have discussed the method of constructing an orthogonality relation of the form (130) for the polynomials $p_n(x)$ in the case when the lattice $\{x_i\}$ is determined by using the equations $p_N(x_i) = 0$. The entire discussion can be carried over if $\{x_i\}$ is determined by using the more general equation $\alpha p_N(x_i) + \beta p_{N-1}(x_i) = 0$, where α and β are real coefficients, not both zero.

As an example we consider an orthogonality relation of the form (130) for the Chebyshev polynomials of the first kind, $T_n(x) = \cos(n \arccos x)$. In this case

$$a_n = \frac{1}{2^{n-1}}, \qquad d_n^2 = \begin{cases} \pi & \text{for } n = 0, \\ \pi/2 & \text{for } n \neq 0; \end{cases}$$

$T_n(x_i) = 0$ for

$$x_i = \cos\left[\frac{\pi}{N}(i + 1/2)\right] \qquad (i = 0, 1, \ldots, N-1),$$

whence

$$D_i^2 = \frac{4}{\pi} T_N'(x_i) T_{N-1}(x_i) = \frac{4N}{\pi}.$$

Hence we can write (130) in the form

$$\sum_{i=0}^{N-1} T_m[x(s_i)] T_n[x(s_i)] \frac{\pi}{4N} = d_n^2 \delta_{mn}, \tag{134}$$

where $x(s) = \cos(\pi s/N)$, $s_i = i + \frac{1}{2}$ $(i = 0, 1, \ldots, N-1)$. The lattice $x_i = \cos(\pi s_i/N)$ is a special case of the lattice (58) with $\omega = \pi/(2N)$. Consequently it is natural to expect that the Chebyshev polynomials $T_n(x)$ coincide, up to a normalizing factor, with the q-analogs of the Racah polynomials $u_n^{(\alpha,\beta)}[x(s)]$, which are orthogonal on the lattice with $x(s) = \cos 2\omega s$, $\omega = \pi/(2N)$, $a = 1/2$, $b = N + \frac{1}{2}$, and some values of α and β.

By comparing (134) and (50), we see that our expectation is fulfilled if

$$\rho(s_i)\Delta x\left(s_i - \frac{1}{2}\right) = \text{const.}\,, \tag{135}$$

where $\rho(s)$ is defined by (106).

Let us verify that (135) is satisfied with $a = 1/2$, $b = N+\frac{1}{2}$, $\alpha = \beta = -\frac{1}{2}$. In fact, (135) will be satisfied if

$$\frac{\rho(s_i+1)}{\rho(s_i)} = \frac{\Delta x(s_i - 1/2)}{\Delta x(s_i + 1/2)}. \tag{135a}$$

Since (see Table 10)

$$\frac{\Delta x(s_i - 1/2)}{\Delta x(s_i + 1/2)} = \frac{\sin(\pi s_i/N)}{\sin(\pi(s_i+1)/N)},$$

$$\frac{\rho(s_i+1)}{\rho(s_i)} = \frac{\sigma(s_i) + \tau(s_i)\Delta x(s_i - 1/2)}{\sigma(s_i+1)}$$

$$= \frac{\psi_q(N - \frac{1}{2} - s_i)\psi_q(s_i)\psi_q(N + s_i)}{\psi_q(N + \frac{1}{2} + s_i)\psi_q(s_i+1)\psi_q(N - 1 - s_i)}$$

$$= \frac{\sin(\pi(N - 1/2 - s_i))\sin(\pi s_i/2N)\sin(\pi(N + s_i))/2N}{\sin(\pi(N + 1/2 + s_i))/2N \sin(\pi(s_i+1))/2N \sin(\pi(N - 1 - s_i))/2N}$$

$$= \frac{\sin(\pi s_i/2N)\cos(\pi s_i/2N)}{\sin(\pi(s_i+1))/2N \cos(\pi s_i(s_i+1))/2N} = \frac{\sin(\pi s_i/N)}{\sin(\pi(s_i+1)/N)}$$

the validity of (135) is established.

Consequently

$$T_n(x) = A_n u_n^{(-1/2,-1/2)}[x(s), q] \tag{150}$$

for $x(s) = \cos 2\omega s$, $\omega = \pi/(2N)$, $a = 1/2, b = N + 1/2$. The constants A_n can be determined by comparing coefficients of the leading terms in (150).

By using (150), one can show that the Chebyshev polynomials of the second kind

$$U_n(x) = \frac{\sin(n+1)\phi}{\sin\phi} \qquad (\phi = \arccos x),$$

coincide, up to a constant multiple, with the q-analogs $u_n^{(1/2,1/2)}(x, q)$ of the Racah polynomials on the lattice $x(s) = \cos 2\omega s$, with $\omega = \pi/(2N)$, for $a = 1$, $b = N$. This follows from the easily verified relation

$$U_n[x_1(s)] = \frac{\Delta T_{n+1}[x(s)]}{\Delta x(s)} \qquad (x_1(s) = x(s+1/2))$$

and the orthogonality of the polynomials

$$\frac{\Delta u_{n+1}^{(-1/2,-1/2)}[x(s), q]}{\Delta x(s)}$$

on the lattice.

Chapter III

Bessel Functions

§ 14 Bessel's differential equation and its solutions

1. Solving the Helmholtz equation in cylindrical coordinates. Bessel functions are perhaps the most frequently used special functions. Typical problems that lead to Bessel functions arise in solving the *Helmholtz equation*

$$\Delta v + \lambda v = 0$$

in cylindrical coordinates. We consider the simplest case, when v is independent of the distance along the axis of the cylinder. Then $v = v(r, \phi)$ and

$$\Delta v + \lambda v = \frac{1}{r}\frac{\partial}{\partial r}\left(r\frac{\partial v}{\partial r}\right) + \frac{1}{r^2}\frac{\partial^2 v}{\partial \phi^2} + \lambda v = 0. \tag{1}$$

So that v will be single-valued, we require that it satisfies the periodicity condition $v(r, \phi + 2\pi) = v(r, \phi)$. Let us expand v in a Fourier series:

$$v(r, \phi) = \sum_{n=-\infty}^{\infty} v_n(r)e^{in\phi},$$

where

$$v_n(r) = \frac{1}{2\pi}\int_{-\pi}^{\pi} v(r, \phi)e^{-in\phi}d\phi. \tag{2}$$

It is easy to obtain a differential equation for $v_n(r)$ by integrating (1) on $(-\pi, \pi)$ with weight $e^{-in\phi}$ and simplifying the terms containing $\partial^2 v/\partial\phi^2$ by integrating twice by parts. Since $v(r,\phi)$ is periodic in ϕ, the integrated terms vanish, and we obtain a differential equation for $u(z) = v_n(r)$ with $z = \sqrt{\lambda}r$:

$$z^2u'' + zu' + (z^2 - n^2)u = 0.$$

We are going to study an equation of somewhat more general form,

$$z^2u'' + zu' + (z^2 - \nu^2)u = 0, \tag{3}$$

where z is a complex variable, and the parameter ν can have any real or complex values.

The solutions of (3) are *Bessel functions of order* ν, or *cylinder functions*, and (3) is *Bessel's equation.*

Many other differential equations can be obtained from Bessel's equation by changes of variable. An example is *Lommel's equation*

$$v'' + \frac{1-2\alpha}{\xi}v' + \left[(\beta\gamma\xi^{\gamma-1})^2 + \frac{\alpha^2 - \nu^2\gamma^2}{\xi^2}\right]v = 0, \tag{4}$$

which is extensively used in applications; its solutions are

$$v(\xi) = \xi^{\alpha}u_{\nu}(\beta\xi^{\gamma})$$

Here $u_\nu(z)$ is a Bessel function of order ν; α, β, γ are constants.

2. Definition of Bessel functions of the first kind and Hankel functions. Bessel's equation (3) is the special case of the generalized equation of hypergeometric type (1.1) for which $\sigma(z) = z$, $\tilde{\tau}(z) = 1$, $\tilde{\sigma}(z) = z^2 - \nu^2$. In reducing (3) to an equation of hypergeometric type, the possible forms of $\phi(z)$ are, as was shown in §1 (see the example given there) $\phi(z) = z^{\pm\nu}e^{\pm iz}$, corresponding to different choices of signs in formula (1.11) for $\pi(z)$ and different values of k. Let us consider, for example, $\phi(z) = z^{\nu}e^{iz}$. Putting $u(z) = \phi(z)y(z)$, we obtain an equation of hypergeometric type,

$$\sigma(z)y'' + \tau(z)y' + \lambda y = 0, \tag{3a}$$

where

$$\sigma(z) = z, \quad \tau(z) = 2iz + 2\nu + 1, \quad \lambda = i(2\nu + 1).$$

By Theorem 1 of §3 a particular solution of (3a) is

$$y(z) = \frac{c_\mu}{\rho(z)}\int_C \frac{\sigma^\mu(s)\rho(s)}{(s-z)^{\mu+1}}ds,$$

where c_μ is a normalizing constant, $\rho(z)$ is a solution of the differential equation

$$[\sigma(z)\rho(z)]' = \tau(z)\rho(z),$$

and μ is a root of the equation

$$\lambda + \mu\tau' + \frac{1}{2}\mu(\mu-1)\sigma'' = 0$$

(we have used formulas (3.2) and (3.3), where in order to avoid confusion we have replaced ν by μ since ν has already been used in the original Bessel equation). The contour C is chosen so that

$$\left.\frac{\sigma^{\mu+1}(s)\rho(s)}{(s-z)^{\mu+2}}\right|_{s_1,s_2} = 0.$$

In the present case,

$$\mu = -\nu - \frac{1}{2}, \quad \rho(z) = z^{2\nu}e^{2iz}.$$

Hence a particular solution of Bessel's equation can be written in the form

$$u_\nu(z) = \phi(z)y(z) = a_\nu z^{-\nu}e^{-iz}\int_C [s(z-s)]^{\nu-\frac{1}{2}}e^{2is}ds, \tag{5}$$

where a_ν is a normalizing constant and C is chosen so that

$$\left. s^{\nu+1/2}(z-s)^{\nu-3/2}e^{2is}\right|_{s_1,s_2} = 0.$$

Let $z > 0$, Re $\nu > 3/2$. Then the ends of the contour can be taken at $s_1 = 0, s_2 = z$. Alternatively, C might go to infinity with $\operatorname{Im} s \to +\infty$. Then C can be one of the contours indicated in Figure 6.

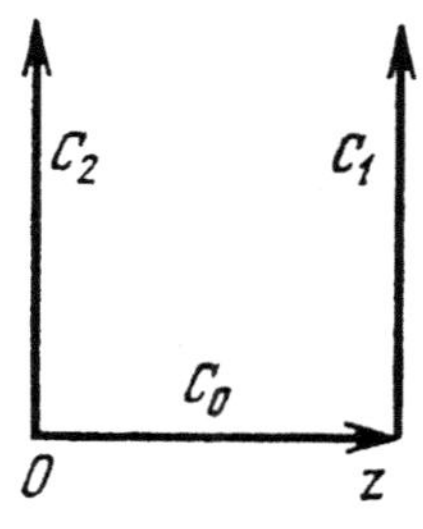

Figure 6.

We thus obtain the following three solutions of Bessel's equation:

$$u_\nu^{(0)}(z) = a_\nu z^{-\nu} e^{-iz} \int\limits_{C_0} [s(z-s)]^{\nu-1/2} e^{2is} ds, \tag{6}$$

$$u_\nu^{(1)}(z) = a_\nu^{(1)} z^{-\nu} e^{-iz} \int\limits_{C_1} [s(z-s)]^{\nu-1/2} e^{2is} ds, \tag{7}$$

$$u_\nu^{(2)}(z) = a_\nu^{(2)} z^{-\nu} e^{-iz} \int\limits_{C_2} [s(z-s)]^{\nu-1/2} e^{2is} ds. \tag{8}$$

In order to have a single-valued branch of the function $[s(z-s)]^{\nu-1/2}$, we take $|\arg s(z-s)| < \pi$. The contours C_0, C_1, C_2 can be parametrized by

$$\begin{array}{ll} s = z(1+t)/2 & (-1 \le t \le 1), \\ s = z(1+it/2) & (0 \le t < \infty), \\ s = izt/2 & (0 \le t < \infty). \end{array}$$

Then (6)–(8) become

$$u_\nu^{(0)}(z) = \frac{a_\nu}{2^{2\nu}} z^\nu \int\limits_{-1}^{1} (1-t^2)^{\nu-1/2} e^{izt} dt = \frac{a_\nu}{2^{2\nu}} z^\nu \int\limits_{-1}^{1} (1-t^2)^{\nu-1/2} \cos zt \, dt, \tag{9}$$

$$u_\nu^{(1)}(z) = -\frac{a_\nu^{(1)}}{\sqrt{2}} \left(\frac{z}{2}\right)^\nu e^{i(z-\pi\nu/2-\pi/4)} \int\limits_0^\infty e^{-zt} t^{\nu-1/2} \left(1+\frac{it}{2}\right)^{\nu-1/2} dt, \tag{10}$$

and

$$u_\nu^{(2)}(z) = \frac{a_\nu^{(2)}}{\sqrt{2}} \left(\frac{z}{2}\right)^\nu e^{-i(z-\pi\nu/2-\pi/4)} \int\limits_0^\infty e^{-zt} t^{\nu-1/2} \left(1-\frac{it}{2}\right)^{\nu-1/2} dt. \tag{11}$$

In accordance with the condition $|\arg s(z-s)| < \pi$, the values of $\arg(1 \pm \frac{1}{2}it)$ in (10) and (11) are taken with the smallest possible absolute values.

If we take the normalizing constants real, and $a_\nu^{(2)} = -a_\nu^{(1)}$, we see from (10) and (11) that when z and ν are real, the functions $u_\nu^{(1)}(z)$ and $u_\nu^{(2)}(z)$ are complex conjugates. It is convenient to introduce a function that is real for real z,

$$u_\nu(z) = \frac{1}{2}[u_\nu^{(1)}(z) + u_\nu^{(2)}(z)]. \tag{12}$$

Let us show that this function is equal to $u_\nu^{(0)}(z)$ if we take

$$a_\nu^{(2)} = -a_\nu^{(1)} = 2a_\nu. \tag{13}$$

To prove this it is enough to apply Cauchy's theorem to the contour C which is the union of C_0, C_1 and C_2 (see Fig. 6). If we close the contour at ∞, Cauchy's theorem yields

$$\int\limits_C [s(z-s)]^{\nu-1/2} e^{2is} ds = -\int\limits_{C_2} [s(z-s)]^{\nu-1/2} e^{2is} ds$$
$$+ \int\limits_{C_0} [s(z-s)]^{\nu-1/2} e^{2is} ds + \int\limits_{C_1} [s(z-s)]^{\nu-1/2} e^{2is} ds = 0$$

(the integral over the part of the contour "at infinity" reduces to zero). Taking account of (13) and using (6)–(8), we obtain

$$u_\nu^{(0)}(z) = \frac{1}{2}[u_\nu^{(1)}(z) + u_\nu^{(2)}(z)], \tag{14}$$

as required.

The function $u_\nu^{(0)}(z)$, with an appropriate choice of a_ν, is the *Bessel function of the first kind*, $J_\nu(z)$; the functions $u_\nu^{(1)}(z)$ and $u_\nu^{(2)}(z)$ with the normalization (13) are the *Hankel functions of the first and second kind*, $H_\nu^{(1)}(z)$ and $H_\nu^{(2)}(z)$. By (14), these functions are connected by the equation

$$J_\nu(z) = \frac{1}{2}[H_\nu^{(1)}(z) + H_\nu^{(2)}(z)]. \tag{15}$$

The integral representations (9)–(11) are useful for investigating the properties of Bessel functions: in particular, the integral representation for $J_\nu(z)$ lets one obtain the power series for $J_\nu(z)$; the integral representations of the Hankel functions are used in obtaining the asymptotic expansions of these functions as $z \to \infty$.

To obtain the power series for $J_\nu(z)$, we replace $\cos zt$ in (9) by its expansion in powers of zt and interchange summation and integration. We obtain

$$J_\nu(z) = \frac{a_\nu}{2^{2\nu}} z^\nu \sum_{k=0}^{\infty} \frac{(-1)^k z^{2k}}{(2k)!} \int\limits_{-1}^{1} (1-t^2)^{\nu-1/2} t^{2k} dt.$$

We can evaluate the coefficients in the series:

$$\int_{-1}^{1}(1-t^2)^{\nu-1/2}t^{2k}dt = 2\int_{0}^{1}(1-t^2)^{\nu-1/2}t^{2k}dt$$

$$= \int_{0}^{1}(1-t)^{\nu-1/2}t^{k-1/2}dt = \frac{\Gamma\left(\nu+\frac{1}{2}\right)\Gamma\left(k+\frac{1}{2}\right)}{\Gamma(\nu+k+1)}$$

$$= \frac{\Gamma\left(\nu+\frac{1}{2}\right)\sqrt{\pi}(2k)!}{2^{2k}k!\,\Gamma(\nu+k+1)}.$$

Here we used the evenness of the integrand, the relation between the beta and gamma functions, and the duplication formula for the gamma function (see Appendix A). Hence we have

$$J_\nu(z) = \frac{a_\nu}{2^\nu}\sqrt{\pi}\Gamma\left(\nu+\frac{1}{2}\right)\sum_{k=0}^{\infty}\frac{(-1)^k(z/2)^{\nu+2k}}{k!\,\Gamma(\nu+k+1)}.$$

The series will have a simpler form if we choose a_ν so that

$$\frac{a_\nu}{2^\nu}\sqrt{\pi}\Gamma\left(\nu+\frac{1}{2}\right) = 1. \tag{16}$$

Finally we obtain

$$J_\nu(z) = \sum_{k=0}^{\infty}\frac{(-1)^k(z/2)^{\nu+2k}}{k!\,\Gamma(\nu+k+1)}. \tag{17}$$

Using the value of a_ν given by (16), we can rewrite (9)–(11) as

$$J_\nu(z) = \frac{(z/2)^\nu}{\sqrt{\pi}\Gamma(\nu+1/2)}\int_{-1}^{1}(1-t^2)^{\nu-1/2}\cos zt\,dt, \tag{18}$$

$$H_\nu^{(1)}(z) = \sqrt{\frac{2}{\pi}}\frac{z^\nu e^{i(z-\pi\nu/2-\pi/4)}}{\Gamma(\nu+1/2)}\int_{0}^{\infty}e^{-zt}t^{\nu-1/2}\left(1+\frac{it}{2}\right)^{\nu-1/2}dt, \tag{19}$$

and

$$H_\nu^{(2)}(z) = \sqrt{\frac{2}{\pi}}\frac{z^\nu e^{-i(z-\pi\nu/2-\pi/4)}}{\Gamma(\nu+1/2)}\int_{0}^{\infty}e^{-zt}t^{\nu-1/2}\left(1-\frac{it}{2}\right)^{\nu-1/2}dt. \tag{20}$$

These are known as *Poisson's integrals* for the Bessel functions.

Other useful representations for the Hankel functions are obtained from (19) and (20) by replacing t by t/z:

$$H_\nu^{(1)}(z) = \sqrt{\frac{2}{\pi z}} \frac{z^\nu e^{i(z-\pi\nu/2-\pi/4)}}{\Gamma(\nu+1/2)} \int_0^\infty e^{-t} t^{\nu-1/2} \left(1+\frac{it}{2z}\right)^{\nu-1/2} dt, \quad (19a)$$

$$H_\nu^{(2)}(z) = \sqrt{\frac{2}{\pi z}} \frac{z^\nu e^{-i(z-\pi\nu/2-\pi/4)}}{\Gamma(\nu+1/2)} \int_0^\infty e^{-t} t^{\nu-1/2} \left(1-\frac{it}{2z}\right)^{\nu-1/2} dt. \quad (20a)$$

§ 15 Basic properties of Bessel functions

1. Recursion relations and differentiation formulas. Recursion relations and differentiation formulas for Bessel functions can be found by the method of §4 from the original integral representation of the functions:

$$u_\nu(z) = a_\nu z^{-\nu} e^{-iz} \int_C [s(z-s)]^{\nu-1/2} e^{2is} ds.$$

As an example, let us find a relation of the form

$$A_1(z)u_\nu'(z) + A_2(z)u_\nu(z) + A_3(z)u_{\nu-1}(z) = 0, \quad (1)$$

where $A_i(z)$ are rational functions of z. We have

$$A_1(z)u_\nu'(z) + A_2(z)u_\nu(z) + A_3(z)u_{\nu-1}(z) = e^{-iz} z^{-\nu-1} \int_C P(s)[s(z-s)]^{\nu-3/2} e^{2is} ds,$$

where

$$P(s) = A_1 a_\nu \left[(-\nu - iz)s(z-s) + \left(\nu - \frac{1}{2}\right) zs\right] + A_2 a_\nu zs(z-s) + A_3 z^2 a_{\nu-1}.$$

For (1) to be satisfied, $A_1(z)$, $A_2(z)$ and $A_3(z)$ are to be determined by the condition

$$P(s)[s(z-s)]^{\nu-3/2} e^{2is} = \frac{d}{ds}\{Q(s)[s(z-s)]^{\nu-1/2} e^{2is}\},$$

where $Q(s)$ is a polynomial. As shown in §4, one coefficient of $Q(s)$ can be chosen arbitrarily. In the present case, $Q(s)$ is a constant and we can take $Q(s) = a_\nu$. Using the condition on $P(s)$ and $Q(s)$, we obtain the equation

$$A_1[(-\nu - iz)s(z-s) + (\nu - 1/2)zs] + A_2 zs(z-s) \\ + A_3 z^2 a_{\nu-1}/a_\nu = 2is(z-s) + (\nu - 1/2)(z-2s).$$

If we use the values of a_ν that correspond to $J_\nu(z)$ and $H_\nu^{(1,2)}(z)$, we obtain $a_{\nu-1}/a_\nu = (\nu - 1/2)/2$. The equation that determines A_i is valid for all s. Hence we can find A_i by taking s to have any convenient value. Taking, for example, $s = 0$, we obtain $A_3 = 2/z$. The value $s = z$ yields $A_1 = -2/z$. The coefficient A_2 is easily found by comparing the coefficients of the highest power of s: $A_2 = -2\nu/z^2$. Letting $u_\nu(z)$ stand for either $J_\nu(z)$ or $H_\nu^{(1,2)}(z)$, we obtain the relation

$$\frac{\nu}{z}u_\nu(z) + u_\nu'(z) = u_{\nu-1}(z). \tag{2}$$

By the same method, we can obtain a recursion involving $u_\nu(z), u_{\nu-1}(z)$ and $u_{\nu-2}(z)$. However, it is easier to differentiate (2) and eliminate $u_\nu''(z), u_\nu'(z)$ and $u_{\nu-1}'(z)$ by using Bessel's equation and (2). We find

$$u_\nu(z) - \frac{2(\nu-1)}{z}u_{\nu-1}(z) + u_{\nu-2}(z) = 0. \tag{3}$$

Equations (2) and (3) can be transformed into the equivalent forms

$$\frac{1}{z}\frac{d}{dz}[z^\nu u_\nu(z) = z^{\nu-1}u_{\nu-1}(z),$$
$$-\frac{1}{z}\frac{d}{dz}[z^{-\nu}u_\nu(z)] = z^{-(\nu+1)}u_{\nu+1}(z).$$

Hence

$$\left(\frac{1}{z}\frac{d}{dz}\right)^n [z^\nu u_\nu(z)] = z^{\nu-n}u_{\nu-n}(z),$$
$$\left(-\frac{1}{z}\frac{d}{dz}\right)^n [z^{-\nu} u_\nu(z)] = z^{-(\nu+n)}u_{\nu+n}(z). \tag{4}$$

2. Analytic continuation and asymptotic formulas. We have defined $J_\nu(z)$, $H_\nu^{(1)}(z)$ and $H_\nu^{(2)}(z)$ only for $z > 0$ and $\operatorname{Re}\nu > 3/2$. Now let z be a point of the complex plane cut along $(-\infty, 0)$, i.e. with $|\arg z| < \pi$. This restriction makes z^ν single-valued when ν is not an integer. By using the integral representations (14.18)–(14.20) we can continue $J_\nu(z)$ and $H_\nu^{(1,2)}(z)$ to larger domains for both z and ν.

The integral for $J_\nu(z)$ converges uniformly in z and ν for Re $\nu \geq -\frac{1}{2}+\delta$, $|z| \leq R$ (δ and R are arbitrary positive numbers), because

$$|(1-t^2)^{\nu-1/2}\cos zt| \leq e^R(1-t^2)^{\delta-1}$$

and the integral $\int_{-1}^{1}(1-t^2)^{\delta-1}dt$ converges. Hence, by Theorem 2, §3, the function $J_\nu(z)$ is an analytic function of z and of ν for $|\arg z| < \pi$, and Re $\nu > -1/2$.

The integrals for $H_\nu^{(1,2)}(z)$,

$$\int_0^\infty e^{-zt}t^{\nu-1/2}\left(1 \pm \frac{1}{2}it\right)^{\nu-1/2} dt,$$

are the Laplace integrals

$$F(z) = \int_0^\infty e^{-zt} f(t)dt,$$

with $f(t) = t^{\nu-1/2}(1 \pm \frac{1}{2}it)^{\nu-1/2}$. The analytic continuation and asymptotic representation of Laplace integrals of the form

$$F(z,p,q) = \int_0^\infty e^{-zt}t^p(1+at)^q dt$$

are discussed in detail in the example for Theorem 1 in Appendix B. In the present case, $p = q = \nu - \frac{1}{2}, a = \pm i/2$, and the results of this example show that the Hankel functions $H_\nu^{(1,2)}(z)$ are analytic in each variable for $z \neq 0$, $|\arg z| < \pi$, Re $\nu > -1/2$. These functions have asymptotic representations as $z \to \infty$ when Re $\nu > -1/2$ and $|\arg z| \leq \pi - \epsilon$:

$$H_\nu^{(1,2)}(z) = \left(\frac{2}{\pi z}\right)^{1/2} e^{\pm i(z-(\pi\nu/2)-\pi/4)}\left[\sum_{k=0}^{n-1} C_k\left(\pm\frac{i}{z}\right)^k + O\left(\frac{1}{z^n}\right)\right]. \quad (5)$$

Here

$$C_k = \frac{\Gamma\left(\nu+\frac{1}{2}+k\right)}{2^k k!\,\Gamma\left(\nu+\frac{1}{2}-k\right)},$$

and the upper signs apply to $H_\nu^{(1)}(z)$, the lower signs to $H_\nu^{(2)}(z)$. If we use the functional equation $\Gamma(z+1) = z\Gamma(z)$, we can simplify the formula for C_k. We have

$$\Gamma\left(\nu+\frac{1}{2}+k\right) = \left(\nu+\frac{1}{2}\right)\left(\nu+\frac{3}{2}\right)\dots\left(\nu+k-\frac{1}{2}\right)\Gamma\left(\nu+\frac{1}{2}\right),$$

$$\Gamma\left(\nu+\frac{1}{2}-k\right) = \frac{\Gamma\left(\nu+\frac{1}{2}\right)}{\left(\nu-\frac{1}{2}\right)\left(\nu-\frac{3}{2}\right)\dots\left(\nu-k+\frac{1}{2}\right)}.$$

Hence

$$C_k = \prod_{l=1}^{k}\left[\frac{4\nu^2-(2l-1)^2}{8l}\right], \quad C_0 = 1.$$

Using the equation

$$J_\nu(z) = \frac{1}{2}[H_\nu^{(1)}(z) + H_\nu^{(2)}(z)]$$

we obtain an asymptotic formula for $J_\nu(z)$:

$$J_\nu(z) = \left(\frac{2}{\pi z}\right)^{1/2}\left\{\sum_{k=0}^{n-1}\frac{C_k}{z^k}\cos\left[z-\frac{\pi}{2}\left(\nu-k+\frac{1}{2}\right)\right] + O\left(\frac{e^{|\mathrm{Im}\, z|}}{z^n}\right)\right\}. \quad (6)$$

We have considered the analytic continuation of Bessel functions for $z \neq 0, |\arg z| < \pi$ and $\mathrm{Re}\ \nu > -1/2$. The condition $\mathrm{Re}\ \nu > -1/2$ is not essential, since when $\mathrm{Re}\ \nu \leq -1/2$ the analytic continuation can be obtained from the recursion relation (3) with ν decreased by 1. By the differentiation formula (2) the derivatives of $J_\nu(z)$ and $H_\nu^{(1,2)}(z)$ are analytic in z and in ν in the same region as the Bessel functions themselves. By the principle of analytic continuation, the analytic continuations of the Bessel functions still satisfy Bessel's equation.

3. Functional equations. The Bessel equation is not changed by replacing ν by $-\nu$. Therefore it not only has $H_\nu^{(1)}(z)$ and $H_\nu^{(2)}(z)$ as solutions, but also $H_{-\nu}^{(1)}(z)$ and $H_{-\nu}^{(2)}(z)$. In finding the formulas that connect $H_\nu^{(1,2)}(z)$ with $H_{-\nu}^{(1,2)}(z)$, we shall suppose for the time being that $|\mathrm{Re}\,\nu| < 1/2$. Then the Hankel functions $H_{\pm\nu}^{(1,2)}(z)$ have the asymptotic representations (5). It is clear from these representations that the functions $H_\nu^{(1,2)}(z)$ have different asymptotic behavior as $z \to \infty$ and therefore are linearly independent solutions of Bessel's equation. Consequently

$$H_{-\nu}^{(1)}(z) = A_\nu H_\nu^{(1)}(z) + B_\nu H_\nu^{(2)}(z), \quad (7)$$

where A_ν and B_ν are constants. If we compare the asymptotic behavior, as $z \to \infty$, of the left-hand and right-hand sides of (7), we find that $A_\nu = e^{i\pi\nu}$ and $B_\nu = 0$, i.e.

$$H_{-\nu}^{(1)}(z) = e^{i\pi\nu} H_\nu^{(1)}(z). \tag{8}$$

Similarly,

$$H_{-\nu}^{(2)}(z) = e^{-i\pi\nu} H_\nu^{(2)}(z). \tag{9}$$

It is easily verified by using (8) and (9) that (5), and therefore also (6), are valid for all values of ν.

We now find the connection between $H_\nu^{(1)}(z), H_\nu^{(2)}(z)$ and $J_\nu(z)$, $J_{-\nu}(z)$. Since

$$\begin{aligned} J_\nu(z) &= \frac{1}{2}[H_\nu^{(1)}(z) + H_\nu^{(2)}(z)], \\ J_{-\nu}(z) &= \frac{1}{2}[H_{-\nu}^{(1)}(z) + H_{-\nu}^{(2)}(z)], \end{aligned} \tag{10}$$

we have

$$\begin{aligned} H_\nu^{(1)}(z) &= \frac{J_{-\nu}(z) - e^{-i\pi\nu} J_\nu(z)}{i \sin \pi\nu}, \\ H_\nu^{(2)}(z) &= \frac{e^{i\pi\nu} J_\nu(z) - J_{-\nu}(z)}{i \sin \pi\nu}. \end{aligned} \tag{11}$$

by (8) and (9).

4. Power series expansions. We have already obtained the power series

$$J_\nu(z) = \sum_{k=0}^{\infty} \frac{(-1)^k (z/2)^{\nu+2k}}{k!\,\Gamma(\nu+k+1)} \tag{12}$$

for real $z > 0$ and $\operatorname{Re} \nu > 3/2$. To establish this expansion for arbitrary values of ν and z we investigate the region of analyticity of (12) by using a theorem of Weierstrass (see [D2], [L3] or [S8]).

Theorem 1. *Let $f_k(z)$ be analytic in a region D and let the series*

$$\sum_{k=0}^{\infty} f_k(z)$$

converge uniformly on every compact subset of D to $f(z)$. Then in D:

1) $f(z)$ is analytic;

2) $f^{(n)}(z) = \sum_{k=0}^{\infty} f_k^{(n)}(z)$;

3) $\sum_{k=0}^{\infty} f_k^{(n)}(z)$ converges uniformly on every compact subset of D.

Remark. The series $\sum_{k=0}^{\infty} f_k(z)$ will converge uniformly in D if there is an m such that for every $z \in D$ and $k > m$ we have

$$\left|\frac{f_k(z)}{f_{k-1}(z)}\right| \leq q < 1,$$

where q is independent of z and $|f_m(z)| \leq C$ for $z \in D$ (C, a constant). This test for the uniform convergence of a series is known as *D'Alembert's test.*

Let us show that (12) converges uniformly for z and ν in the regions $0 < \delta \leq |z| \leq R$, $|\nu| \leq N$, where R and N are arbitrarily large fixed numbers. It will be sufficient to use the following estimate of the ratio of two successive terms of the series:

$$\left|\frac{u_k(z)}{u_{k-1}(z)}\right| = \frac{|z|^2}{4k|k+\nu|} \leq \frac{R^2}{4k(k-N)} \leq \frac{1}{4}$$

where $k \geq \max(R^2, N+1)$. Since the terms of the series are analytic functions of z and ν for $\delta \leq |z| \leq R$, $|\arg z| < \pi$, and $|\nu| \leq N$, the series (12) represents an analytic function of z and ν for all ν and $|\arg z| < \pi$.

Consequently both sides of (12) are analytic functions of each of z and ν for all ν and $|\arg z| < \pi$. By the principle of analytic continuation, (12) is valid in the specified domain of z and ν.

If $\nu \neq 0, 1, 2, \ldots$, the functions $J_\nu(z)$ and $J_{-\nu}(z)$ are linearly independent, since they behave differently as $z \to 0$:

$$J_\nu(z) \approx \frac{(z/2)^\nu}{\Gamma(\nu+1)}, \quad J_{-\nu}(z) \approx \frac{(z/2)^{-\nu}}{\Gamma(-\nu+1)}.$$

It follows that when $\nu \neq n (n = 0, 1, 2, \ldots)$ the general solution of Bessel's equation can be written

$$u(z) = C_1 J_\nu(z) + C_2 J_{-\nu}(z).$$

From (12) and (11) we can obtain the power series for $H_\nu^{(1,2)}(z)$. This presents no difficulty for $\nu \neq n$; therefore we consider only the case $\nu = n$. The points $\nu = n$ are removable singular points on the right-hand sides of (11), since the left-hand sides are analytic functions of ν and consequently approach limits as $\nu \to n$. The denominators in (11) vanish at $\nu = n$; hence if the limits are to exist, the numerators must vanish at $\nu = n$, i.e.

$$J_{-n}(z) = (-1)^n J_n(z).$$

It follows that when $\nu = n$ the solutions $J_n(z)$ and $J_{-n}(z)$ are linearly dependent. Taking limits as $\nu \to n$ and using L'Hospital's rule, we find

$$H_n^{(1,2)}(z) = J_n(z) \pm \frac{i}{\pi}[a_n(z) + (-1)^n a_{-n}(z)], \tag{13}$$

where $a_\nu(z) = \partial J_\nu(z)/\partial\nu$ (the plus sign corresponds to $H_n^{(1)}(z)$).

Since the series for $J_\nu(z)$ converges uniformly in the region we are considering, and its terms are analytic functions of ν, we may, by the theorem of Weierstrass, differentiate the series termwise in order to evaluate $a_\nu(z)$. We obtain

$$a_\nu(z) = J_\nu(z)\ln(z/2) - \sum_{k=0}^{\infty} \frac{(-1)^k (z/2)^{\nu+2k}}{k!\Gamma(k+\nu+1)}\psi(k+\nu+1),$$

where $\psi(z)$ is the logarithmic derivative of the gamma function (see Appendix A). Since

$$\frac{\psi(z)}{\Gamma(z)} \to (-1)^{n+1} n!, \quad z \to -n$$

(see formula (27) in Appendix A), we have

$$\begin{aligned}(-1)^n a_{-n}(z) &= (-1)^n J_{-n}(z)\ln(z/2) \\ &\quad - (-1)^n \left\{ \sum_{k=0}^{n-1} \frac{(-1)^k (z/2)^{-n+2k}}{k!}(-1)^{n-k}(n-k-1)! \right. \\ &\qquad \left. + \sum_{k=n}^{\infty} \frac{(-1)^k (z/2)^{-n+2k}\psi(k-n+1)}{k!\Gamma(k-n+1)} \right\} \\ &= J_n(z)\ln(z/2) - \sum_{k=0}^{n-1} \frac{(n-k-1)!}{k!}\left(\frac{z}{2}\right)^{2k-n} \\ &\qquad - \sum_{k=0}^{\infty} \frac{(-1)^k (z/2)^{n+2k}}{k!(n+k)!}\psi(k+1).\end{aligned}$$

Therefore

$$\begin{aligned}H_n^{(1,2)}(z) = J_n(z) \pm \frac{i}{\pi}\left\{ 2J_n(z)\ln\frac{z}{2} - \sum_{k=0}^{n-1} \frac{(n-k-1)!}{k!}\left(\frac{z}{2}\right)^{2k-n} \right. \\ \left. - \sum_{k=0}^{\infty} \frac{(-1)^k (z/2)^{n+2k}}{k!(n+k)!}[\psi(n+k+1) + \psi(k+1)] \right\}.\end{aligned} \tag{14}$$

When $n = 0$ the first sum is to be taken as zero. The values of $\psi(x)$ for integral x can be found from formula (16), Appendix A.

It follows from (11) and (14) that $H_\nu^{(1,2)}(z)$ have algebraic singular points of type $z^{\pm\nu}$ at $z = 0$ when Re $\nu \neq 0$, and logarithmic singular points when $\nu = 0$.

§ 16 Sommerfeld's integral representations

1. Sommerfeld's integral representation for Bessel functions. The Poisson integral representations of $J_\nu(z)$ and $H_\nu^{(1,2)}(z)$ are useful in discussing properties of the solutions of Bessel's equation. There is a different integral representation which is useful, for example, in diffraction problems. It is obtained in the following way. As we showed in §14, the function

$$u_n(z) = \frac{1}{2\pi}\int_{-\pi}^{\pi} v(r,\phi)e^{-in\phi}d\phi,$$

with $z = \sqrt{\lambda}r$, is a Bessel function of order n if v satisfies the equation $\Delta v + \lambda v = 0$. The simplest solution of $\Delta v + \lambda v = 0$ when $\lambda = k^2 > 0$ is a plane wave $v = e^{i\mathbf{k}\cdot\mathbf{r}}$, where $\mathbf{k}$ is the wave vector. If the y axis is taken in the direction of $\mathbf{k}$, then

$$v(r,\phi) = e^{ikr\sin\phi}.$$

Hence we obtain the following integral representation for the Bessel function $u_n(z)$:

$$u_n(z) = \frac{1}{2\pi}\int_{-\pi}^{\pi} e^{iz\sin\phi - in\phi}d\phi. \tag{1}$$

There is a similar representation for Bessel functions of arbitrary order ν. To obtain it, it is natural to look for a solution of Bessel's equation for arbitrary ν in the form

$$u_\nu(z) = \int_C e^{iz\sin\phi - i\nu\phi}d\phi.$$

We shall show that $u_\nu(z)$ is a solution of Bessel's equation if the contour C is properly chosen. We start, as in the derivation of (1), from the fact that $v(r,\phi) = e^{ikr\sin\phi}$ is a solution of the Helmholtz equation

$$\frac{1}{r}\frac{\partial v}{\partial r}\left(r\frac{\partial v}{\partial r}\right) + \frac{1}{r^2}\frac{\partial^2 v}{\partial \phi^2} + k^2 v = 0. \tag{2}$$

It is easily verified that (2) remains valid for arbitrary complex values of r and ϕ.

We can obtain an equation for $v_\nu(r) = \int_C v(r,\phi)e^{-i\nu\phi}d\phi$ by starting from (2), integrating over C with weight $e^{-i\nu\phi}$, and simplifying the term in $\partial^2 v/\partial\phi^2$ by integrating by parts twice. If we require that the integrated terms, namely

$$e^{-i\nu\phi}\left(\frac{\partial v}{\partial \phi} + i\nu v\right)\Bigg|_{\phi_2}^{\phi_1} = e^{ikr\sin\phi - i\nu\phi}(kr\cos\phi + \nu)\Bigg|_{\phi_1}^{\phi_2}$$

(where ϕ_1 and ϕ_2 are the endpoints of C), are zero, we obtain Bessel's equation for $u_\nu(z) = v_\nu(r)$ with $z = kr$.

We have therefore shown that the function

$$u_\nu(z) = \int_C e^{iz\sin\phi - i\nu\phi}d\phi \tag{3}$$

is actually a solution of Bessel's equation provided that

$$e^{iz\sin\phi - i\nu\phi}(z\cos\phi + \nu)\Big|_{\phi_1}^{\phi_2} = 0. \tag{4}$$

Since $\cos\phi = (e^{i\phi} + e^{-i\phi})/2$, condition (4) will evidently be satisfied if

$$e^{iz\sin\phi - i\nu\phi}\Big|_{\phi=\phi_1,\phi_2} = 0. \tag{5}$$

for every ν.

A representation of the form (3) is known as a *Sommerfeld representation*.

2. Sommerfeld's integral representations for Hankel functions and Bessel functions of the first kind. The contour C in the integral representation for $u_\nu(z)$ can, for example, be chosen as a contour that extends to infinity in such a way that

$$\mathrm{Re}(iz\sin\phi - i\nu\phi) = \mathrm{Re}\left[\frac{1}{2}|z|e^{i\theta}(e^{i\phi} - e^{-i\phi}) - i\nu\phi\right] \to -\infty \tag{6}$$

as $\phi \to \infty$, where $\theta = \arg z$.

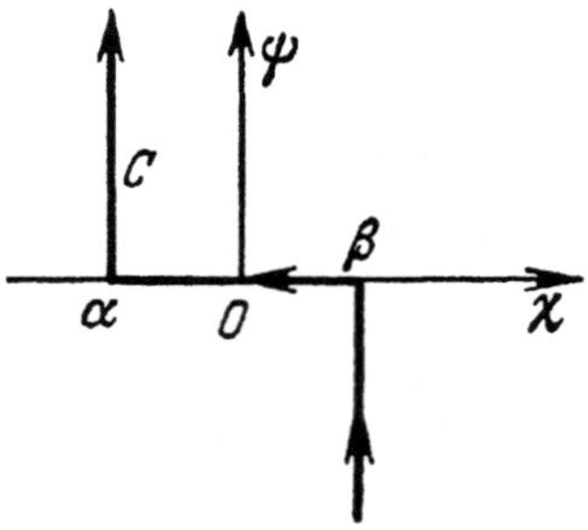

Figure 7.

Let us consider the contour C indicated in Figure 7, where $\phi = \chi + i\psi$. We need to see what conditions on α and β will make the contour have the required properties.

Let $\chi = \alpha$, $\psi \to +\infty$. In this case the terms in ϕ and $e^{i\phi}$ in (6) can be neglected in comparison with $e^{-i\phi}$. The condition on the contour becomes

$$\operatorname{Re} e^{i(\theta-\phi)} \to +\infty, \qquad \psi \to +\infty.$$

It is satisfied if $\cos(\theta - \alpha) > 0$. We may therefore suppose that

$$\theta - \pi/2 < \alpha < \theta + \pi/2. \tag{7}$$

Now let $\chi = \beta$, $\psi \to -\infty$. We find similarly that it is enough to require that $\cos(\theta + \beta) < 0$. This is satisfied if $\beta = -\alpha \pm \pi$. We denote the corresponding contours by C_+ and C_-.

There is a certain amount of freedom in the choice of the contours. Let C' be determined by numbers α' and β' that satisfy

$$\cos(\theta - \alpha') > 0, \quad \cos(\theta + \beta') < 0.$$

We can easily show by Cauchy's theorem that C' can be replaced by any other contour C'' determined by numbers α'' and β'' provided that the inequalities $\cos(\theta-\alpha) > 0$, $\cos(\theta+\beta) < 0$ are satisfied for all $\alpha \in [\alpha', \alpha'']$ and $\beta \in [\beta', \beta'']$. Consequently it is clear, in particular, that the contour C in the Sommerfeld representation can be replaced by a contour that has been shifted by an amount less than π, without affecting the value of the Sommerfeld integral.

Since $u_\nu(z)$ satisfies Bessel's equation, it can be represented in the form

$$u_\nu(z) = C_\nu H_\nu^{(1)}(z) + D_\nu H_\nu^{(2)}(z). \tag{8}$$

We can find C_ν and D_ν by using the asymptotic behavior of $H_\nu^{(1,2)}(z)$. Let us first consider the case when C is taken to be C_+. Let $|z| \to \infty$ and

$\arg z = \pi/2$. Then we can take $\alpha = \beta = \pi/2$; that is, take $\phi = \pi/2 + i\psi$, where $-\infty < \psi < \infty$, in the formula for $u_\nu(z)$. This yields

$$u_\nu(z) = ie^{-i\pi\nu/2} \int_{-\infty}^{\infty} e^{-|z|\cosh\psi} e^{\nu\psi} d\psi$$

$$= 2ie^{-i\pi\nu/2} \int_{0}^{\infty} e^{-|z|\cosh\psi} \cosh\nu\psi d\psi.$$

To find the asymptotic behavior of $u_\nu(z)$ as $z \to \infty$ we can use Watson's lemma (see Appendix B), after first making the change of variable $\cosh\psi = 1 + t$. In fact, after this substitution we obtain

$$u_\nu(z) = 2i\exp(-i\pi\nu/2 - |z|) \int_{0}^{\infty} e^{-z|t|} f(t) dt,$$

where

$$f(t) = \frac{1}{\sqrt{t(2+t)}} \cosh[\nu\ln(1 + t + \sqrt{t(2+t)})].$$

Since $f(t) = (2t)^{-1/2}[1 + O(t)]$ as $t \to 0$, Watson's lemma yields

$$u_\nu(z) = 2i\exp(-i\pi\nu/2 - |z|) \frac{\Gamma\left(\frac{1}{2}\right)}{(2|z|)^{1/2}} \left[1 + O\left(\frac{1}{z}\right)\right]$$

$$= i(2\pi/|z|)^{1/2} \exp(-i\pi\nu/2 - |z|) \left[1 + O\left(\frac{1}{z}\right)\right]$$

as $z \to \infty$. Comparing the leading terms on the two sides of (8), we find $D_\nu = 0$, $C_\nu = -\pi$. Therefore

$$H_\nu^{(1)}(z) = -\frac{1}{\pi} \int_{C_+} e^{iz\sin\phi - i\nu\phi} d\phi. \tag{9}$$

In a similar way we obtain

$$H_\nu^{(2)}(z) = \frac{1}{\pi} \int_{C_-} e^{iz\sin\phi - i\nu\phi} d\phi \tag{10}$$

by using the contour C_-. Hence

$$J_\nu(z) = \frac{1}{2}[H_\nu^{(1)}(z) + H_\nu^{(2)}(z)] = \frac{1}{2\pi} \int_{C_1} e^{iz\sin\phi - i\nu\phi} d\phi, \tag{11}$$

where C_1 is indicated in Figure 8. When $\nu = n$, the integral over C_1 reduces to an integral over $(-\alpha - \pi, -\alpha + \pi)$ since the integrand is periodic.

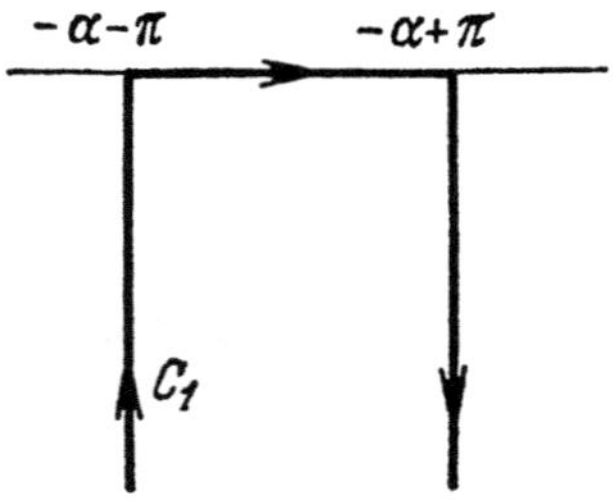

Figure 8.

Since the integral of a periodic function over an interval of length equal to a period is independent of the location of the interval, we have

$$J_n(z) = \frac{1}{2\pi} \int_{-\pi}^{\pi} e^{iz\sin\phi - in\phi} d\phi, \tag{11a}$$

i.e. the functions $J_n(z)$ are the Fourier coefficients of $e^{iz\sin\phi}$. Therefore

$$e^{iz\sin\phi} = \sum_{n=-\infty}^{\infty} J_n(z) e^{in\phi}. \tag{12}$$

By the principle of analytic continuation, (12) is valid for all complex ϕ.

We can simplify (11a) by using the formula

$$e^{iz\sin\phi - in\phi} = \cos(z\sin\phi - n\phi) + i\sin(z\sin\phi - n\phi)$$

and the evenness or oddness (in ϕ) of $\cos(z\sin\phi - n\phi)$ and $\sin(z\sin\phi - n\phi)$. Hence we obtain *Bessel's integral*

$$J_n(z) = \frac{1}{\pi} \int_0^{\pi} \cos(z\sin\phi - n\phi) d\phi.$$

§ 17 Special classes of Bessel functions

1. Bessel functions of the second kind. In practice we often deal with solutions of Bessel's equation for real ν and positive z. It is not always convenient to use the Hankel functions since they take complex values. However, $H_\nu^{(2)}(z) = \overline{H_\nu^{(1)}(z)}$ in the present case (the bar denotes the complex conjugate), and

$$J_\nu(z) = \frac{1}{2}[H_\nu^{(1)}(z) + H_\nu^{(2)}(z)] = \text{Re}\, H_\nu^{(1)}(z).$$

This suggests taking the second linearly independent solution of Bessel's equation to be Im $H_\nu^{(1)}(z)$, i.e.

$$Y_\nu(z) = \frac{1}{2i}[H_\nu^{(1)}(z) - H_\nu^{(2)}(z)]. \tag{1}$$

The functions $Y_\nu(z)$ are known as *Bessel functions of the second kind.**

We can consider $Y_\nu(z)$ as defined by (1) for arbitrary complex values of ν and z. It is analytic in ν except for $\nu = n (n = 0, \pm 1, \pm 2, \ldots)$ and analytic in z for $z \neq 0, |\arg z| < \pi$.

We list the basic properties of $Y_\nu(z)$, which follow from the corresponding properties of the Hankel functions.

a) *$Y_\nu(z)$ expressed in terms of $J_\nu(z)$ and $J_{-\nu}(z)$:*

$$Y_\nu(z) = \frac{\cos \pi\nu J_\nu(z) - J_{-\nu}(z)}{\sin \pi\nu} \qquad (\nu \neq n).$$

b) *Series expansion of $Y_\nu(z)$ for $\nu = n$:*

$$Y_n(z) = \frac{1}{\pi}\left\{2J_n(z)\ln\frac{z}{2} - \sum_{k=0}^{n-1}\frac{(n-k-1)!}{k!}\left(\frac{z}{2}\right)^{2k-n}\right.$$
$$\left. - \sum_{k=0}^{\infty}\frac{(-1)^k(z/2)^{n+2k}}{k!(n+k)!}[\psi(n+k+1)+\psi(k+1)]\right\}.$$

c) *Asymptotic formula for $Y_\nu(z)$ as $z \to \infty$:*

$$Y_\nu(z) = \left(\frac{2}{\pi z}\right)^{1/2}\left[\sin\left(z - \frac{\pi\nu}{2} - \frac{\pi}{4}\right) + O\left(\frac{e^{|\text{Im}\, z|}}{z}\right)\right].$$

---

* They are also called *Weber functions* or *Neumann functions*, and denoted by $N_\nu(z)$. We note that the Hankel functions are also called *Bessel functions of the third kind.*

d) *Recursion relation and differentiation formula:*

$$Y_{\nu-1}(z)+Y_{\nu+1}(z)=(2\nu/z)Y_\nu(z),$$
$$Y_{\nu-1}(z)-Y_{\nu+1}(z)=2Y_\nu'(z).$$

Graphs of $J_\nu(x)$ and $Y_\nu(x)$ for some integral values of ν and $x>0$ are given in Figures 9 and 10.

2. Bessel functions whose order is half an odd integer. Bessel polynomials. The Bessel functions of order half an odd integer* form a distinct class. They are remarkable for being expressible in terms of elementary functions. To establish this, we first use (19a) and (20a) of §14 to show that

$$H_{1/2}^{(1,2)}(z)=\left(\frac{2}{\pi z}\right)^{1/2}e^{\pm i(z-\pi/2)},$$

whence

$$J_{1/2}(z)=\left(\frac{2}{\pi z}\right)^{1/2}\sin z,\quad Y_{1/2}(z)=-\left(\frac{2}{\pi z}\right)^{1/2}\cos z.$$

Moreover, according to (15.8) and (15.9),

$$H_{-1/2}^{(1)}(z)=e^{i\pi/2}H_{1/2}^{(1)}=\left(\frac{2}{\pi z}\right)^{1/2}e^{iz},$$
$$H_{-1/2}^{(2)}(z)=e^{-i\pi/2}H_{1/2}^{(2)}(z)=\left(\frac{2}{\pi z}\right)^{1/2}e^{-iz}.$$

Hence

$$J_{-1/2}(z)=\left(\frac{2}{\pi z}\right)^{1/2}\cos z,\quad Y_{-1/2}(z)=\left(\frac{2}{\pi z}\right)^{1/2}\sin z.$$

Taking $\nu=-1/2$ in formulas (15.4), we obtain

$$H_{n-1/2}^{(1,2)}(z)=\left(\frac{2}{\pi z}\right)^{1/2}z^n\left(-\frac{1}{z}\frac{d}{dz}\right)^n e^{\pm iz}, \tag{2}$$

* These functions include, for example, the solutions of the Helmholtz equation that are obtained by separating variables in spherical coordinates.

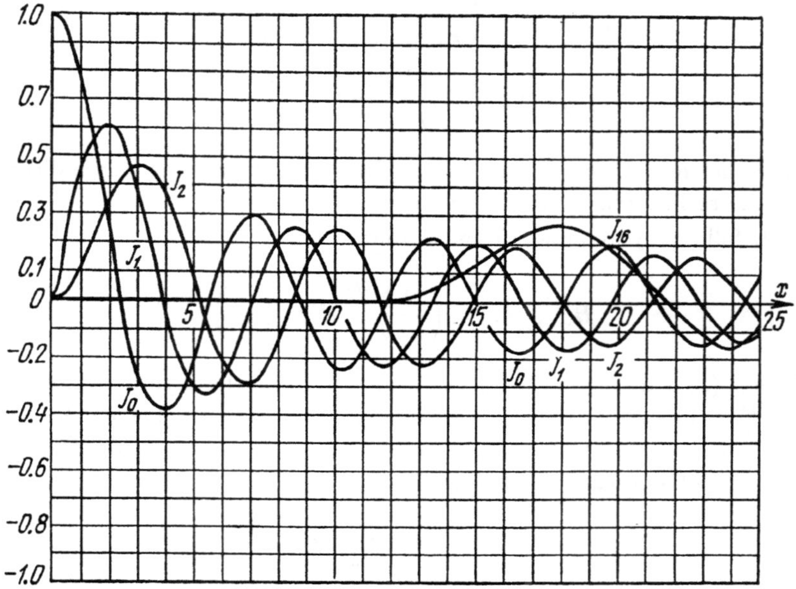

Figure 9.

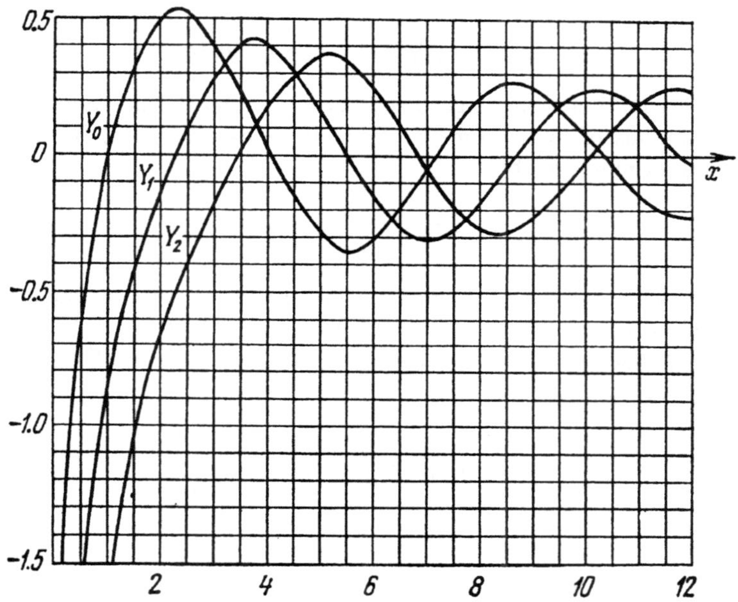

Figure 10.

$$J_{n-1/2}(z) = \left(\frac{2}{\pi z}\right)^{1/2} z^n \left(-\frac{1}{z}\frac{d}{dz}\right)^n \cos z, \tag{3}$$

$$Y_{n-1/2}(z) = \left(\frac{2}{\pi z}\right)^{1/2} z^n \left(-\frac{1}{z}\frac{d}{dz}\right)^n \sin z. \tag{4}$$

It was shown by Liouville that halves of odd integers are the only indices for which Bessel functions are elementary.

It follows by induction from (2) that

$$H^{(1)}_{n+1/2}(z) = \left(\frac{2}{\pi z}\right)^{1/2} e^{iz} p_n\left(\frac{1}{iz}\right),$$

where $p_n(s)$ is a polynomial in s of degree n. From the asymptotic behavior of $H^{(1)}_{n+1/2}(z)$ as $z \to \infty$ it follows that $p_n(0) = (-i)^{n+1}$. Let us show that $p_n(s)$ is a polynomial of hypergeometric type and can be expressed in terms of Bessel polynomials (see §5, Part 1):

$$y_n(z) = 2^{-n} e^{2/z} \frac{d^n}{dz^n}(z^{2n} e^{-2/z}).$$

In fact, from the differential equation for the Hankel functions $H^{(1)}_{n+1/2}(z)$ we can obtain a differential equation for $p_n(s)$:

$$s^2 p_n''(s) + 2(s+1)p_n'(s) - n(n+1)p_n(s) = 0.$$

This is an equation of hypergeometric type, and so the polynomials $p_n(s)$ are polynomials of hypergeometric type. If we express $p_n(s)$ by using the Rodrigues formula, we obtain

$$p_n(s) = B_n e^{2/s} \frac{d^n}{ds^n}(s^{2n} e^{-2/s}).$$

It is then clear that the polynomials $p_n(s)$ are, up to a normalizing factor, the Bessel polynomials $y_n(s)$. Since $p_n(0) = (-i)^{n+1}$, $y_n(0) = 1$, we finally obtain the following formula connecting the Hankel functions $H^{(1)}_{n+1/2}(z)$ with the Bessel polynomials:

$$H^{(1)}_{n+1/2}(z) = (-i)^{n+1} \left(\frac{2}{\pi z}\right)^{1/2} e^{iz} y_n\left(\frac{1}{iz}\right).$$

Similarly,

$$H^{(2)}_{n+1/2}(z) = i^{n+1} \left(\frac{2}{\pi z}\right)^{1/2} e^{-iz} y_n\left(-\frac{1}{iz}\right).$$

3. Modified Bessel functions. We have discussed the Bessel equation

$$z^2u'' + zu' + (z^2 - \nu^2)u = 0$$

for complex z. In the most important applications, z is positive. However, in many problems we are also interested in solutions of the equation

$$z^2u'' + zu' - (z^2 + \nu^2)u = 0 \tag{5}$$

for $z > 0$. This equation is the Bessel equation with z replaced by iz, and hence its solutions are known as *Bessel functions with imaginary argument*, or *modified Bessel functions.* Evidently $J_\nu(iz)$ and $H_\nu^{(1)}(iz)$ are linearly independent solutions of (5). The first solution is bounded as $z \to 0$ if $\nu > 0$, and the second, as $z \to \infty$.

It is customary to use

$$I_\nu(z) = e^{-i\pi\nu/2}J_\nu(iz), \tag{6}$$

and

$$K_\nu(z) = \frac{1}{2}\pi e^{i\pi(\nu+1)/2}H_\nu^{(1)}(iz) \tag{7}$$

instead of $J_\nu(iz)$ and $H_\nu^{(1)}(iz)$. These functions are real when $z > 0$ and ν is real, as follows from the formulas (see (14.7) and (14.19a))

$$I_\nu(z) = \sum_{k=0}^{\infty} \frac{(z/2)^{\nu+2k}}{k!\Gamma(k+\nu+1)},$$

$$K_\nu(z) = \frac{\pi}{2}\frac{I_{-\nu}(z) - I_\nu(z)}{\sin \pi\nu}.$$

These are corollaries of the power series expansion of $J_\nu(iz)$ and the equation (15.11) that connects $H_\nu^{(1)}(iz)$ with $J_\nu(iz)$ and $J_{-\nu}(iz)$. The function $K_\nu(z)$ is known as *Macdonald's function.*

We list the basic properties of $I_\nu(z)$ and $K_\nu(z)$; these follow from their relationship to $J_\nu(iz)$ and $H_\nu^{(1)}(iz)$.

1. *Poisson integral representations.* It follows from (18) and (19a), §14, that

$$I_\nu(z) = \frac{(z/2)^\nu}{\pi^{1/2}\Gamma(\nu+\frac{1}{2})}\int_{-1}^{1}(1-t^2)^{\nu-1/2}\cosh zt\,dt$$

$$K_\nu(z) = \left(\frac{\pi}{2z}\right)^{1/2} e^{-z}\frac{\int_0^\infty e^{-t}t^{\nu-1/2}\left(1+t/(2z)\right)^{\nu-1/2}dt}{\Gamma(\nu+1/2)}.$$

2) *Series expansions:*

$$I_\nu(z) = \sum_{k=0}^{\infty} \frac{(z/2)^{\nu+2k}}{k!\Gamma(k+\nu+1)},$$

$$K_\nu(z) = \frac{\pi}{2}\frac{I_{-\nu}(z) - I_\nu(z)}{\sin \pi\nu} \quad (\nu \neq n),$$

$$K_n(z) = (-1)^{n+1} I_n(z)\ln(z/2) + \frac{1}{2}\sum_{k=0}^{n-1} \frac{(-1)^k(n-k-1)!}{k!}\left(\frac{z}{2}\right)^{2k-n}$$

$$+ \frac{1}{2}(-1)^n \sum_{k=0}^{\infty} \frac{(z/2)^{2k+n}}{k!(k+n)!}[\psi(n+k+1) + \psi(k+1)] \tag{8}$$

(when $n = 0$ the first sum is to be taken to be zero).

It is evident from the expansion of $I_\nu(z)$ that when $z > 0$ and $\nu \geq 0$ the function $I_\nu(z)$ is positive and monotone increasing as z increases (see Figure 11).

3) *Connections between $K_\nu(z)$ and $K_{-\nu}(z)$, $I_n(z)$ and $I_{-n}(z)$:*

$$I_{-n}(z) = I_n(z),$$
$$K_{-\nu}(z) = K_\nu(z). \tag{9}$$

4) *Asymptotic behavior as $z \to +\infty$:*

$$I_\nu(z) = \frac{e^z}{\sqrt{2\pi z}}[1 + O(1/z)],$$

$$K_\nu(z) = \sqrt{\pi/(2z)}e^{-z}[1 + O(1/z)].$$

5) *Recursion relations and differentiation formulas:*

$$I_{\nu-1}(z) - I_{\nu+1}(z) = \frac{2\nu}{z} I_\nu(z),$$

$$I_{\nu-1}(z) + I_{\nu+1}(z) = 2I'_\nu(z),$$

$$K_{\nu-1}(z) - K_{\nu+1}(z) = -\frac{2\nu}{z} K_\nu(z),$$

$$K_{\nu-1}(z) + K_{\nu+1}(z) = -2K'_\nu(z),$$

in particular

$$I'_0(z) = I_1(z), \qquad K'_0(z) = -K_1(z).$$

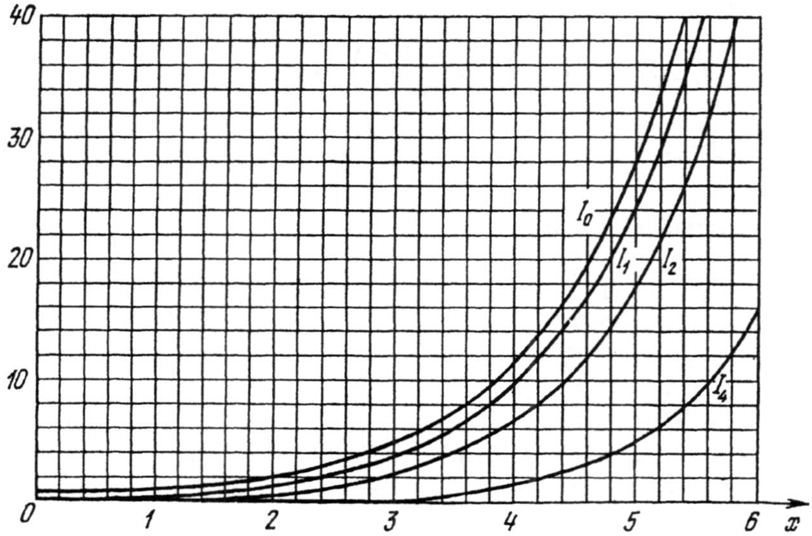

Figure 11.

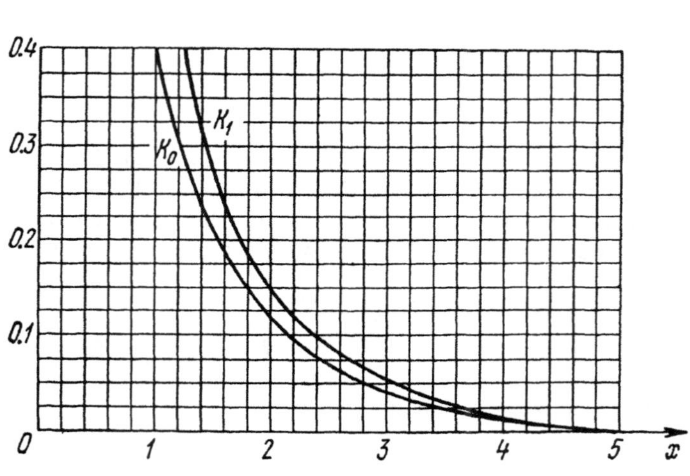

Figure 12.

6) *$I_\nu(z)$ and $K_\nu(z)$ expressed as elementary functions when ν is half an odd integer*:

$$I_{n-1/2}(z) = \left(\frac{2}{\pi z}\right)^{1/2} z^n \left(\frac{1}{z}\frac{d}{dz}\right)^n \cosh z \quad (n = 0, 1, \ldots),$$

$$K_{n-1/2}(z) = \left(\frac{\pi}{2z}\right)^{1/2} z^n \left(-\frac{1}{z}\frac{d}{dz}\right)^n e^{-z} \quad (n = 0, 1, \ldots).$$

7) *Sommerfeld integral for $K_\nu(z)$ for $z > 0$*:

$$K_\nu(z) = \frac{1}{2}\int_{-\infty}^{\infty} e^{-z\cosh\psi + \nu\psi} d\psi = \int_0^\infty e^{-z\cosh\psi} \cosh \nu\psi d\psi. \tag{10}$$

To derive (10) we took $\alpha = \pi/2$, $\phi = \pi/2 + i\psi$, $-\infty < \psi < \infty$, in (16.9). It is evident from (10) that when $z > 0$ and ν is real, $K_\nu(z)$ is positive and monotone decreasing as z increases (see Figure 12, p. 225).

If we make the substitution $\frac{1}{2}ze^{-\psi} = t$ in (10) for $z > 0$, we obtain the Sommerfeld integral:

$$K_\nu(z) = \frac{1}{2}\left(\frac{z}{2}\right)^\nu \int_0^\infty \exp\left(-z - \frac{z^2}{4t}\right) t^{-\nu-1} dt. \tag{11}$$

Remark. It follows from the properties of $I_\nu(z)$ and $K_\nu(z)$ that when $\nu \geq 0$ and $z \geq 0$, the general solution of (5) has the form

$$u(z) = AI_\nu(z) + BK_\nu(z),$$

where $B = 0$ if $u(z)$ is bounded at $z = 0$; if $u(z)$ is bounded as $z \to +\infty$ then $A = 0$.

We have considered several useful special classes of Bessel functions. There are other classes of functions that are related to Bessel functions and are convenient in special problems. They include the real and imaginary parts of $u_\nu(z)$ for $\operatorname{Im} \nu = 0$, $\arg z = \pm\pi/4$, $\pm 3\pi/4$; and the Airy function

$$Ai(z) = \begin{cases} (|z|/3\pi)^{1/2} K_{1/3}\left(\frac{2}{3}|z|^{3/2}\right) & \text{for } z < 0, \\ \frac{1}{3}(\pi z)^{1/2}\left[I_{-1/3}\left(\frac{2}{3}z^{3/2}\right) + I_{1/3}\left(\frac{2}{3}z^{3/2}\right)\right] & \text{for } z > 0. \end{cases}$$

The Airy function is a solution of

$$u'' + zu = 0$$

(see (14.4)).

§ 18 Addition theorems

The addition formulas for Bessel functions have the form

$$u_\nu(R) = F(r,\rho,\theta)\sum_{n=0}^{\infty} f_n(r)g_n(\rho)h_n(\theta), \tag{1}$$

where r, ρ, R are the lengths of the sides of a triangle, θ is the angle between r and ρ (Figure 13), and $F(r,\rho,\theta)$ is an elementary function of simple form.

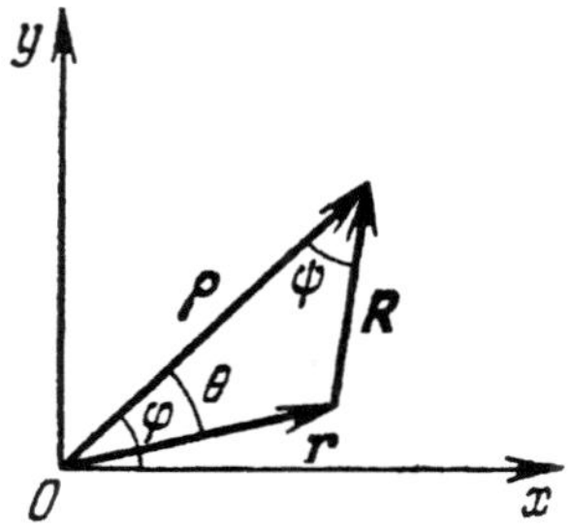

Figure 13.

These formulas provide series expansions of Bessel functions $u_\nu(R)$ of order ν, with terms that are products of the function $F(r,\rho,\theta)$, which is independent of the summation index, by factors each of which depends on only one of r, ρ and θ. Formulas of this kind are important in mathematical physics and other applications of Bessel functions.

1. Graf's addition theorem. Let $u_\nu(z)$ be one of $J_\nu(z)$, $H_\nu^{(1)}(z)$, $H_\nu^{(2)}(z)$. For the simplest addition theorem we use Sommerfeld's integral for $u_\nu(R)$:

$$u_\nu(R) = A\int_C \exp\{iR\sin\phi - i\nu\phi\}d\phi \tag{2}$$

(A is a normalizing constant, which is independent of ν in the present case.)

Consider the triangle in Figure 13. Projecting the vector equation $\mathbf{R} = \boldsymbol{\rho} - \mathbf{r}$ on the y axis, we have

$$R\sin(\phi+\psi) = \rho\sin\phi - r\sin(\phi-\theta).$$

It is clear, by the principle of analytic continuation, that this equation remains valid for complex ϕ.

The contour C can, as we showed in §16, be chosen so that a shift by $\psi < \pi$ does not affect the value of the integral. If we replace ϕ in (2) by $\phi+\psi$, we obtain

$$u_\nu(R)e^{i\nu\psi} = A\int_C \exp\{iR\sin(\phi+\psi) - i\nu\phi\}d\phi$$
$$= A\int_C \exp\{i\rho\sin\phi + ir\sin(\theta-\phi) - i\nu\phi\}d\phi.$$

Since

$$e^{ir\sin(\theta-\phi)} = \sum_{n=-\infty}^{\infty} J_n(r)e^{in(\theta-\phi)}$$

by (16.12), we have

$$u_\nu(R)e^{i\nu\psi} = \sum_{n=-\infty}^{\infty} e^{in\theta}J_n(r)A\int_C e^{i\rho\sin\phi - i(\nu+n)\phi}d\phi$$
$$= \sum_{n=-\infty}^{\infty} J_n(r)u_{\nu+n}(\rho)e^{in\theta}.$$

We can interchange summation and integration when $r < \rho$. Hence we obtain

$$u_\nu(R)e^{i\nu\psi} = \sum_{n=-\infty}^{\infty} J_n(r)u_{\nu+n}(\rho)e^{in\theta}.$$

Since the substitutions $R \to kR$, $r \to kr$, $\rho \to k\rho$ do not change θ and ψ, we can write the formula as

$$u_\nu(kR)e^{i\nu\psi} = \sum_{n=-\infty}^{\infty} J_n(kr)u_{\nu+n}(k\rho)e^{in\theta}. \tag{3}$$

This is *Graf's addition theorem.*

2. Gegenbauer's addition theorem. There is a different addition formula when $F(r,\rho,\theta) = R^\nu$. To obtain it, we consider the function

$$R^{-\nu}u_\nu(R) = v(R).$$

Suppose for definiteness that $r < \rho$. In this case $R \neq 0$ and $v(R)$ is bounded as $r \to 0$.

The function $v(R)$ satisfies the equation

$$Rv'' + (2\nu + 1)v' + Rv = 0 \tag{4}$$

(see Lommel's equation, (14.4)). It is easy to deduce a partial differential equation in r and $\mu = \cos\theta$ for fixed ρ. Since $R = (r^2 + \rho^2 - 2r\rho\mu)^{1/2}$, we have

$$\frac{\partial v}{\partial r} = \frac{dv}{dR}\frac{r - \rho\mu}{R}, \quad \frac{\partial v}{\partial \mu} = -\frac{dv}{dR}\frac{r\rho}{R},$$

$$\frac{\partial^2 v}{\partial r^2} = \frac{d^2 v}{dR^2}\left(\frac{r - \rho\mu}{R}\right)^2 + \frac{dv}{dR}\left[\frac{1}{R} - \frac{(r - \rho\mu)^2}{R^3}\right],$$

$$\frac{\partial^2 v}{\partial \mu^2} = \frac{d^2 v}{dR^2}\left(\frac{r\rho}{R}\right)^2 - \frac{dv}{dR}\frac{(r\rho)^2}{R^3}.$$

If we eliminate ρ, we obtain

$$\frac{1}{R}\frac{dv}{dR} = \frac{1}{r}\frac{\partial v}{\partial r} - \frac{\mu}{r^2}\frac{\partial v}{\partial \mu}.$$

Since $R^2 = (r - \rho\mu)^2 + \rho^2(1 - \mu^2)$, we have

$$\frac{d^2 v}{dR^2} = \frac{\partial^2 v}{\partial r^2} + \frac{1 - \mu^2}{r^2}\frac{\partial^2 v}{\partial \mu^2}.$$

Substituting the formulas for dv/dR and d^2v/dR^2 into (4), we obtain the partial differential equation

$$r^2\frac{\partial^2 v}{\partial r^2} + (2\nu + 1)r\frac{\partial v}{\partial r} + r^2 v + (1 - \mu^2) - \frac{\partial^2 v}{\partial \mu^2}(2\nu + 1)\mu\frac{\partial v}{\partial \mu} = 0. \tag{5}$$

In the present case the addition formula (1) has the form

$$v(R) = \sum_{n=0}^{\infty} f_n(r)g_n(\rho)h_n(\mu) \qquad (\mu = \cos\theta). \tag{6}$$

We can determine the forms of $f_n(r)$, $g_n(\rho)$ and $h_n(\mu)$ by the requirement that each term of (6) satisfies (5). For this purpose we look for particular bounded solutions of (5) by the method of separation of variables, taking

$$v = f(r)h(\mu). \tag{7}$$

Substituting (7) into (5), we obtain

$$\frac{r^2 f'' + (2\nu + 1)rf' + r^2 f}{f} = \frac{-(1 - \mu^2)h'' + (2\nu + 1)\mu h'}{h} = \lambda, \tag{5a}$$

where λ is a constant. Hence we obtain an equation of hypergeometric type for $h(\mu)$,

$$(1-\mu^2)h'' - (2\nu+1)\mu h' + \lambda h = 0.$$

Its solutions for $\lambda = n(n+2\nu)$ are the Jacobi polynomials $P_n^{(\nu-1/2,\nu-1/2)}(\mu)$. Hence it is reasonable to take

$$h_n(\mu) = P_n^{(\nu-1/2,\nu-1/2)}(\mu)$$

in (6). Then (6) becomes the expansion of $v(R)$ in a series of Jacobi polynomials:

$$v(R) = \sum_{n=0}^{\infty} a_n(r,\rho) P_n^{(\nu-1/2,\nu-1/2)}(\mu). \tag{8}$$

Since $v(R)$ satisfies the hypotheses of the theorem on expansions in series of the Jacobi polynomials $P_n^{(\nu-1/2,\nu-1/2)}(\mu)$ for $\nu > -1/2$ (see §8), we have

$$\begin{aligned} a_n(r,\rho) &= \frac{1}{d_n^2}\int_{-1}^{1} v(R)(1-\mu^2)^{\nu-1/2} P_n^{(\nu-1/2,\nu-1/2)}(\mu)\,d\mu \\ &= \frac{1}{d_n^2}\int_{-1}^{1} \frac{u_\nu(R)}{R^\nu}(1-\mu^2)^{\nu-1/2} P_n^{(\nu-1/2,\nu-1/2)}(\mu)\,d\mu, \end{aligned}$$

where d_n^2 is the squared norm of the Jacobi polynomial.

It remains to show that the coefficients $a_n(r,\rho)$ can be represented in the form

$$a_n(r,\rho) = f_n(r) g_n(\rho).$$

To establish this, we integrate (5) over $(-1,1)$ after first multiplying by $(1-\mu^2)^{\nu-1/2} P_n^{(\nu-1/2,\nu-1/2)}(\mu)$, and eliminate the terms in $\partial^2 v/\partial\mu^2$ and $\partial v/\partial\mu$ by integration by parts. Since

$$\left[(1-\mu^2)\frac{\partial^2 v}{\partial\mu^2} - (2\nu+1)\mu\frac{\partial v}{\partial\mu}\right](1-\mu^2)^{\nu-1/2} = \frac{\partial}{\partial\mu}\left[(1-\mu^2)^{\nu+1/2}\frac{\partial v}{\partial\mu}\right],$$

we have

$$\int_{-1}^{1}\left[(1-\mu^2)\frac{\partial^2 v}{\partial\mu^2}-(2\nu+1)\mu\frac{\partial v}{\partial\mu}\right](1-\mu^2)^{\nu-1/2}P_n^{(\nu-1/2,\nu-1/2)}(\mu)d\mu$$

$$=(1-\mu^2)^{\nu+1/2}\frac{\partial v}{\partial\mu}P_n^{(\nu-1/2,\nu-1/2)}(\mu)\Big|_{-1}^{1}$$

$$-\int_{-1}^{1}\frac{\partial v}{\partial\mu}(1-\mu^2)^{\nu+1/2}\frac{d}{d\mu}P_n^{(\nu-1/2,\nu-1/2)}(\mu)d\mu$$

$$=(1-\mu^2)^{\nu+1/2}\left[\frac{\partial v}{\partial\mu}P_n^{(\nu-1/2,\nu-1/2)}(\mu)-v\frac{d}{d\mu}P_n^{(\nu-1/2,\nu-1/2)}(\mu)\right]\Big|_{-1}^{1}$$

$$+\int_{-1}^{1}v\frac{d}{d\mu}\left[(1-\mu^2)^{\nu+1/2}\frac{d}{d\mu}P_n^{(\nu-1/2,\nu-1/2)}(\mu)\right]d\mu.$$

Since $\nu+\frac{1}{2}>0$, the integrated terms vanish. Moreover, it follows from the equation of the Jacobi polynomials that

$$\frac{d}{d\mu}\left[(1-\mu^2)^{\nu+1/2}\frac{d}{d\mu}P_n^{(\nu-1/2,\nu-1/2)}(\mu)\right]$$
$$=-n(n+2\nu)(1-\mu^2)^{\nu-1/2}P_n^{(\nu-1/2,\nu-1/2)}(\mu).$$

Hence we obtain a differential equation for $a_n(r,\rho)$:

$$\frac{\partial^2 a_n}{\partial r^2}+\frac{2\nu+1}{r}\frac{\partial a_n}{\partial r}+\left[1-\frac{n(n+2\nu)}{r^2}\right]a_n=0.$$

As we would expect, this agrees with (5a) when $\lambda=n(n+2\nu)$.

The last equation is a special case of Lommel's equation (14.4). The only solution that is bounded as $r\to 0$ is, up to a factor independent of r, the function $r^{-\nu}J_{\nu+n}(r)$, i.e.

$$a_n(r,\rho)=r^{-\nu}J_{\nu+n}(r)g_n(\rho).$$

Consequently

$$a_n(r,\rho)=r^{-\nu}J_{\nu+n}(r)g_n(\rho)$$
$$=\frac{1}{d_n^2}\int_{-1}^{1}R^{-\nu}u_\nu(R)(1-\mu^2)^{\nu-1/2}P_n^{(\nu-1/2,\nu-1/2)}(\mu)d\mu, \tag{9}$$

where d_n^2 is the squared norm of the Jacobi polynomial. To find $g_n(\rho)$ we calculate the integral on the right-hand side of (9) by using the Rodrigues formula for the Jacobi polynomials,

$$P_n^{(\nu-1/2,\nu-1/2)}(\mu) = \frac{(-1)^n}{2^n n!}\frac{1}{(1-\mu^2)^{\nu-1/2}}\frac{d^n}{d\mu^n}[(1-\mu^2)^{n+\nu-1/2}],$$

and integrating by parts n times:

$$\int_{-1}^{1} R^{-\nu}u_\nu(R)(1-\mu^2)^{\nu-1/2}P_n^{(\nu-1/2,\nu-1/2)}(\mu)d\mu$$

$$= \frac{1}{2^n n!}\int_{-1}^{1}(1-\mu^2)^{n+\nu-1/2}\frac{\partial^n}{\partial\mu^n}[R^{-\nu}u_\nu(R)]d\mu.$$

The integrated terms vanish at ± 1 since the factor $(1-\mu^2)$ enters with positive exponents.

On the other hand,

$$\frac{\partial}{\partial\mu}v(R) = -\frac{r\rho}{R}\frac{dv}{dR}$$

for every function $v(R)$, whence

$$\frac{\partial^n}{\partial\mu^n}\left[\frac{u_\nu(R)}{R^\nu}\right] = (r\rho)^n\left(-\frac{1}{R}\frac{d}{dr}\right)^n\left[\frac{u_\nu(R)}{R^\nu}\right].$$

By the differentiation formula (4) of §15 we have

$$\left(-\frac{1}{R}\frac{d}{dR}\right)^n\left[\frac{u_\nu(R)}{R^\nu}\right] = \frac{u_{\nu+n}(R)}{R^{\nu+n}}.$$

We therefore obtain

$$\int_{-1}^{1} R^{-\nu}u_\nu(R)(1-\mu^2)^{\nu-1/2}P_n^{(\nu-1/2,\nu-1/2)}(\mu)d\mu$$

$$= \frac{1}{2^n n!}(r\rho)^n\int_{-1}^{1} R^{-(\nu+n)}u_{\nu+n}(R)(1-\mu^2)^{n+\nu-1/2}d\mu.$$

Hence, by (9),

$$g_n(\rho)\frac{J_{\nu+n}(r)}{r^{\nu+n}} = \frac{\rho^n}{2^n n! d_n^2}\int_{-1}^{1}\frac{u_{\nu+n}(R)}{R^{\nu+n}}(1-\mu^2)^{n+\nu-1/2}d\mu. \tag{10}$$

Let $r \to 0$. Then $R \to \rho$ and consequently

$$\frac{g_n(\rho)}{2^{\nu+n}\Gamma(\nu+n+1)} = \frac{u_{\nu+n}(\rho)}{\rho^\nu}\frac{1}{2^n n! d_n^2}\int_{-1}^{1}(1-\mu^2)^{n+\nu-1/2}d\mu.$$

Since (see §5,Table 1)

$$d_n^2 = \frac{2^{2\nu-1}\Gamma^2(n+\nu+1/2)}{n!(n+\nu)\Gamma(n+2\nu)},$$

and

$$\int_{-1}^{1}(1-\mu^2)^{n+\nu-1/2}d\mu = 2\int_{0}^{1}(1-\mu^2)^{n+\nu-1/2}d\mu$$

$$= \int_0^1 (1-t)^{n+\nu-1/2}t^{-1/2}dt = \frac{\Gamma(n+\nu+1/2)\Gamma(1/2)}{\Gamma(n+\nu+1)},^*$$

we finally obtain

$$g_n(\rho) = \frac{\sqrt{\pi}}{2^{\nu-1}}\frac{(n+\nu)\Gamma(n+2\nu)u_{\nu+n}(\rho)}{\Gamma(n+\nu+1/2)\rho^\nu}.$$

Then (8) takes the form

$$\frac{u_\nu(R)}{R^\nu} = \frac{\sqrt{\pi}}{2^{\nu-1}}\sum_{n=0}^{\infty}\frac{(n+\nu)\Gamma(n+2\nu)}{\Gamma(n+\nu+1/2)}\frac{J_{\nu+n}(r)}{r^\nu}\frac{u_{\nu+n}(\rho)}{\rho^\nu}P_n^{(\nu-1/2,\nu-1/2)}(\mu). \tag{11}$$

If we use the Gegenbauer polynomials

$$C_n^\nu(\mu) = \frac{(2\nu)_n}{(\nu+1/2)_n}P_n^{(\nu-1/2,\nu-1/2)}(\mu)$$

---

* The integrand is even; the substitution $t=\mu^2$ and the connection between the beta and gamma functions (Appendix A) yield the value of the integral.

instead of the Jacobi polynomials, the expansion (11) takes the simpler form

$$\frac{u_\nu(R)}{R^\nu} = 2^\nu \Gamma(\nu) \sum_{n=0}^{\infty} (\nu + n) \frac{J_{\nu+n}(r)}{r^\nu} \frac{u_{\nu+n}(\rho)}{\rho^\nu} C_n^\nu(\mu). \tag{12}$$

We recall that the formula was obtained for $\nu > -1/2$, $r < \rho$.

Evidently (12) remains valid if we make the substitutions $R \to kR$, $r \to kr$, $\rho \to k\rho$, i.e.

$$\frac{u_\nu(kR)}{(kR)^\nu} = 2^\nu \Gamma(\nu) \sum_{n=0}^{\infty} (\nu + n) \frac{J_{\nu+n}(kr)}{(kr)^\nu} \frac{u_{k+n}(k\rho)}{(k\rho)^\nu} C_n^\nu(\mu). \tag{13}$$

Formula (13) is *Gegenbauer's addition formula.*

The Graf and Gegenbauer formulas were obtained under certain restrictions on the parameters. However, (3) and (13) can be extended to a wider range of parameters by the principle of analytic continuation.

3. Expansion of spherical and plane waves in series of Legendre polynomials. Let us consider some corollaries of Gegenbauer's addition formula. These are often applied, for example in quantum-mechanical scattering theory, for solving diffraction problems.

1) In (13) put $\nu = 1/2$, $u_\nu(z) = H_\nu^{(1)}(z)$, and use the explicit expression for $H_{1/2}^{(1)}(z)$. We obtain

$$\frac{e^{ikR}}{r} = i\pi \sum_{n=0}^{\infty} (n + 1/2) \frac{J_{n+1/2}(kr)}{r^{1/2}} \frac{H_{n+1/2}^{(1)}(k\rho)}{\rho^{1/2}} P_n(\mu).$$

We have used the identity $C_n^{1/2}(\mu) = P_n(\mu)$, where $P_n(\mu)$ is the Legendre polynomial.

2) An interesting limiting form of the addition theorem is obtained from (13) with $u_\nu(z) = H_\nu^{(2)}(z)$ by letting $\rho \to \infty$. We have

$$R = \rho\left(1 - \frac{2r\mu}{\rho} + \frac{r^2}{\rho^2}\right)^{1/2} = \rho - r\mu + O(1/\rho),$$

$$\lim_{\rho\to\infty} \frac{(k\rho)^{-\nu} H_{\nu+n}^{(2)}(k\rho)}{(kR)^{-\nu} H_\nu^{(2)}(kR)} = \lim_{\rho\to\infty} (R/\rho)^{\nu+1/2} i^n e^{-ik(\rho-R)} = i^n e^{-ikr\mu}.$$

Hence we find from (13) that

$$e^{ikr\mu} = 2^\nu \Gamma(\nu) \sum_{n=0}^{\infty} i^n (\nu + n) \frac{J_{\nu+n}(kr)}{(kr)^\nu} C_n^\nu(\mu).$$

When $\nu = 1/2$, it is then easy to obtain the expansion of a plane wave $e^{i\mathbf{k}\cdot\mathbf{r}}$ in Legendre polynomials:

$$e^{i\mathbf{k}\cdot\mathbf{r}} = \left(\frac{2\pi}{kr}\right)^{1/2} \sum_{n=0}^{\infty} i^n \left(n + \frac{1}{2}\right) J_{n+1/2}(kr) P_n(\mu). \tag{14}$$

Here $\mathbf{k}$ is the wave vector, $\mu = \cos\theta$, and θ is the angle between $\mathbf{k}$ and $\mathbf{r}$.

§ 19 Semiclassical approximation (WKB method)

The transition from the classical physics of the late nineteenth century to the quantum mechanics of the early twentieth century is exemplified by the problem of finding uniformly asymptotic solutions of differential equations of the form

$$[k(x)y']' + \lambda r(x)y = 0 \tag{1}$$

as $\lambda \to +\infty$. We describe such approximate solutions as *semiclassical approximations* [L1]. The initial investigations of Wentzel, Kramers and Brillouin were subsequently extended by Langer and many others. Semiclassical approximations are useful in many problems of mathematical physics.

1. Semiclassical approximation for the solutions of equations of second order. Let us study the behavior of the solutions of an equation of the form

$$[k(x)y']' + \lambda r(x)y = 0 \tag{1}$$

as $\lambda \to +\infty$. We see from simple examples with $k(x) = \text{const.}$ or $r(x) = \text{const.}$ that the behavior of the solutions depends in an essential way on the signs of $k(x)$ and $r(x)$. Consequently we are going to discuss (1) in regions where $k(x)$ and $r(x)$ have constant signs. We first consider the case when $k(x)$ and $r(x)$ have the same sign, say $k(x) > 0$ and $r(x) > 0$, and we suppose that these functions have continuous first and second derivatives.

1°. Let us try to reduce (1) to a simpler form by the substitutions

$$y(x) = \phi(x)u(s), \qquad s = s(x) \tag{2}$$

If we substitute (2) into (1), we see that (1) becomes

$$u'' + f(s)u' + [\lambda g(s) - q(s)]u = 0, \tag{1a}$$

where

$$f(s) = \frac{2\,k(x)s'(x)\,\phi'(x) + [k(x)s'(x)]'\phi(x)}{k(x)\,\phi(x)[s'(x)]^2},$$

$$g(s) = \frac{r(x)}{k(x)[s'(x)]^2}, \qquad q(s) = -\frac{[k(x)\,\phi'(x)]'}{k(x)\,\phi(x)[s'(x)]^2}.$$

It will be convenient to choose the functions $s(x)$ and $\phi(x)$ so that $g(s) \to 1$ and $f(s) \to 0$ as $\lambda \to +\infty$, i.e.

$$[s'(x)]^2 = \frac{r(x)}{k(x)},$$
$$\frac{\phi'(x)}{\phi(x)} = -\frac{1}{2}\frac{[k(x)s'(x)]'}{k(x)s'(x)} = -\frac{1}{4}\frac{[k(x)r(x)]'}{k(x)r(x)}. \tag{3}$$

Then equation (1) assumes the standard form

$$u'' + [\lambda - q(s)]u = 0. \tag{4}$$

Here

$$q(s) = -\frac{[k(x)\phi'(x)]'}{r(x)\phi(x)}.$$

If we use (3) for $\phi(x)$, we can write

$$q(s) = \frac{k}{4r}\left[\left(\frac{k'}{k} + \frac{r'}{r}\right)' + \left(\frac{3}{4}\frac{k'}{k} - \frac{1}{4}\frac{r'}{r}\right)\left(\frac{k'}{k} + \frac{r'}{r}\right)\right]. \tag{5}$$

If $k(x)$ and $r(x)$ have different signs on (a, b), (2) takes (1) to a form similar to (4), namely

$$u''(s) - [\lambda + q(s)]u(s) = 0. \tag{6}$$

Since the behavior of the solutions of (6) as $\lambda \to +\infty$ can be studied by the same methods as for (4), we shall consider only the case when (1) has been carried into (4) by the substitutions (2).

In (3) we have

$$s(x) = \int_{x_0}^{x} [r(t)/k(t)]^{1/2} dt \quad (a < x_0 < b), \quad \phi(x) = [k(x)r(x)]^{-1/4}.$$

Let $s(a) = c$ ($c < 0$) and $s(b) = d$ ($d > 0$). Then $s(x)$ is continuous and monotone increasing on (a, b). Hence it has an inverse $x = x(s)$ which is monotone increasing and continuous on (c, d), and $q(s)$ is continuous on (c, d).

2°. It is natural to expect that as $\lambda \to +\infty$ the solutions of (4) will agree in the limit with the solutions of the simplified equation

$$u'' + \lambda u = 0,$$

i.e. that as $\lambda \to +\infty$ we will have the approximate equation

$$u(s) \approx A \cos \mu s + B \sin \mu s,$$

where $\mu = \sqrt{\lambda}$ and A and B are constants.

We can verify this conjecture by a method that was proposed by Steklov [S5]. We solve the equation

$$u'' + \mu^2 u = q(s)u \tag{3a}$$

by variation of parameters, treating the right-hand side as known. We obtain

$$u(s) = \bar{u}(s) + R_\mu(s), \tag{7}$$

where

$$\bar{u}(s) = A \cos \mu s + B \sin \mu s,$$

$$R_\mu(s) = \frac{1}{\mu} \int_0^s \sin \mu(s - s') q(s') u(s') ds'.$$

Let us show that when $c < c_1 \leq s \leq d_1 < d$ ($c_1 < 0, d_1 > 0$) we can neglect $R_\mu(s)$ in (7) as $\mu \to \infty$, i.e.

$$\lim_{\mu \to \infty} \frac{R_\mu(s)}{\bar{M}(\mu)} = 0, \tag{8}$$

where $\bar{M}(\mu) = \max_{c_1 \leq s \leq d_1} |\bar{u}(s)|$. It is clear from the formula for $R_\mu(s)$ that

$$|R_\mu(s) \leq \frac{1}{\mu} L M(\mu), \tag{9}$$

where

$$L = \int_{c_1}^{d_1} |q(s')| ds', \quad M(\mu) = \max_{c_1 \leq s \leq d_1} |u(s)|.$$

We estimate $M(\mu)$ as $\mu \to \infty$. From (7) and (9) we have

$$|u(s)| \le \bar{M}(\mu) + \mu^{-1} L M(\mu),$$

whence

$$M(\mu) \le \bar{M}(\mu) + \mu^{-1} L M(\mu).$$

If we solve this inequality for $M(\mu)$ and use (9) for $\mu > L$, we obtain

$$\frac{R_\mu(s)}{\bar{M}(\mu)} \le \frac{L}{\mu - L},$$

which establishes (8).

Returning to the original variables, we see that when $k(x) > 0$ and $r(x) > 0$ on (a, b) the solutions of (1) have the representation

$$y(x) \approx \frac{1}{\sqrt{k(x)p(x)}}[A \cos \xi(x) + B \sin \xi(x)], \tag{10}$$

as $\lambda \to +\infty$, on every interval $[a_1, b_1] \subset (a, b)$; here

$$p(x) = \left(\lambda \frac{r(x)}{k(x)}\right)^{1/2}, \quad \xi(x) = \int_{x_0}^{x} p(t)dt.$$

The semiclassical method of solving (1) consists of replacing the solution of (1) by the approximate solution (10).

When $k(x) > 0$ and $r(x) < 0$, we find similarly that

$$y(x) \approx \frac{1}{\sqrt{k(x)p(x)}}[Ae^{\xi(x)} + Be^{-\xi(x)}], \tag{10a}$$

where

$$p(x) = \left(\lambda \left|\frac{r(x)}{k(x)}\right|\right)^{1/2}, \quad \xi(x) = \int_{x_0}^{x} p(t)dt.$$

In replacing the exact solution by an approximate solution, what is significant is only the inequality $|q(s)| \ll \mu$ in (4). Consequently the approximate solutions (10) and (10a) can be used not only in cases when λ is large, but also when $\lambda \sim 1$ provided that $|q(s)| \ll 1$. As we see from (5), this is the case when $k(x)$ and $r(x)$ have small derivatives, i.e. when the coefficients of (1) vary slowly and smoothly. We note that logarithmic derivatives of the functions $k(x)$ and $r(x)$, as well as derivatives of the corresponding logarithmic derivatives, enter into the right-hand side of formula (5). Here the summand

$$\frac{k}{4r}\left(\frac{3}{4}\frac{k'}{k}-\frac{1}{4}\frac{r'}{r}\right)\left(\frac{k'}{k}+\frac{r'}{r}\right)$$

will seem to dominate. Hence the smallness condition

$$\left|\frac{k}{4r}\left(\frac{3}{4}\frac{k'}{k}-\frac{1}{4}\frac{r'}{r}\right)\left(\frac{k'}{k}+\frac{r'}{r}\right)\right| \ll 1 \tag{5a}$$

must first be fulfilled. Condition (5a) is somewhat cruder than the condition $|q(s)| \ll 1$. However, if we consider the Schrödinger equation in the form

$$\frac{d^2\psi}{dx^2}+p^2(x)\psi=0, \quad p^2(x)=2[E-U(x)],$$

then the condition (5a) for this equation will coincide with that for applicability of the semiclassical approximation $|p'(x)/p^2(x)| \ll 1$, which is widely used in quantum mechanics. For this it is sufficient to set $k(x) = 1$ and $r(x) = p^2(x)$ in (5a) provided that $\lambda \approx 1$.

3°. It is also of practical interest to find an approximate solution of (1) that is valid up to the endpoints of (a, b) as $\lambda \to +\infty$. Consider, for example, the problem of an approximate representation of (1) for $a \leq x < b$. If $k(a) > 0$ and $r(a) > 0$, all the reasoning that led to (10) remains valid. Hence we need to consider the case when at least one of the functions $k(x)$ and $r(x)$ is zero at $x = a$. Let, for example,

$$k(x) = (x-a)^m\, k_0(x), \quad r(x) = (x-a)^l r_0(x),$$

where $k_0(a) > 0$, $r_0(a) > 0$ and $k_0(x)$ and $r_0(x)$ have continuous second derivatives for $a \leq x < b$. We assume that $(l-m)/2 > -1$ so that $s(a)$ will be finite. In this case the expressions for $s(x)$ and $q(s)$ are

$$s(x) = \int_a^x \left(\frac{r_0(t)}{k_0(t)}\right)^{1/2} (t-a)^{(l-m)/2} dt, \tag{11}$$

$$\begin{aligned} q(s) = (x-a)^{m-l-2}\frac{k_0(x)}{4r_0(x)}\Bigg\{&\frac{(l+m)(3m-l-4)}{4} \\ &+\frac{x-a}{2}\left[(3m+l)\frac{k_0'}{k_0}+(m-l)\frac{r_0'}{r_0}\right] \\ &+(x-a)^2\left[\left(\frac{k_0'}{k_0}+\frac{r_0'}{r_0}\right)'+\left(\frac{3}{4}\frac{k_0'}{k_0}-\frac{1}{4}\frac{r_0'}{r_0}\right)\left(\frac{k_0'}{k_0}+\frac{r_0'}{r_0}\right)\right]\Bigg\}. \end{aligned} \tag{12}$$

If $x \approx a$, we have

$$s(x) \approx \left(\frac{r_0(a)}{k_0(a)}\right)^{1/2} \frac{(x-a)^{(l-m+2)/2}}{\frac{1}{2}(l-m+2)}$$

and consequently we can represent $q(s)$ in the form

$$q(s) = \frac{\nu^2 - \frac{1}{4}}{s^2} + s^{\gamma-2} f(s),$$

where

$$\gamma = \frac{2}{l-m+2} > 0, \qquad \nu = \frac{|m-1|}{l-m+2},$$

and $f(s)$ is continuous for $0 \le s < s(b)$. We see that in the present case $q(s)$ has a singular point at $s = 0$. Hence in order to apply Steklov's method we have to separate out the singularity of $q(s)$, i.e. we write (3) in the form

$$u'' + \left(\mu^2 - \frac{\nu^2 - \frac{1}{4}}{s^2}\right) u = s^{\gamma-2} f(s) u \quad (\mu = \sqrt{\lambda}) \tag{13}$$

and solve this equation by variation of parameters, treating the right-hand side of (13) as known. Since the equation

$$u'' + \left(\mu^2 - \frac{\nu^2 - \frac{1}{4}}{s^2}\right) u = 0$$

is a Lommel equation (14.4) with solution

$$u = A v_\nu(\mu s) + B v_{-\nu}(\mu s),$$

where $v_\nu(z) = \sqrt{z} J_\nu(z)$ and A and B are constants, we obtain a solution of (13) in the form

$$u(s) = A v_\nu(\mu s) + B v_{-\nu}(\mu s) + R_\mu(s), \tag{14}$$

where

$$R_\mu(s) = \int_{s_0}^{s} K_\mu(s, s')(s')^{\gamma-2} f(s') u(s') ds',$$

$$K_\mu(s, s') = \frac{\pi}{2\mu \sin \pi\nu} [v_\nu(\mu s) v_{-\nu}(\mu s') - v_\nu(\mu s') v_{-\nu}(\mu s)].$$

It can be shown that $R_\mu(s)$ can be neglected in (14) as $\mu \to +\infty$. In estimating $R_\mu(s)$ it is convenient to take $s_0 > 0$ when $B \neq 0$ and $s_0 = 0$ when $B = 0$. The estimates can be carried out along the same lines as before, but they are rather more complicated technically because to estimate the functions $v_{\pm\nu}(\mu s)$, which appear in place of $\cos \mu s$ and $\sin \mu s$, it is necessary to consider small and large values of μs separately:

$$|v_{\pm\nu}(\mu s)| \leq \begin{cases} C(\mu s)^{\pm\nu+1/2} & \text{for } \mu s \leq 1, \\ C & \text{for } \mu s > 1 \end{cases}$$

(where C is a constant).

Returning to the original variables, we find that, in the case when $k(x) = (x-a)^m k_0(x)$ and $r(x) = (x-a)^l r_0(x)$, with $l - m + 2 > 0$, and $k_0(x)$ and $r_0(x)$ are positive and have continuous second derivatives for $a \leq x < b$, the solutions of (1) can be approximately represented, as $\lambda \to +\infty$, $a \leq x \leq b_1 < b$, in the form

$$\begin{gathered} y(x) \approx \left(\frac{\xi(x)}{k(x)p(x)}\right)^{1/2} \{AJ_\nu[\xi(x)] + BJ_{-\nu}[\xi(x)]\}, \\ p(x) = \left(\lambda \frac{r(x)}{k(x)}\right)^{1/2}, \quad \xi(x) = \int_a^x p(t)dt; \quad \nu \neq 0, 1, \ldots \end{gathered} \tag{15}$$

When ν is an integer we have to replace $J_{-\nu}(\xi)$ in (15) by $Y_\nu(\xi)$. We may observe that when $\xi(x) \gg 1$, replacing the Bessel functions in (15) by the first terms of their asymptotic expansions leads to a formula equivalent to (10). If $k(x) > 0$ and $r(x) > 0$, (15) is replaced by

$$\begin{gathered} y(x) \approx \left(\frac{\xi(x)}{k(x)p(x)}\right)^{1/2} \{AI_\nu[\xi(x)] + BK_\nu[\xi(x)]\}, \\ p(x) = \left(\lambda \left|\frac{r(x)}{k(x)}\right|\right)^{1/2}, \quad \xi(x) = \int_a^x p(t)dt. \end{gathered} \tag{16}$$

Similar formulas can be obtained for $a < x \leq b$ if $k(x)$ and $r(x)$ have the forms

$$\begin{gathered} k(x) = (b-x)^m k_0(x), \quad r(x) = (b-x)^l r_0(x), \\ k_0(x) > 0, \quad r_0(x) > 0. \end{gathered}$$

We have described a method of obtaining an asymptotic formula for the solutions of (1) as $\lambda \to +\infty$. We now turn to applications of this formula to the problems with which we are directly concerned.

2. Asymptotic formulas for classical orthogonal polynomials for large values of n. Let us obtain an approximate formula for the Jacobi polynomials $P_n^{(\alpha,\beta)}(x)$ for large n when $\alpha \geq 0$, $\beta \geq 0$ and $x \in [-1,1]$. The function $y(x) = P_n^{(\alpha,\beta)}(x)$ satisfies the differential equation (1) with

$$k(x) = (1-x)^{\alpha+1}(1+x)^{\beta+1}, \quad r(x) = (1-x)^{\alpha}(1+x)^{\beta},$$
$$\lambda = n(n+\alpha+\beta+1).$$

In the present case $m = \beta+1$, $l = \beta$, $\nu = \beta$. If $n \to \infty$, we have $\lambda \to +\infty$. When $-1 \leq x \leq 1-\delta$, we have

$$y(x) \approx \frac{\sqrt{\xi}}{(1-x)^{\alpha/2+1/4}(1+x)^{\beta/2+1/4}}[AJ_\beta(\xi) + BJ_{-\beta}(\xi)],$$

where

$$\xi = \xi(x) = \mu \int_{-1}^{x} \frac{dt}{\sqrt{1-t^2}} = \mu \arccos(-x), \quad \mu = \sqrt{\lambda}.$$

Since the limit

$$\lim_{x\to -1} y(x) = P_n^{(\alpha,\beta)}(-1) = (-1)^n \frac{\Gamma(n+\beta+1)}{\Gamma(\beta+1)n!}$$

exists, we have

$$B = 0,$$
$$A = \lim_{x\to -1} \frac{(1-x)^{(\alpha/2)+(1/4)}(1+x)^{(\beta/2)+(1/4)}y(x)}{\sqrt{\xi}J_\beta(\xi)}$$
$$= 2^{(2\alpha+1)/4} P_n^{(\alpha,\beta)}(-1)\, 2^{\beta}\Gamma(\beta+1) \lim_{x\to -1}\left(\frac{\sqrt{1+x}}{\xi}\right)^{\beta+1/2}.$$

By L'Hospital's rule,

$$\lim_{x\to -1} \frac{\sqrt{1+x}}{\xi(x)} = \lim_{x\to -1} \frac{1}{2\sqrt{1+x}\,\xi'(x)} = \frac{1}{\sqrt{2}}.$$

Therefore

$$A = (-1)^n \frac{2^{(\alpha+\beta)/2}\Gamma(n+\beta+1)}{n!\mu^{\beta+1/2}}, \quad \mu = \sqrt{n(n+\alpha+\beta+1)}.$$

Putting $x = -\cos\theta$, we have

$$P_n^{(\alpha,\beta)}(-\cos\theta) \approx \frac{(-1)^n\Gamma(n+\beta+1)\sqrt{\theta/2}}{n!\mu^\beta(\cos(\theta/2))^{\alpha+1/2}(\sin(\theta/2))^{\beta+1/2}} J_\beta(\mu\theta) \tag{17}$$

for $0 \le \theta \le \pi - \delta$.

From (17) we can easily deduce an approximate formula for the Jacobi polynomials $P_n^{(\alpha,\beta)}(x)$ for $-1 < x \le 1$ if we use the equation

$$P_n^{(\alpha,\beta)}(x) = (-1)^n P_n^{(\beta,\alpha)}(-x).$$

If $x \in [-1+\delta, 1-\delta]$, an approximate formula for $P_n^{(\alpha,\beta)}(x)$ can be found by using the asymptotic representation of $J_\beta(\mu\theta)$ as $\mu\theta \to +\infty$ and the asymptotic representation of $\Gamma(z)$ as $z \to \infty$ (see Appendix A):

$$P_n^{(\alpha,\beta)}(\cos\theta) \approx \frac{\cos\{[n+(\alpha+\beta+1)/2]\theta - (2\alpha+1)\pi/4\}}{\sqrt{\pi n}(\sin(\theta/2))^{\alpha+1/2}(\cos(\theta/2))^{\beta+1/2}} \qquad (0 < \delta \le \theta \le \pi - \delta). \tag{18}$$

When $\alpha = \beta = 0$, this becomes an asymptotic formula for the Legendre polynomials:

$$P_n(\cos\theta) \approx \left(\frac{2}{\pi n}\right)^{1/2} \frac{\cos[(n+1/2)\theta - \pi/4]}{\sqrt{\sin\theta}}.$$

Similarly we can obtain an approximate formula for the Laguerre polynomials $L_n^\alpha(x)$ for $x > 0$ and large n. In particular, we have

$$L_n^\alpha(x) \approx \pi^{-1/2} e^{x/2} x^{-\alpha/2-1/4} n^{\alpha/2-1/4} \cos\left[2\sqrt{nx} - (2\alpha+1)\frac{\pi}{4}\right] \tag{19}$$

for $0 < \delta \le x \le N < \infty$. If $\alpha = \pm 1/2$, formula (19) is valid down to $x = 0$, since in this case $\nu = \pm 1/2$ and (13) does not have a singular point at $s = 0$.

A corresponding formula for the Hermite polynomials $H_n(x)$ can be obtained from (19) by using formulas (6.14) and (6.15), which express the Hermite polynomials in terms of Laguerre polynomials:

$$H_n(x) \approx \sqrt{2}\left(\frac{2n}{e}\right)^{n/2} e^{x^2/2} \cos\left(\sqrt{2n}\, x - \frac{1}{2}\pi n\right) \qquad (|x| \le N < \infty). \tag{20}$$

Remark 1. Inequalities (20a), (27a) and (28a) of §7 (p. 54), which were obtained there by rather complicated calculations, are easily deducible from the estimates (18)–(20).

Remark 2. We derived (18) for $\alpha \geq 0$ and $\beta \geq 0$. However, it remains valid for all α and β. We can prove this by induction. Suppose that (18) holds for $P_n^{(\alpha+1,\beta+1)}(\cos\theta)$ and $P_n^{(\alpha+2,\beta+2)}(\cos\theta)$. Applying the differential equation of the Jacobi polynomials and the differentiation formula (5.6), we obtain

$$P_n^{(\alpha,\beta)}(x) = -\frac{1}{\lambda_n}\left[\tau(x)\frac{n+\alpha+\beta+1}{2}P_{n-1}^{(\alpha+1,\beta+1)}(x)\right.$$
$$\left.+\sigma(x)\frac{(n+\alpha+\beta+1)(n+\alpha+\beta+2)}{4}P_{n-2}^{(\alpha+2,\beta+2)}(x)\right],$$

where

$$\lambda_n = (n+\alpha+\beta+1), \quad \tau(x) = \beta-\alpha-(\alpha+\beta+2)x, \quad \sigma(x) = 1-x^2.$$

Hence

$$P_n^{(\alpha,\beta)}(\cos\theta)$$
$$= -\left[\frac{\beta-\alpha-(\alpha+\beta+2)\cos\theta}{2n}P_{n-1}^{(\alpha+1,\beta+1)}(\cos\theta)\right.$$
$$\left.+\frac{\sin^2\theta}{4}\left(1+\frac{\alpha+\beta+2}{n}\right)P_{n-2}^{(\alpha+2,\beta+2)}(\cos\theta)\right].$$

Substitute the asymptotic representations for $P_{n-1}^{(\alpha+1,\beta+1)}(\cos\theta)$ and $P_{n-2}^{(\alpha+2,\beta+2)}(\cos\theta)$, as obtained from (18), into the right-hand side, retaining only the principal terms, and we find

$$P_n^{(\alpha,\beta)}(\cos\theta) \approx -\frac{\sin^2\theta}{4}\frac{\cos\{[n-2+(\alpha+\beta+5)/2]\theta-(2\alpha+5)\pi/4\}}{\sqrt{\pi n}(\sin(\theta/2))^{\alpha+5/2}(\cos(\theta/2))^{\beta+5/2}}$$
$$= \frac{\cos\{[n+(\alpha+\beta+1)/2]\theta-(2\alpha+1)\pi/4\}}{\sqrt{\pi n}(\sin(\theta/2))^{\alpha+1/2}(\cos(\theta/2))^{\beta+1/2}}.$$

This agrees with (18). Similarly we can establish the validity of (19) for all real α.

3. Semiclassical approximation for equations with singular points. The central field. In discussing the motion of a particle in a central field it is useful to obtain a semiclassical approximation for an equation of the form

$$u'' + r(x)u = 0, \tag{21}$$

where $x^2r(x)$ is continuous together with its first and second derivatives for $0 \le x \le b$. The previous approximation cannot be used for (21) in a neighborhood of $x = 0$. However, the change of variables $x = e^z, u = e^{z/2}v(z)$ transforms the equation into

$$v''(z) + r_1(z)v = 0, \tag{22}$$

where

$$r_1(z) = -\frac{1}{4} + x^2 r(x)\Big|_{x = e^z}.$$

As $z \to -\infty$ (which corresponds to $x \to 0$), $r_1(z)$ differs little from a constant, namely $-\frac{1}{4}+\lim_{x\to 0} x^2r(x)$. Moreover, $\lim_{z\to-\infty} r_1^{(k)}(z) = 0\ (k = 1,2)$. Hence $r_1(z)$ and its derivatives vary slowly for negative z of large modulus, and we can apply the semiclassical approximation to (22). If the conditions of applicability of this method are satisfied for all required values of z, we can return to the original variables and obtain an approximate solution of (21) of the previous form, but with $r(x)$ replaced by $r(x) - 1/(4x^2)$.

For example, consider the solution of the Schrödinger equation in spherical coordinates for the radial part of the wave function $R(r)$,

$$-\frac{1}{2}R'' + \left[U(r) + \frac{l(l+1)}{2r^2}\right] R = ER,$$

where r is distance from the origin, $U(r)$ is the potential energy, E is the total energy of the particle, and $l = 0, 1, 2, \ldots$ are the orbital quantum numbers. In the semiclassical approximation we obtain

$$R(r) = \begin{cases} (\xi/p)^{1/2}[AJ_{1/3}(\xi) + BJ_{-1/3}(\xi)] & (r \ge r_0), \\ (\xi/p)^{1/2}[CI_{1/3}(\xi) + DK_{1/3}(\xi)] & (r \le r_0), \end{cases}$$

where

$$p = p(r) = \left|2[E - U(r)] - \frac{(l+1/2)^2}{r^2}\right|^{1/2},$$

$$\xi = \xi(r) = \left|\int_{r_0}^{r} p(r')dr'\right|,$$

and r_0 is a root (supposed simple) of the equation $p(r) = 0$.

Since $R(r)$ must be bounded as $r \to 0$, i.e. as $\xi \to \infty$, we have $C = 0$. Since $R(r)$ and $R'(r)$ must agree at $r = r_0$, we can express A and B in terms of D. If we expand the function under the square root sign in the formula for $p(r)$ in powers of $(r - r_0)$, we easily see that $p(r)/|r - r_0|^{1/2}$ and $\xi(r)/|r - r_0|^{3/2}$ and their first derivatives are continuous at $r = r_0$. Hence the joining conditions for $R(r)$ and $R'(r)$ at $r = r_0$ imply similar joining conditions for

$$\Phi(r) = (\xi/2)^{1/3}(p/\xi)^{1/2}R(r)$$

and its derivative. We have

$$\Phi(r) = \begin{cases} \dfrac{A(\xi/2)^{2/3}}{\Gamma(4/3)} + \dfrac{B}{\Gamma(2/3)} + O[(r - r_0)^3] & (r \geq r_0), \\ \dfrac{\pi D}{2\sin(\pi/3)}\left[\dfrac{1}{\Gamma(2/3)} - \dfrac{(\xi/2)^{2/3}}{\Gamma(4/3)}\right] + O[(r - r_0)^3] & (r \leq r_0). \end{cases}$$

Since $\xi^{2/3}/|r - r_0|$ is continuous at $r = r_0$, comparing coefficients of powers of $(r - r_0)$ yields

$$A = B = \frac{\pi}{\sqrt{3}}D.$$

4. Asymptotic formulas for Bessel functions of large order. Langer's formulas. The method explained above can be used to obtain asymptotic formulas for Bessel functions whose order ν is large. Let us transform the Bessel equation

$$x^2y'' + xy' + (x^2 - \nu^2)y = 0$$

into the form (21) by the substitution $u(x) = \sqrt{x}y(\nu x)$ (see the Lommel equation (14.4)). Then $u(x)$ satsifies

$$u'' + r(x)u = 0, \quad r(x) = \nu^2 - \frac{\nu^2 - 1/4}{x^2}.$$

Here we may use the reasoning of Part 3, putting $x = e^z, u = e^{z/2}v(z)$. Then we obtain

$$v'' + r_1(z)v = 0, \quad r_1(z) = \nu^2(e^{2z} - 1). \tag{23}$$

Since $\nu \gg 1$, we can apply a semiclassical approximation to (23). In the original variables the approximation for $u(x)$ is

$$u(x) = \left(\frac{\xi}{p}\right)^{1/2} \begin{cases} AI_{1/3}(\xi) + BK_{1/3}(\xi) & (x \leq 1), \\ CH^{(1)}_{1/3}(\xi) + DH^{(2)}_{1/3}(\xi) & (x \geq 1). \end{cases} \tag{24}$$

Here

$$p = p(x) = \nu s/x, \quad s = |1 - x^2|^{1/2},$$

$$\xi = \xi(x) = \left| \int_1^x p(t)dt \right| = \begin{cases} \nu(\tanh^{-1} s - s) & (x \leq 1), \\ \nu(s - \tan^{-1} s) & (x \geq 1). \end{cases}$$

For example, put $u(x) = \sqrt{x} H_\nu^{(1)}(\nu x)$. To determine C and D we compare the principal terms of the asymptotic formulas (for $x \to \infty$) for the two sides of (24). Since

$$s(x) = x + O(1/x), \quad \xi(x) = \nu(x - \pi/2) + O(1/x),$$

as $x \to \infty$ for fixed ν, we have

$$\sqrt{x} H_\nu^{(1)}(\nu x) \approx \left(\frac{2}{\pi\nu}\right)^{1/2} \exp[i(\nu x - \pi\nu/2 - \pi/4)]$$

$$= \left(\frac{2}{\pi\nu}\right)^{1/2} \left\{ C e^{i[\nu(x-\pi/2)-\pi/6-\pi/4]} + D e^{-i[\nu(x-\pi/2)-\pi/6-\pi/4)]} \right\},$$

from which it follows that

$$D = 0, \quad C = e^{i\pi/6}.$$

We determine A and B by the joining conditions for $u(x)$ and $u'(x)$ at $x = 1$, as in the previous example. We find

$$A = -2i, \quad B = (2/\pi)e^{-i\pi/3},$$

i.e. in the semiclassical approximation, for large ν,

$$H_\nu^{(1)}(\nu x) = \begin{cases} 2\sqrt{\dfrac{\tanh^{-1} s}{s} - 1} \left[-iI_{1/3}(\xi) + \dfrac{e^{-i\pi/3}}{\pi} K_{1/3}(\xi) \right] & (x \leq 1), \\ \sqrt{1 - \dfrac{\tan^{-1} s}{s}} e^{i\pi/6} H_{1/3}^{(1)}(\xi) & (x \geq 1). \end{cases} \tag{25}$$

If we compare real parts in (25), we obtain the semiclassical approximation for $J_\nu(\nu x)$ for large ν:

$$J_\nu(\nu x) = \begin{cases} \dfrac{1}{\pi}\sqrt{\dfrac{\tanh^{-1} s}{s} - 1}\, K_{1/3}(\xi) & (x \le 1), \\ \dfrac{1}{\sqrt{3}}\sqrt{1 - \dfrac{\tan^{-1} s}{s}}\, [J_{-1/3}(\xi) + J_{1/3}(\xi)] & (x \ge 1). \end{cases} \tag{26}$$

Formulas (25) and (26) are *Langer's formulas* [L2]. More precise estimates show that they provide uniform approximations to the Bessel functions with error $O(\nu^{-4/3})$ [L2]. It is interesting that (26) gives the behavior of $J_\nu(\nu x)$ correctly as $x \to 0$, even though it was derived by using the asymptotic formulas for Bessel functions as $x \to \infty$.

5. Finding the energy eigenvalues for the Schrödinger equation in the semiclassical approximation. The Bohr-Sommerfeld formula. The solution of the Schrödinger equation

$$-\frac{1}{2}\psi''(x) + U(x)\psi(x) = E\psi(x) \quad (-\infty < x < \infty) \tag{27}$$

describing the motion of a particle in a field with potential $U(x)$ can be found explicitly for only a few special forms of $U(x)$ (E is the total energy of the particle; we use a system of units in which the mass m and Planck's constant $\hbar$ are both 1). This can be done, for example, when (27) can be reduced to a generalized equation of hypergeometric type (see the theorem of §9, part 2). However, it is more useful to have methods for solving (27) approximately for any potential $U(x)$.

Let us find the energy eigenvalues of (27) in the semiclassical approximation. We are to find the values of E for which $E - U(x) < 0$ as $x \to \pm\infty$, and $\psi(x)$ satisfies the normalizing condition

$$\int_{-\infty}^{\infty} |\psi(x)|^2 dx = 1. \tag{28}$$

Let

$$E - U(x) \ge 0 \text{ for } x_1 \le x \le x_2$$

(this is the interval of classical motion) and let

$$E - U(x) < 0 \quad \text{for } x < x_1 \quad \text{and} \quad x > x_2,$$

where x_1 and x_2 are simple roots of the equation $E = U(x)$ (called turning points in quantum mechanics).

In solving the problem, we suppose that the integrals

$$\int_{-\infty}^{x_1} p(x)dx, \qquad \int_{x_2}^{\infty} p(x)dx$$

diverge ($p(x) = (2|E-U(x)|)^{1/2}$) and that $\int_{x_1}^{x_2} p(x)dx$ is sufficiently large. In the semiclassical approximation we have, for $-\infty < x \leq x_2$,

$$\psi(x) = \begin{cases} (\xi/p)^{1/2}[A_1 I_{1/3}(\xi) + B_1 K_{1/3}(\xi)] & \text{for } x \leq x_1, \\ (\xi/p)^{1/2}[A_2 J_{-1/3}(\xi) + B_2 J_{1/3}(\xi)] & \text{for } x_1 \leq x \leq x_2, \end{cases}$$
$$\xi = \xi(x) = \left|\int_{x_1}^{x} p(s)ds\right|. \tag{29}$$

We put $A_1 = 0$ so that $\int_{-\infty}^{\infty} |\psi(x)|^2 dx$ will converge. We can express A_2 and B_2 in terms of B_1 by using the joining conditions for $\psi(x)$ and $\psi'(x)$ at $x = x_1$ (see the example in Part 3). This yields $A_2 = B_2 = \pi B_1/\sqrt{3}$. Consequently

$$\psi(x) = A_2(\xi/p)^{1/2}[J_{-1/3}(\xi) + J_{1/3}(\xi)] \tag{30}$$

for $x_1 \leq x \leq x_2$.

We are considering values of x which make $\xi(x)$ sufficiently large. For such values of x we can use the asymptotic formula for $J_{\pm 1/3}(z)$:

$$J_{\pm 1/3}(z) \approx \sqrt{\frac{2}{\pi z}} \cos\left(z \mp \frac{\pi}{6} - \frac{\pi}{4}\right).$$

We obtain

$$\psi(x) = \frac{c_1}{\sqrt{p(x)}} \cos\left[\int_{x_1}^{x} p(s)ds - \frac{\pi}{4}\right] \tag{31}$$

(c_1, a constant).

In a similar way we can obtain an expression for $\psi(x)$ for $x_1 < x \leq x_2$, by starting from the behavior of the function as $x \to \infty$. If we consider values of x for which $\int_x^{x_2} p(s)ds$ is sufficiently large, we obtain

$$\psi(x) = \frac{c_2}{\sqrt{p(x)}} \cos\left[\int_x^{x_2} p(s)ds - \frac{\pi}{4}\right] \tag{32}$$

(c_2, a constant).

Let us consider an x for which both $\int_{x_1}^{x} p(s)ds$ and $\int_x^{x_2} p(s)ds$ are large. In this case we have the two expressions (31) and (32) for $\psi(x)$. These expressions and their derivatives coincide at x only if

$$\int_{x_1}^{x} p(s)ds + \int_{x}^{x_2} p(s)ds = \int_{x_1}^{x_2} p(s)ds = \pi\left(n + \frac{1}{2}\right), \tag{33}$$

where $n = 0, 1, 2, \ldots$. It is easy to see that n is the number of zeros of $\psi(x)$ (which can have zeros only for $x_1 < x < x_2$). Moreover, $c_2 = (-1)^n c_1$. Therefore, in the semiclassical approximation, the energy values $E = E_n$ ($n = 0, 1, 2, \ldots$) in the discrete spectrum must be subject to (33). This is the Bohr-Sommerfeld condition in quantum mechanics.

We can also obtain the Bohr-Sommerfeld condition for a particle in a central field $U(r)$. Repeating the preceding reasoning and using the results of Part 3, we obtain the Bohr-Sommerfeld condition for the energy eigenvalues E_{nl} of the discrete spectrum in the form

$$\begin{gathered} \int_{r_1(E)}^{r_2(E)} p(r)dr = \pi\left(n + \frac{1}{2}\right), \\ p(r) = \left\{2\left|E - U(r) - \frac{(l+1/2)^2}{2r^2}\right|\right\}^{1/2}, \quad p(r)|_{r=r_1,r_2} = 0. \end{gathered} \tag{34}$$

Example 1. Find, in the semiclassical approximation, the energy levels of a particle in the field $u(x) = \frac{1}{2}\mu\omega^2 x^2$ (the linear harmonic oscillator).

We obtained the exact solution in §9. If we use the same units as in §9, the Schrödinger equation (27) becomes

$$-\frac{1}{2}\psi'' + \frac{1}{2}x^2\psi = \mathcal{E}\psi \quad (E = \hbar\omega\mathcal{E}).$$

In the present case

$$p(x) = \sqrt{2\mathcal{E} - x^2}, \quad x_1 = -\sqrt{2\mathcal{E}}, \quad x_2 = \sqrt{2\mathcal{E}}.$$

The energy levels are found from the Bohr-Sommerfeld condition

$$\int_{-\sqrt{2\mathcal{E}}}^{\sqrt{2\mathcal{E}}} \sqrt{2\mathcal{E} - x^2}\,dx = \pi\left(n + \frac{1}{2}\right). \tag{35}$$

Since

$$\int \sqrt{\alpha x^2 + \beta}\,dx = \frac{1}{2}x\sqrt{\alpha x^2 + \beta} + \frac{1}{2}\beta \int \frac{dx}{\sqrt{\alpha x^2 + \beta}},$$

we have

$$\int_{x_1}^{x_2} \sqrt{2\mathcal{E} - x^2}\,dx = \mathcal{E} \int_{x_1}^{x_2} \frac{dx}{\sqrt{2\mathcal{E} - x^2}} = \mathcal{E} \arcsin \frac{x}{\sqrt{2\mathcal{E}}}\Bigg|_{x_1}^{x_2} = \pi\mathcal{E},$$

From (35) we find

$$\mathcal{E} = \mathcal{E}_n = n + \frac{1}{2},$$

which agrees with the exact solution even when the condition $\int_{x_1}^{x_2} p(x)dx \gg 1$ is not satisfied.

Example 2. Find, in the semiclassical approximation, the energy levels in the field $U(r) = -Z/r$ (we are using atomic units).

In the present case we put $U(r) = -Z/r$ in (34). After integrating by parts, we find that

$$\int_{r_1}^{r_2} p(r)dr = rp(r)\Bigg|_{r_1}^{r_2} - \int_{r_1}^{r_2} \frac{r\{(-Z/r^2) + (l + \frac{1}{2})^2/r^3\}}{\{2[E + (Z/r) - \frac{1}{2}(l + \frac{1}{2})^2/r^2]\}^{1/2}}dr$$

$$= z\int_{r_1}^{r_2} \frac{dr}{\{2[Er^2 + Zr - \frac{1}{2}(l + \frac{1}{2})^2]\}^{1/2}}$$

$$= \int_{x_1}^{x_2} \frac{(l + \frac{1}{2})^2 dx}{\{2[E + Zx - \frac{1}{2}(l + \frac{1}{2})^2 x^2]\}^{1/2}}$$

$$= \pi\left(\frac{Z}{\sqrt{-2E}} - l - \frac{1}{2}\right) \qquad (x = 1/r,\ E < 0).$$

Substituting this expression for the integral in (34), we obtain

$$E = E_{nl} = -\frac{Z^2}{2(n + l + 1)^2},$$

which agrees with the exact value for all n and l.

Chapter IV

Hypergeometric functions

In Chapters II and III we discussed properties of the classical orthogonal polynomials and of Bessel functions. Those functions satisfy differential equations which are special cases of the generalized equation of hypergeometric type

$$u'' + \frac{\tilde{\tau}(z)}{\sigma(z)}u' + \frac{\tilde{\sigma}(z)}{\sigma^2(z)}u = 0. \tag{1}$$

Here $\sigma(z)$ and $\tilde{\sigma}(z)$ are polynomials of degree at most 2, and $\tilde{\tau}(z)$ is a polynomial of degree at most 1.

By using the results of Chapter I, we can investigate the properties of the solutions of any generalized equation of hypergeometric type. By the substitution $u = \phi(z)y$ with a suitable chosen $\phi(z)$ we can reduce an equation of the form (1) to the equation of hypergeometric type

$$\sigma(z)y'' + \tau(z)y' + \lambda y = 0, \tag{2}$$

where $\tau(z)$ is a polynomial of degree at most 1, and λ is a constant (see §1). In §3 we gave a method for constructing particular solutions of (2). In the present chapter we apply this method to make a more detailed study of the particular solutions.

§ 20 The equations of hypergeometric type and their solutions

1. Reduction to canonical form. We are going to transform (2) to a canonical form by a linear change of independent variable. There are three cases, according to the degree of $\sigma(z)$.

1) Let $\sigma(z)$ be of degree 2: $\sigma(z) = (z-a)(b-z), a \neq b$.* Under the substitution $z = a + (b-a)s$, we obtain

$$s(1-s)y'' + \frac{1}{b-a}\tau[a+(b-a)s]y' + \lambda y = 0.$$

It is always possible to choose the parameters α, β and γ so that this equation can be written in the form

$$s(1-s)y'' + [\gamma - (\alpha+\beta+1)s]y' - \alpha\beta y = 0.$$

This is the *hypergeometric equation*, also often called Gauss's equation.

2) Let $\sigma(z)$ be of degree 1: $\sigma(z) = z - a$. Putting $z = a + bs$, we transform (2) to

$$sy'' + \tau(a+bs)y' + \lambda by = 0. \tag{3}$$

If $\tau'(z) = 0$, then (3) is the Lommel equation (14.4) for any b, and its solutions can be expressed in terms of Bessel functions. If $\tau'(z) \neq 0$, then, with $b = -1/\tau'(z)$,

$$\tau(a+bs) = \tau(a) + \tau'(a)bs = \tau(a) - s.$$

Set $\gamma = \tau(a), \alpha = -\lambda b$. Then (3) reduces to

$$sy'' + (\gamma - s)y' - \alpha y = 0.$$

This is the *confluent hypergeometric equation.*

3) If $\sigma(z)$ is independent of z, we may take $\sigma(z) = 1$. When $\tau'(z) = 0$, equation (2) is a homogeneous linear equation with constant coefficients. If $\tau'(z) \neq 0$, we can write (2) in the form

$$y'' + b\tau(a+bs)y' + \lambda b^2 y = 0$$

after the substitution $z = a + bs$. By choosing a, b and ν in this equation we can put it in the form

$$y'' - 2sy' + 2\nu y = 0,$$

which is the *Hermite equation* (when $\nu = n$ it is the equation for the Hermite polynomials).

---

* If $\sigma(z)$ has a double zero, (2) can be reduced to an equation of hypergeometric type for which $\sigma(z)$ is of degree 1(see §1).

2. Construction of particular solutions. Particular solutions of the hypergeometric and confluent hypergeometric equations, and of the Hermite equation, can be found by the method explained in §3. The number of such solutions can be increased by using the transformation of the original equation into another equation of the same form by the process suggested in §3. Let us consider the corresponding transformations.

The equation of hypergeometric type (2) is the special case of (1) with $\tilde{\tau}(z) = \tau(z)$, $\tilde{\sigma}(z) = \lambda\sigma(z)$. Therefore it can be transformed into an equation of the same type by the substitution $u = \phi(z)y$ (see §1), provided that $\phi(z)$ satisfies

$$\phi'/\phi = \pi(z)/\sigma(z),$$

where

$$\pi(z) = \frac{\sigma' - \tau}{2} \pm \left\{\left(\frac{\sigma' - \tau}{2}\right)^2 - \kappa\sigma\right\}^{1/2} \qquad (\kappa = \lambda - k)$$

is a polynomial of degree 1 at most. The constant κ is determined by the condition that the discriminant of the quadratic under the square root sign is zero.

1°. For the hypergeometric equation

$$z(1-z)u'' + [\gamma - (\alpha+\beta+1)z]u' - \alpha\beta u = 0 \tag{4}$$

we have

$$\left(\frac{\sigma' - \tau}{2}\right)^2 - \kappa\sigma = \left[\frac{1-\gamma+(\alpha+\beta-1)z}{2}\right]^2 - \kappa z(1-z).$$

Setting the discriminant of this quadratic equal to zero, we obtain two possible values for κ:

$$\kappa_1 = (1-\gamma)(\alpha+\beta-\gamma), \quad \kappa_2 = 0.$$

When $\kappa_1 = (1-\gamma)(\alpha+\beta-\gamma)$, there are the following possibilities for $\pi(z)$ and $\phi(z)$:

$$\begin{aligned} &a) \quad \pi(z) = (1-\gamma)(1-z), \quad \phi(z) = z^{1-\gamma};\\ &b) \quad \pi(z) = (\alpha+\beta-\gamma)z, \quad \phi(z) = (1-z)^{\gamma-\alpha-\beta}. \end{aligned}$$

Substituting $u = \phi(z)y$ with $\phi(z) = z^{1-\gamma}$ we arrive at the following equation for $y(z)$:

$$z(1-z)y'' + [2-\gamma-(\alpha+\beta-2\gamma+3)z]y' - (\alpha-\gamma+1)(\beta-\gamma+1)y = 0.$$

This can be written in the canonical form

$$z(1-z)y'' + [\gamma' - (\alpha' + \beta' + 1)z]y' - \alpha'\beta' y = 0, \tag{4a}$$

by taking $\alpha' = \alpha - \gamma + 1$, $\beta' = \beta - \gamma + 1$, $\gamma' = 2 - \gamma$. Similarly, when $\phi(z) = (1-z)^{\gamma-\alpha-\beta}$ the substitution $u = \phi(z)y$ leads to (4a) with $\alpha' = \gamma - \alpha$, $\beta' = \gamma - \beta$, $\gamma' = \gamma$.

Let $u(z) = f(\alpha, \beta, \gamma, z)$ be a particular solution of (4). Then $y(z) = u(z)/\phi(z)$ satisfies the hypergeometric equation with parameters α', β', γ'. Hence $u(z) = \phi(z) f(\alpha', \beta', \gamma', z)$ is also a solution of (4). Hence we have the following particular solutions of (4):

$$\begin{aligned} u_1(z) &= f(\alpha, \beta, \gamma, z), \\ u_2(z) &= z^{1-\gamma} f(\alpha - \gamma + 1, \beta - \gamma + 1, 2 - \gamma, z), \\ u_3(z) &= (1-z)^{\gamma-\alpha-\beta} f(\gamma - \alpha, \gamma - \beta, \gamma, z). \end{aligned} \tag{5}$$

We have considered transformations of (4) into an equation (4a) of the same type for the case when $\kappa = (1-\gamma)(\alpha + \beta - \gamma)$. Similar transformations corresponding to $\kappa = 0$ are not of interest, since they can be obtained by applying two successive transformations of the preceding types.

Equation (4) is unchanged if α and β are interchanged. Hence there are also solutions obtainable from (5) by this operation: $u_4(z) = f(\beta, \alpha, \gamma, z)$.

For the confluent hypergeometric equation

$$zu'' + (\gamma - z)u' - \alpha u = 0 \tag{6}$$

a solution $u_1(z) = f(\alpha, \gamma, z)$ leads to the solutions

$$\begin{aligned} u_2(z) &= z^{1-\gamma} f(\alpha - \gamma + 1, 2 - \gamma, z), \\ u_3(z) &= e^z f(\gamma - \alpha, \gamma, -z). \end{aligned} \tag{7}$$

For the Hermite equation,

$$u'' - 2zu' + 2\nu u = 0, \tag{8}$$

a solution $u_1(z) = f_\nu(z)$ generates the solution

$$u_2(z) = e^{-z^2} f_{-\nu-1}(iz).$$

Since (8) is also invariant under replacement of z by $-z$, we also have the solutions

$$u_3(z) = f_\nu(-z), \quad u_4(z) = e^{-z^2} f_{-\nu-1}(-iz).$$

2°. We now construct the solutions of (4), (6) and (8) explicitly. As we showed in §3, the equation

$$\sigma(z)u'' + \tau(z)u' + \lambda u = 0$$

of hypergeometric type has particular solutions of the form

$$u(z) = \frac{C_\nu}{\rho(z)} \int_C \frac{\sigma^\nu(s)\rho(s)}{(s-z)^{\nu+1}} ds. \tag{9}$$

Here $\rho(z)$ is a solution of $(\sigma\rho)' = \tau\rho$, where ν is a root of the equation $\lambda + \nu\tau' + \frac{1}{2}\nu(\nu-1)\sigma'' = 0$, and C satisfies

$$\left. \frac{\sigma^{\nu+1}(s)\rho(s)}{(s-z)^{\nu+2}} \right|_{s_1, s_2} = 0 \tag{10}$$

(s_1 and s_2 are the endpoints of C).

We shall first construct solutions when $z > 0$. In addition, for (4) we suppose that $z < 1$.

For (4),

$$\sigma(z) = z(1-z), \quad \rho(z) = z^{\gamma-1}(1-z)^{\alpha+\beta-\gamma}, \quad \nu = -\alpha \quad (\text{or } \nu = -\beta);$$

for (6),

$$\sigma(z) = z, \quad \rho(z) = z^{\gamma-1}e^{-z}, \quad \nu = -\alpha;$$

and for (8),

$$\sigma(z) = 1, \quad \rho(z) = e^{-z^2}.$$

For (4), condition (10) takes the form

$$s^{\gamma-\alpha}(1-s)^{\beta-\gamma+1}(s-z)^{\alpha-2}|_{s_1,s_2} = 0.$$

Under certain restrictions on α, β and γ, this condition can be satisfied if the endpoints of C are taken at $s = 0, 1, z$ or ∞. To construct solutions that have simple behavior in neighborhoods of $z = 0, 1$, or ∞, we can conveniently take C to be a straight line connecting $s = 0, 1$ or ∞ with $s = z$. Assuming that $0 \le t \le 1$, we may take parametric equations of these lines in the forms

$$s = zt, \qquad \operatorname{Re}\gamma > \operatorname{Re}\alpha > 2;$$

$$s = 1 - (1-z)t, \qquad \operatorname{Re}\gamma < \operatorname{Re}\beta + 1, \qquad \operatorname{Re}\alpha > 2;$$

$$s = z/t, \qquad \operatorname{Re}\beta > 1, \qquad \operatorname{Re}\alpha > 2.$$

Similarly we are led to the following contours for solutions of (6) and (8):

a) for the confluent hypergeometric equation (6):

$$s = zt, \quad \mathrm{Re}\,\gamma > \mathrm{Re}\,\alpha > 2 \qquad (0 \le t \le 1),$$
$$s = z(1+t), \quad \mathrm{Re}\,\alpha > 2 \qquad (0 \le t < \infty);$$

b) for the Hermite equation (8):

$$s = z + t \qquad (0 \le t < \infty).$$

3°. For (4) and (6), the contour $s = zt$ leads to the following solutions:

$$u_1(z) = F(\alpha,\beta,\gamma,z)$$
$$= C(\alpha,\beta,\gamma)(1-z)^{\gamma-\alpha-\beta}\int_0^1 t^{\gamma-\alpha-1}(1-t)^{\alpha-1}(1-zt)^{-\beta}dt, \tag{11}$$

$$u_1(z) = F(\alpha,\gamma,z) = C(\alpha,\gamma)e^z\int_0^1 t^{\gamma-\alpha-1}(1-t)^{\alpha-1}e^{-zt}dt. \tag{12}$$

The functions $F(\alpha,\beta,\gamma,z)$ and $F(\alpha,\gamma,z)$ are the *hypergeometric* and *confluent hypergeometric functions*, respectively. The corresponding normalizing constants $C(\alpha,\beta,\gamma)$ and $C(\alpha,\gamma)$ are chosen so that

$$F(\alpha,\beta,\gamma,0) = F(\alpha,\gamma,0) = 1;$$

we then have

$$C(\alpha,\beta,\gamma) = C(\alpha,\gamma) = \frac{1}{\mathrm{B}(\alpha,\gamma-\alpha)} = \frac{\Gamma(\gamma)}{\Gamma(\alpha)\Gamma(\gamma-\alpha)}. \tag{13}$$

Here $\Gamma(z)$ is the gamma function, and $\mathrm{B}(u,v)$, the beta function (see Appendix A).

For $F(\alpha,\beta,\gamma,z)$ and $F(\alpha,\gamma,z)$, condition (10) is satisfied only under certain restrictions on the parameters. It will be shown in the next part that (11) and (12) allow $F(\alpha,\beta,\gamma,z)$ and $F(\alpha,\gamma,z)$ to be continued analytically in z and each parameter into the region $\mathrm{Re}\,\gamma > \mathrm{Re}\,\alpha > 0$. An analytic continuation of $F(\alpha,\beta,\gamma,z)$ or $F(\alpha,\gamma,z)$ must, in the first place, satisfy (4) or (6), respectively. In order to have $F(\alpha,\beta,\gamma,z)$ single-valued in (11), we must require that $|\arg(1-zt)| < \pi$; to ensure this, we make a cut in the z plane along the real axis for $z \ge 1$.

We also obtain solutions of (4) by taking $f(\alpha,\beta,\gamma,z) = F(\alpha,\beta,\gamma,z)$ in (5):

$$u_2(z) = z^{1-\gamma}F(\alpha-\gamma+1,\beta-\gamma+1,2-\gamma,z),$$
$$u_3(z) = (1-z)^{\gamma-\alpha-\beta}F(\gamma-\alpha,\gamma-\beta,\gamma,z).$$

There are also the solutions obtained by interchanging α and β. In particular,

$$u_4(z) = F(\beta,\alpha,\gamma,z)$$

is a solution of the hypergeometric equation. The integral representations defining these four solutions exist simultaneously provided that $0 < \text{Re}\,\alpha < 1$ and $0 < \text{Re}\,(\gamma-\alpha) < 1$. Since the hypergeometric equation has only two linearly independent solutions, there must be a linear relation among the $u_i(z)$. For $\gamma \neq 1$ the functions $u_1(z)$ and $u_2(z)$ are linearly independent, since they behave differently as $z \to 0$. Hence for $\gamma \neq 1$ both $u_3(z)$ and $u_4(z)$ must be linear combinations of $u_1(z)$ and $u_2(z)$. By comparing the behavior of these functions as $z \to 0$, we find that

$$u_3(z) = u_1(z), \quad u_4(z) = u_1(z) \qquad (\text{Re}\,\gamma > 1),$$

i.e., for Re $\gamma > 1$,

$$F(\alpha,\beta,\gamma,z) = (1-z)^{\gamma-\alpha-\beta}F(\gamma-\alpha,\gamma-\beta,\gamma,z), \tag{14}$$
$$F(\alpha,\beta,\gamma,z) = F(\beta,\alpha,\gamma,z). \tag{15}$$

By using the principle of analytic continuation, we can drop the restriction on γ. In (14) we may take $(1-z)^{\gamma-\alpha-\beta}$ to be the branch that is 1 at $z=0$, i.e. $|\arg(1-z)| < \pi$.

In a similar way we obtain the two linearly independent solutions

$$\left.\begin{aligned} u_1(z) &= F(\alpha,\gamma,z), \\ u_2(z) &= z^{1-\gamma}F(\alpha-\gamma+1,2-\gamma,z) \end{aligned}\right. \tag{16}$$

of the confluent hypergeometric equation, and the functional equation

$$F(\alpha,\gamma,z) = e^z F(\gamma-\alpha,\gamma,-z). \tag{17}$$

By using (14) and (17) we can replace (11)–(13) by the simpler integral representations

$$F(\alpha,\beta,\gamma,z) = \frac{\Gamma(\gamma)}{\Gamma(\alpha)\Gamma(\gamma-\alpha)}\int_0^1 t^{\alpha-1}(1-t)^{\gamma-\alpha-1}(1-zt)^{-\beta}dt, \tag{18}$$

$$F(\alpha,\gamma,z) = \frac{\Gamma(\gamma)}{\Gamma(\alpha)\Gamma(\gamma-\alpha)} \int_0^1 t^{\alpha-1}(1-t)^{\gamma-\alpha-1} e^{zt}\,dt. \tag{19}$$

Other contours for the hypergeometric equation lead to the following pairs of linearly independent solutions:

1) the contour $s = 1-(1-z)t \quad (0 \le t \le 1)$:

$$\begin{aligned} u_1(z) &= F(\alpha,\beta,\alpha+\beta-\gamma+1,1-z),\\ u_2(z) &= (1-z)^{\gamma-\alpha-\beta} F(\gamma-\alpha,\gamma-\beta,\gamma-\alpha-\beta+1,1-z); \end{aligned} \tag{20}$$

2) the contour $s = z/t \quad (0 \le t \le 1)$:

$$\begin{aligned} u_1(z) &= z^{-\alpha} F(\alpha,\alpha-\gamma+1,\alpha-\beta+1,1/z),\\ u_2(z) &= z^{-\beta} F(\beta,\beta-\gamma+1,\beta-\alpha+1,1/z). \end{aligned} \tag{21}$$

For the confluent hypergeometric equation the contour $s = z(1+t)$ leads to the solution

$$u_1(z) = G(\alpha,\gamma,z) = C(\alpha,\gamma) \int_0^\infty e^{-zt} t^{\alpha-1}(1+t)^{\gamma-\alpha-1}\,dt.$$

The function $G(\alpha,\gamma,z)$ is *the confluent hypergeometric function of the second kind*. The integral defining $G(\alpha,\gamma,z)$ is a Laplace integral and therefore by Watson's lemma (see Appendix B)

$$\lim_{z\to\infty} z^\alpha G(\alpha,\gamma,z) = C(\alpha,\gamma)\Gamma(\alpha).$$

It is convenient to take $C(\alpha,\gamma) = 1/\Gamma(\alpha)$ so that the limit will be 1. Then

$$G(\alpha,\gamma,z) = \frac{1}{\Gamma(\alpha)} \int_0^\infty e^{-zt} t^{\alpha-1}(1+t)^{\gamma-\alpha-1}\,dt \tag{22}$$

$$(\operatorname{Re}\alpha > 0, |\arg z| \le \pi).$$

By using $u_1(z) = G(\alpha,\gamma,z)$ as $f(\alpha,\gamma,z)$ in (7), we can construct a second linearly independent solution

$$u_2(z) = e^z G(\gamma-\alpha,\gamma,-z)$$

and obtain the functional equation

$$G(\alpha,\gamma,z) = z^{1-\gamma}G(\alpha-\gamma+1, 2-\gamma, -z). \tag{23}$$

For (8), we can obtain similarly

$$\begin{aligned} u_1(z) &= H_\nu(z) = C_\nu \int_0^\infty e^{-t^2-2zt} t^{-\nu-1}\,dt, \\ u_2(z) &= e^{-z^2} H_{-\nu-1}(iz), \\ u_3(z) &= H_\nu(-z), \\ u_4(z) &= e^{-z^2} H_{-\nu-1}(-iz). \end{aligned} \tag{24}$$

With $C_\nu = 1/\Gamma(-\nu)$, the functions $H_\nu(z)$ are the *Hermite functions*.*

4°. We collect some of the simplest properties of functions of hypergeometric type, those that follow immediately from the integral representations (18), (19), (22) and (24). It was shown in §2 that all derivatives of functions of hypergeometric type are also functions of hypergeometric type. We can use the integral representations of $F(\alpha,\beta,\gamma,z)$, $F(\alpha,\gamma,z)$, $G(\alpha,\gamma,z)$ and $H_\nu(z)$ to make these general statements specific:

$$\begin{aligned} \frac{d}{dz}F(\alpha,\beta,\gamma,z) &= \frac{\alpha\beta}{\gamma}F(\alpha+1,\beta+1,\gamma+1,z), \\ \frac{d}{dz}F(\alpha,\gamma,z) &= \frac{\alpha}{\gamma}F(\alpha+1,\gamma+1,z), \\ \frac{d}{dz}G(\alpha,\gamma,z) &= -\alpha G(\alpha+1,\gamma+1,z), \\ \frac{d}{dz}H_\nu(z) &= 2\nu H_{\nu-1}(z). \end{aligned} \tag{25}$$

Combining the *differentiation formulas* (25) with (4), (6) and (8), we obtain *recursion relations*:

$$\begin{aligned} \phi(\alpha,\beta,\gamma,z) = {}&(\alpha+1)(\beta+1)z(1-z)\phi(\alpha+2,\beta+2,\gamma+2,z) \\ &+[\gamma-(\alpha+\beta+1)z]\phi(\alpha+1,\beta+1,\gamma+1,z); \end{aligned} \tag{26}$$

$$\phi(\alpha,\gamma,z) = (\alpha+1)z\phi(\alpha+2,\gamma+2,z)+(\gamma-z)\phi(\alpha+1,\gamma+1,z); \tag{27}$$

* The constant C_ν is chosen so that analytic continuation of $H_\nu(z)$ with respect to ν leads to the Hermite polynomial $H_n(z)$ for $\nu=n$ (see §22).

$$G(\alpha,\gamma,z)=(\alpha+1)zG(\alpha+2,\gamma+2,z)-(\gamma-z)G(\alpha+1,\gamma+1,z); \quad (28)$$

$$H_\nu(z)=2zH_{\nu-1}(z)-(2\nu-2)H_{\nu-2}(z). \quad (29)$$

Here

$$\phi(\alpha,\beta,\gamma,z)=\frac{1}{\Gamma(\gamma)}F(\alpha,\beta,\gamma,z),$$

$$\phi(\alpha,\gamma,z)=\frac{1}{\Gamma(\gamma)}F(\alpha,\gamma,z).$$

If we replace α and β in (26) by $\gamma-\alpha$ and $\gamma-\beta$ and use (14), we obtain the recursion relation

$$\phi(\alpha,\beta,\gamma,z)=(\gamma-\alpha+1)(\gamma-\beta+1)\frac{z}{1-z}\phi(\alpha,\beta,\gamma+2,z)$$

$$+[\gamma-(2\gamma-\alpha-\beta+1)z]\frac{1}{1-z}\phi(\alpha,\beta,\gamma+1,z). \quad (30)$$

Similarly we can obtain another recursion relation for the confluent hypergeometric function:

$$\phi(\alpha,\gamma,z)=-(\gamma-\alpha+1)z\phi(\alpha,\gamma+2,z)+(\gamma+z)\phi(\alpha,\gamma+1,z). \quad (31)$$

The recursion relations (26)–(31) are useful for obtaining analytic continuations of functions of hypergeometric type. The general problem of finding recursion relations is discussed in the next section.

3. Analytic continuation. Let us consider analytic continuations of the functions $F(\alpha,\beta,\gamma,z)$, $F(\alpha,\gamma,z)$, $G(\alpha,\gamma,z)$ and $H_\nu(z)$. In the first place, we shall determine the largest domains of z and of the parameters into which these functions can be continued by using their integral representations and the theorem on the analyticity of functions defined by integrals that depend on parameters (Theorem 2, §3).

We shall show that the hypergeometric function $F(\alpha,\beta,\gamma,z)$ defined by equation (18)

$$F(\alpha,\beta,\gamma,z)=\frac{\Gamma(\gamma)}{\Gamma(\alpha)\Gamma(\gamma-\alpha)}\int_0^1 t^{\alpha-1}(1-t)^{\gamma-\alpha-1}(1-zt)^{-\beta}dt,$$

is analytic in each of α,β,γ, and z for $\mathrm{Re}\,\gamma>\mathrm{Re}\,\alpha>0, |\arg(1-z)|<\pi$. To find the domain of analyticity we need to find the domain in which (18) converges uniformly with respect to z and the corresponding parameters. We have

$$t^{\alpha-1}(1-t)^{\gamma-\alpha-1}(1-zt)^{-\beta}=t^{\delta-1}(1-t)^{\delta-1}\psi(t),$$

where

$$\psi(t) = t^{\alpha-\delta}(1-t)^{\gamma-\alpha-\delta}(1-zt)^{-\beta}.$$

For each $\delta > 0$ the function $\psi(t)$ is continuous in all the variables collectively in the closed region $0 \le t \le 1$, $\delta \le \mathrm{Re}\,\alpha \le N$, $\delta \le \mathrm{Re}\,(\gamma-\alpha) \le N$, $|\beta| \le N$, $|z| \le N$, $|\arg(1-\delta-z)| \le \pi - \delta$, and hence is bounded there:

$$|t^{\alpha-\delta}(1-t)^{\gamma-\alpha-\delta}(1-zt)^{-\beta}| \le C$$

(C, some constant). The restriction $|\arg(1-\delta-z)| \le \pi-\delta$ is imposed so that the region under consideration will not contain the singular points of $(1-zt)^{-\beta}$, i.e. the points $z = t^{-1}$ $(0 \le t \le 1)$. Consequently

$$|t^{\alpha-1}(1-t)^{\gamma-\alpha-1}(1-zt)^{-\beta}| \le Ct^{\delta-1}(1-t)^{\delta-1}$$

in the region under consideration. Since the integral $\int_0^1 t^{\delta-1}(1-t)^{\delta-1}dt$ converges, the integral (18) that defines $F(\alpha,\beta,\gamma,z)$ converges uniformly in the same region and is therefore an analytic function of each argument.

Since δ and N are arbitrary, the function $F(\alpha,\beta,\gamma,z)$ is analytic in each argument for $\mathrm{Re}\,\gamma > \mathrm{Re}\,\alpha > 0$, $|\arg(1-z)| < \pi$. The last condition means that there is a cut in the z plane along the real axis for $z \ge 1$.

Similarly we can show that $F(\alpha,\gamma,z)$ is analytic in each argument for $\mathrm{Re}\,\gamma > \mathrm{Re}\,\alpha > 0$ and all z.

The integral defining $G(\alpha,\gamma,z)$ is a Laplace integral, which is discussed in the example for Theorem 1 in Appendix B. It follows from this discussion that $G(\alpha,\gamma,z)$ is analytic in each variable for $|\arg z| < 3\pi/2$, $z \ne 0$, $\mathrm{Re}\,\alpha > 0$. It has the following asymptotic representation for $z \to \infty$ with $\mathrm{Re}\,\alpha > 0$ and $|\arg z| \le (3\pi/2) - \epsilon$ $(\epsilon > 0)$:

$$G(\alpha,\gamma,z) = \frac{\Gamma(\gamma-\alpha)}{\Gamma(\alpha)z^\alpha}\left[\sum_{k=0}^{n-1}\frac{\Gamma(\alpha+k)}{k!\,\Gamma(\gamma-\alpha-k)}\frac{1}{z^k} + O\left(\frac{1}{z^n}\right)\right]. \tag{32}$$

To make $G(\alpha,\gamma,z)$ single-valued, it is enough to introduce a cut along the real axis for $z < 0$ and suppose that $-\pi < \arg z \le \pi$. The asymptotic formula (32) will be valid in this region.

To determine the region of analyticity of

$$H_\nu(z) = \frac{1}{\Gamma(-\nu)}\int_0^\infty e^{-t^2-2zt}t^{-\nu-1}dt$$

we need to find a region in which the integral converges uniformly in z and ν. This takes place in a region $\mathrm{Re}\,z \ge -N$, $\delta - 1 \le -\mathrm{Re}\,\nu - 1 \le N$

$(N > 0, \delta > 0)$ because

$$|e^{-t^2-2zt}t^{-\nu-1}| < e^{-t^2+2Nt}(t^{\delta-1} + t^N)$$

and the integral

$$\int_0^\infty e^{-t^2+2Nt}(t^{\delta-1} + t^N)dt$$

converges. Since N and δ are arbitrary, $H_\nu(z)$ will be analytic in each variable for $\mathrm{Re}\,\nu < 0$.

We can continue $F(\alpha,\beta,\gamma,z)$, $F(\alpha,\gamma,z)$, $G(\alpha,\gamma,z)$ and $H_\nu(z)$ analytically under certain restrictions on the parameters. These restrictions can be removed by using the recursion relations (26)–(31). Since (26) and (27) involve functions $\phi(\alpha,\beta,\gamma,z)$ and $\phi(\alpha,\gamma,z)$ for which the difference $\gamma-\alpha$ is the same, if we then decrease α by 1 in (26) and (27) we can continue the functions

$$\phi(\alpha,\beta,\gamma,z) = \frac{1}{\Gamma(\gamma)}F(\alpha,\beta,\gamma,z),$$

and

$$\phi(\alpha,\gamma,z) = \frac{1}{\Gamma(\gamma)}F(\alpha,\gamma,z)$$

to arbitrary values of α under the additional condition $\mathrm{Re}\,(\gamma-\alpha) > 0$. The analytic continuation of $\phi(\alpha,\beta,\gamma,z)$ and $\phi(\alpha,\gamma,z)$ for $\mathrm{Re}\,(\gamma-\alpha) \leq 0$ can be obtained by repeatedly decreasing γ by 1 in (30) and (31). In a similar way, we can continue $G(\alpha,\gamma,z)$ and $H_\nu(z)$ by using (28) and (29).

Because of the differentiation formula

$$\frac{d}{dz}\phi(\alpha,\beta,\gamma,z) = \alpha\beta\phi(\alpha+1,\beta+1,\gamma+1,z),$$

which follows from (25), the derivative of $\phi(\alpha,\beta,\gamma,z)$ will be analytic in z and the parameters α,β, and γ in the same region as $\phi(\alpha,\beta,\gamma,z)$ itself. By the principle of analytic continuation, $\phi(\alpha,\beta,\gamma,z)$ will therefore satisfy the hypergeometric equation (4) in the same region. Similar considerations apply to $\phi(\alpha,\gamma,z)$, $G(\alpha,\gamma,z)$ and $H_\nu(z)$.

§ 21 Basic properties of functions of hypergeometric type

Our integral representations of functions of hypergeometric type make it possible to obtain the basic properties of these functions: recursion relations, power series expansions, functional equations, asymptotic formulas. We shall make extensive use of the results of Chapter I.

1. Recursion relations. We can show by the method of §4 that any three hypergeometric functions $F(\alpha_i, \beta_i, \gamma_i, z)(i = 1, 2, 3)$, for which the differences $\alpha_i - \alpha_k$, $\beta_i - \beta_k$, and $\gamma_i - \gamma_k$ are integers, are connected by a linear relation

$$\sum_{i=1}^{3} C_i(z) F(\alpha_i, \beta_i, \gamma_i, z) = 0,$$

where $C_i(z)$ are polynomials. To prove this, we consider the expression

$$\sum_i C_i(z) F(\alpha_i, \beta_i, \gamma_i, z).$$

Let us show that the coefficients $C_i = C_i(z)$ can be chosen so that this combination is zero. For a given z, if we use the integral representation (20.18) we have

$$\sum_i C_i F(\alpha_i, \beta_i, \gamma_i, z) = \int_0^1 t^{\alpha_0 - 1}(1-t)^{\gamma_0 - \alpha_0 - 1}(1-zt)^{-\beta_0} P(t)dt.$$

Here $\alpha_0, \gamma_0 - \alpha_0$, and $-\beta_0$ are the numbers $\alpha_i, \gamma_i - \alpha_i$, and $-\beta_i$ with the smallest real parts; $P(t)$ is a polynomial. The coefficients $C_i = C_i(z)$ are determined by the condition

$$\begin{aligned} &t^{\alpha_0 - 1}(1-t)^{\gamma_0 - \alpha_0 - 1}(1-zt)^{-\beta_0} P(t) \\ &\qquad = \frac{d}{dt}[t^{\alpha_0}(1-t)^{\gamma_0 - \alpha_0}(1-zt)^{1-\beta_0} Q(t)], \end{aligned} \tag{1}$$

where $Q(t)$ is a polynomial. We obtain

$$\sum_i C_i F(\alpha_i, \beta_i, \gamma_i, z) = t^{\alpha_0}(1-t)^{\gamma_0 - \alpha_0}(1-zt)^{1-\beta_0} Q(t)|_0^1.$$

Since $\text{Re}\,(\gamma_0 - \alpha_0) = \min \text{Re}\,(\gamma_i - \alpha_i) > 0$ and $\text{Re}\,\alpha_0 = \min \text{Re}\,\alpha_i > 0$, the integrated terms reduce to zero. By selecting the coefficients $C_i = C_i(z)$ in this way, we shall have the linear equation

$$\sum_i C_i F(\alpha_i, \beta_i, \gamma_i, z) = 0.$$

It is easy to see, by the same reasoning as in §4, that the $C_i (i = 1, 2, 3)$ are polynomials, determined up to a constant multiple. In a similar way we can deduce a recursion relation for the confluent hypergeometric functions $F(\alpha, \gamma, z)$ by using (20.19),

$$\sum_i C_i F(\alpha_i, \gamma_i, z) = \int_0^1 t^{\alpha_0 - 1}(1-t)^{\gamma_0 - \alpha_0 - 1} e^{zt} P(t)\,dt,$$

where $P(t)$ is a polynomial. Here the $C_i(z)$ satisfy

$$t^{\alpha_0 - 1}(1-t)^{\gamma_0 - \alpha_0 - 1} e^{zt} P(t) = \frac{d}{dt}[t^{\alpha_0}(1-t)^{\gamma_0 - \alpha_0} e^{zt} Q(t)], \tag{2}$$

where $Q(t)$ is a polynomial. Since

$$\sum_i C_i F(\alpha_i, \gamma_i, z) = t^{\alpha_0}(1-t)^{\gamma_0 - \alpha_0} e^{zt} Q(t)|_0^1 = 0$$

for Re $\gamma_i >$ Re $\alpha_i > 0$, we obtain the required relation.

If we replace t by $-t$ in the integral representation (20.22), we obtain an analog of the representation (20.19) for $F(\alpha, \gamma, z)$:

$$G(\alpha, \gamma, z) = \frac{1}{\Gamma(\alpha)} \int_{-\infty}^0 (-t)^{\alpha - 1}(1-t)^{\gamma - \alpha - 1} e^{zt}\,dt.$$

This differs from the integral representation of $F(\alpha, \gamma, z)$ only by a numerical factor and the limits of integration. Repeating the preceding discussion, we can easily see that the functions

$$G(\alpha, \gamma, z) \qquad \text{and} \qquad e^{i\pi\alpha} \frac{\Gamma(\gamma - \alpha)}{\Gamma(\gamma)} F(\alpha, \gamma, z)$$

satisfy the same recursion relations.

As examples, we deduce the recursion relations that connect $F(\alpha, \gamma, z)$ and $F(\alpha \pm 1, \gamma, z)$. In this case

$$\alpha_1 = \alpha - 1, \quad \alpha_2 = \alpha, \quad \alpha_3 = \alpha + 1, \quad \alpha_0 = \alpha - 1,$$
$$\gamma_0 - \alpha_0 = \gamma - \alpha - 1.$$

Up to factors independent of t, the polynomial $P(t)$ has the form

$$P(t) = C_1 \alpha(\alpha - 1)(1-t)^2 + C_2 \alpha(\gamma - \alpha) t(1-t) + C_3(\gamma - \alpha)(\gamma - \alpha - 1) t^2. \tag{3}$$

The degree of $Q(t)$ is zero; hence we may take $Q(t) = 1$. Then (2) becomes

$$e^{zt}t^{\alpha-2}(1-t)^{\gamma-\alpha-2}P(t) = \frac{d}{dt}[e^{zt}t^{\alpha-1}(1-t)^{\gamma-\alpha-1}].$$

Hence

$$P(t) = zt(1-t) + (\alpha-1)(1-t) - (\gamma-\alpha-1)t.$$

Substituting this into (3) and comparing coefficients on the two sides of the equation, we obtain

$$C_1 = \frac{1}{\alpha}, \quad C_2 = \frac{2\alpha-\gamma+z}{\alpha(\gamma-\alpha)}, \quad C_3 = -\frac{1}{\gamma-\alpha}.$$

Finally we have

$$(\gamma-\alpha)F(\alpha-1,\gamma,z) + (2\alpha-\gamma+z)F(\alpha,\gamma,z) - \alpha F(\alpha+1,\gamma,z) = 0.$$

2. Power series. We can obtain series for $F(\alpha,\beta,\gamma,z)$ and $F(\alpha,\gamma,z)$ by using (20.18) and (20.19) and expanding $(1-zt)^{-\beta}$ and e^{zt} in their power series:

$$\begin{aligned} e^{zt} &= \sum_{n=0}^{\infty} \frac{(zt)^n}{n!}, \\ (1-zt)^{-\beta} &= \sum_{n=0}^{\infty} \frac{(\beta)_n (zt)^n}{n!}, \qquad |zt| < 1, \end{aligned} \tag{4}$$

where

$$(\beta)_0 = 1, \quad (\beta)_n = \beta(\beta+1)\cdots(\beta+n-1) = \frac{\Gamma(\beta+n)}{\Gamma(\beta)}.$$

If $|z| < 1$, the series (4) converges uniformly for $0 \le t \le 1$, and therefore we can interchange summation and integration in the integral representation. For $\operatorname{Re}\gamma > \operatorname{Re}\alpha > 0$ we obtain

$$\begin{aligned} F(\alpha,\beta,\gamma,z) &= \frac{\Gamma(\gamma)}{\Gamma(\alpha)\Gamma(\gamma-\alpha)} \sum_{n=0}^{\infty} \frac{(\beta)_n}{n!} z^n \int_0^1 t^{n+\alpha-1}(1-t)^{\gamma-\alpha-1}\,dt \\ &= \sum_{n=0}^{\infty} \frac{(\alpha)_n(\beta)_n}{(\gamma)_n n!} z^n. \end{aligned} \tag{5}$$

Similarly we obtain

$$F(\alpha,\gamma,z)=\sum_{n=0}^{\infty}\frac{(\alpha)_n}{(\gamma)_n n!}z^n \tag{6}$$

for all z.

Series (6) differs from (5) only by the omission of the factor $(\beta)_n$ from each term. Series (5) is the ***hypergeometric series*** and (6) is the ***confluent hypergeometric series***.

By D'Alembert's test (see §15, p. 212), the series (5) and (6) converge uniformly in all parameters in every compact subset of their domain, not containing negative integral or zero values of γ, where in (5) we must also require in addition that $|z|\le q<1$. Hence, by Weierstrass's theorem (§15, part 4) these series represent analytic functions in all variables for $\gamma\ne -k$ ($k=0,1,2,\ldots$), and (for (5)) under the additional restriction $|z|<1$. By the principle of analytic continuation, equations (5) and (6) remain valid throughout the specified domains.

If $\alpha=-m$ ($m=0,1,\ldots$), the hypergeometric series (5) terminates and $F(\alpha,\beta,\gamma,z)$ becomes a polynomial of degree m in z. This polynomial is well-defined for $\gamma=-k$ if $m\le k$, since $(\gamma)_n=(-k)_n\ne 0$ for $n\le m$. Since $F(\alpha,\beta,\gamma,z)=F(\beta,\alpha,\gamma,z)$, this is evidently true also when $\beta=-m$. Similar remarks can also be made about (6).

Series expansions of the confluent hypergeometric function $G(\alpha,\gamma,z)$ of the second kind and of the Hermite function $H_\nu(z)$ can be obtained from the formulas that express these functions in terms of $F(\alpha,\gamma,z)$ (see part 3, below).

In solving special problems, one sometimes needs the generalized hypergeometric function ${}_pF_q(\alpha_1,\alpha_2,\ldots,\alpha_p;\beta_1,\beta_2,\ldots,\beta_q;z)$, whose power series is a generalization of (5) and (6) [L6]:

$${}_pF_q(\alpha_1,\alpha_2,\ldots,\alpha_p;\beta_1,\beta_2,\ldots,\beta_q;z)=\sum_{n=0}^{\infty}\frac{(\alpha_1)_n(\alpha_2)_n\cdots(\alpha_p)_n}{(\beta_1)_n(\beta_2)_n\cdots(\beta_q)_n}\frac{z^n}{n!}.$$

The series for these functions can converge only for $p\le q+1$; for $p=q+1$ the series converges only for $|z|<1$. In the present notation

$$F(\alpha,\beta,\gamma,z)={}_2F_1(\alpha,\beta;\gamma;z),$$

$$F(\alpha,\gamma,z)={}_1F_1(\alpha;\gamma;z).$$

3. Functional equations and asymptotic formulas. The equations

$$F(\alpha,\beta,\gamma,z)=(1-z)^{\gamma-\alpha-\beta}F(\gamma-\alpha,\gamma-\beta,\gamma,z), \tag{7}$$

$$F(\alpha,\beta,\gamma,z)=F(\beta,\alpha,\gamma,z) \tag{8}$$

that we obtained above are examples of functional equations that connect hypergeometric functions of the same argument z. Hypergeometric functions also possess a whole series of functional equations that connect functions with different arguments. In part 2, §20, we obtained some pairs of linearly independent solutions of the hypergeometric equation, containing functions of arguments $z, 1-z$, and $1/z$. Since the hypergeometric equation has only two linearly independent solutions, every solution $u(z)$ must be representable as a linear combination of any two linearly independent solutions $u_1(z)$ and $u_2(z)$:

$$u(z)=C_1u_1(z)+C_2u_2(z). \tag{9}$$

Let us notice a simple property of the coefficients C_1 and C_2: *if $u(z)$, $u_1(z)$ and $u_2(z)$ and their z-derivatives are analytic functions of α,β,γ, and z in some domain, then the coefficients $C_1=C_1(\alpha,\beta,\gamma)$ and $C_2=C_2(\alpha,\beta,\gamma)$ are also analytic in each parameter, in the same domain.*

This follows from the explicit form of C_1 and C_2:

$$C_1=W(u,u_2)/W(u_1,u_2),\quad C_2=W(u_1,u)/W(u_1,u_2). \tag{10}$$

Here

$$W(f,g)=f(z)g'(z)-f'(z)g(z)$$

is the Wronskian, which is different from zero if the functions are linearly independent. Hence in finding C_1 and C_2, it is sufficient to find them under restrictions on the parameters, and then apply the principle of analytic continuation.

1°. Let $u(z)=F(\alpha,\beta,\gamma,z)$. To determine the coefficients C_1 and C_2 in (9), we shall use (7) and (8), and the values of $u(z)$ at 0, 1 and ∞. When $\operatorname{Re}\gamma>\operatorname{Re}\alpha>0$ and $\operatorname{Re}\beta<0$, we have, from the representation (20.18),

$$\begin{aligned}\lim_{z\to1}F(\alpha,\beta,\gamma,z)&=\frac{\Gamma(\gamma)}{\Gamma(\alpha)\Gamma(\gamma-\alpha)}\int_0^1 t^{\alpha-1}(1-t)^{\gamma-\alpha-\beta-1}dt\\&=\frac{\Gamma(\gamma)\Gamma(\gamma-\alpha-\beta)}{\Gamma(\gamma-\alpha)\Gamma(\gamma-\beta)},\end{aligned}$$

$$\lim_{z\to\infty}\frac{F(\alpha,\beta,\gamma,z)}{(-z)^{-\beta}}=\frac{\Gamma(\gamma)}{\Gamma(\alpha)\Gamma(\gamma-\alpha)}\int_0^1 t^{\alpha-\beta-1}(1-t)^{\gamma-\alpha-1}dt$$
$$=\frac{\Gamma(\gamma)\Gamma(\alpha-\beta)}{\Gamma(\alpha)\Gamma(\gamma-\beta)}.$$

Here $|\arg(-z)|<\pi$. Moreover,

$$F(\alpha,\beta,\gamma,0)=1.$$

We obtain expansions of $F(\alpha,\beta,\gamma,z)$ in terms of hypergeometric functions of $1-z$ and $1/z$:

$$\begin{aligned}F(\alpha,\beta,\gamma,z)&=C_1(\alpha,\beta,\gamma)F(\alpha,\beta,\alpha+\beta-\gamma+1,1-z)\\&\quad+C_2(\alpha,\beta,\gamma)(1-z)^{\gamma-\alpha-\beta}F(\gamma-\alpha,\gamma-\beta,\gamma-\alpha-\beta+1,1-z),\end{aligned}\tag{11}$$

$$\begin{aligned}&F(\alpha,\beta,\gamma,z)\\&\quad=D_1(\alpha,\beta,\gamma)(-z)^{-\alpha}F(\alpha,\alpha-\gamma+1,\alpha-\beta+1,1/z)\\&\qquad+D_2(\alpha,\beta,\gamma)(-z)^{-\beta}F(\beta,\beta-\gamma+1,\beta-\alpha+1,1/z).\end{aligned}\tag{12}$$

If $\operatorname{Re}\gamma>\operatorname{Re}\alpha>0$ and $\operatorname{Re}\beta<0$, we can take limits in (11) as $z\to1$, and in (12) as $z\to\infty$, and find

$$C_1(\alpha,\beta,\gamma)=\frac{\Gamma(\gamma)\Gamma(\gamma-\alpha-\beta)}{\Gamma(\gamma-\alpha)\Gamma(\gamma-\beta)},\qquad D_2(\alpha,\beta,\gamma)=\frac{\Gamma(\gamma)\Gamma(\alpha-\beta)}{\Gamma(\alpha)\Gamma(\gamma-\beta)}.$$

To determine $C_2(\alpha,\beta,\gamma)$ and $D_1(\alpha,\beta,\gamma)$, we can use (7) in (11) and (8) in (12). We obtain

$$C_2(\alpha,\beta,\gamma)=C_1(\gamma-\alpha,\gamma-\beta,\gamma)=\frac{\Gamma(\gamma)\Gamma(\alpha+\beta-\gamma)}{\Gamma(\alpha)\Gamma(\beta)},$$
$$D_1(\alpha,\beta,\gamma)=D_2(\beta,\alpha,\gamma)=\frac{\Gamma(\gamma)\Gamma(\beta-\alpha)}{\Gamma(\beta)\Gamma(\gamma-\alpha)}.$$

Therefore

$$\begin{aligned}F(\alpha,\beta,\gamma,z)&=\frac{\Gamma(\gamma)\Gamma(\gamma-\alpha-\beta)}{\Gamma(\gamma-\alpha)\Gamma(\gamma-\beta)}F(\alpha,\beta,\alpha+\beta-\gamma+1,1-z)\\&+\frac{\Gamma(\gamma)\Gamma(\alpha+\beta-\gamma)}{\Gamma(\alpha)\Gamma(\beta)}(1-z)^{\gamma-\alpha-\beta}F(\gamma-\alpha,\gamma-\beta,\gamma-\alpha-\beta+1,1-z),\end{aligned}\tag{13}$$

$$F(\alpha,\beta,\gamma,z)=\frac{\Gamma(\gamma)\Gamma(\beta-\alpha)}{\Gamma(\beta)\Gamma(\gamma-\alpha)}(-z)^{-\alpha}F\left(\alpha,\alpha-\gamma+1,\alpha-\beta+1,\frac{1}{z}\right)$$
$$+\frac{\Gamma(\gamma)\Gamma(\alpha-\beta)}{\Gamma(\alpha)\Gamma(\gamma-\beta)}(-z)^{-\beta}F\left(\beta,\beta-\gamma+1,\beta-\alpha+1,\frac{1}{z}\right)\quad (|\arg(-z)|<\pi). \tag{14}$$

By the principle of analytic continuation, (13) and (14) are valid for all α,β and γ.

By using (13) and (14) it is easy to find the behavior of $F(\alpha,\beta,\gamma,z)$ as $z\to 1$ or as $z\to\infty$, by using the expansions of the functions in terms of $1-z$ or $1/z$. By combining (13), (14) and (7) we can evidently obtain still more functional equations which make it possible to express $F(\alpha,\beta,\gamma,z)$ in terms of hypergeometric functions of

$$1/(1-z),\qquad 1-1/z,\qquad \text{and}\qquad 1/(1-1/z)=z/(z-1)$$

(see the section on Basic Formulas, p. 410).

2°. Similarly we can obtain functional equations for confluent hypergeometric functions. We have

$$G(\alpha,\gamma,z)=C_1(\alpha,\gamma)F(\alpha,\gamma,z)+C_2(\alpha,\gamma)z^{1-\gamma}F(\alpha-\gamma+1,2-\gamma,z). \tag{15}$$

To find $C_2(\alpha,\gamma)$ we suppose temporarily that $\operatorname{Re}\gamma-1>\operatorname{Re}\alpha>0$ and that $z>0$, and take limits in (15) and in (22) of §20 as $z\to 0$. This yields

$$C_2(\alpha,\gamma)=\lim_{z\to 0} z^{\gamma-1}G(\alpha,\gamma,z)$$
$$=\lim_{z\to 0}\frac{z^{\gamma-1}}{\Gamma(\alpha)}\int_0^\infty e^{-zt}(1+t)^{\gamma-\alpha-1}dt$$
$$=\lim_{z\to 0}\frac{1}{\Gamma(\alpha)}\int_0^\infty e^{-s}s^{\alpha-1}(z+s)^{\gamma-\alpha-1}ds=\frac{1}{\Gamma(\alpha)}\int_0^\infty e^{-s}s^{\gamma-2}ds=\frac{\Gamma(\gamma-1)}{\Gamma(\alpha)}.$$

We can determine $C_1(\alpha,\gamma)$ by using (23) of §20:

$$G(\alpha,\gamma,z)=z^{1-\gamma}G(\alpha-\gamma+1,2-\gamma,z).$$

This yields

$$C_1(\alpha,\gamma)=C_2(\alpha-\gamma+1,2-\gamma)=\frac{\Gamma(1-\gamma)}{\Gamma(\alpha-\gamma-1)}.$$

Therefore

$$G(\alpha,\gamma,z)=\frac{\Gamma(1-\gamma)}{\Gamma(\alpha-\gamma+1)}F(\alpha,\gamma,z)+\frac{\Gamma(\gamma-1)}{\Gamma(\alpha)}z^{1-\gamma}F(\alpha-\gamma+1,2-\gamma,z). \tag{16}$$

This makes it possible to expand $G(\alpha,\gamma,z)$ in powers of z (see formula (6)). By using (16), and (17) of §20, we can obtain a representation of $F(\alpha,\gamma,z)$ as a linear combination of $G(\alpha,\gamma,z)$ and $e^z G(\gamma-\alpha,\gamma,-z)$. We have

$$\begin{aligned} e^z G(\gamma-\alpha,\gamma,-z) &= \frac{\Gamma(1-\gamma)}{\Gamma(1-\alpha)}e^z F(\gamma-\alpha,\gamma,-z) \\ &\quad+\frac{\Gamma(\gamma-1)}{\Gamma(\gamma-\alpha)}(-z)^{1-\gamma}e^z F(1-\alpha,2-\gamma,-z) \qquad (16a)\\ &= \frac{\Gamma(1-\gamma)}{\Gamma(1-\alpha)}F(\alpha,\gamma,z)+\frac{\Gamma(\gamma-1)}{\Gamma(\gamma-\alpha)}(-z)^{1-\gamma}F(\alpha-\gamma+1,2-\gamma,z). \end{aligned}$$

Here $0<\arg(-z)\le\pi$, from which it follows that

$$(-z)^{1-\gamma}=z^{1-\gamma}e^{\mp i\pi(1-\gamma)}=-z^{1-\gamma}e^{\pm i\pi\gamma}$$

for $-\pi<\arg z\le\pi$ (the plus sign corresponds to $0<\arg z\le\pi$).

If we eliminate the function $z^{1-\gamma}F(\alpha-\gamma+1,2-\gamma,z)$ from (16) and (16a) and use the addition formula for the gamma function, we obtain the functional equation

$$\begin{aligned} F(\alpha,\gamma,z) &= \frac{\Gamma(\gamma)}{\Gamma(\gamma-\alpha)}e^{\pm i\pi\alpha}G(\alpha,\gamma,z) \\ &\quad+\frac{\Gamma(\gamma)}{\Gamma(\alpha)}e^{\pm i\pi(\alpha-\gamma)}e^z G(\gamma-\alpha,\gamma,-z) \end{aligned} \tag{17}$$

(the plus sign corresponds to $0<\arg z\le\pi$; the minus, to $-\pi<\arg z\le 0$). From (17) we can obtain an asymptotic formula for $F(\alpha,\gamma,z)$ as $z\to\infty$ (see (33), §20):

$$\begin{aligned} F(\alpha,\gamma,z) &= \Gamma(\gamma)(-z)^{-\alpha}\left[\sum_{k=0}^{n-1}\frac{(\alpha)_k}{k!\Gamma(\gamma-\alpha-k)}\frac{1}{z^k}+O\left(\frac{1}{z^n}\right)\right] \\ &\quad+\Gamma(\gamma)e^z z^{\alpha-\gamma}\left[\sum_{k=0}^{n-1}\frac{(\gamma-\alpha)_k}{k!\Gamma(\gamma-k)}\frac{1}{z^k}+O\left(\frac{1}{z^n}\right)\right]. \end{aligned} \tag{18}$$

Here $(-z)^{-\alpha}$ and $z^{\alpha-\gamma}$ are to be interpreted with $-\pi<\arg(-z)\le\pi$ and $-\pi<\arg z\le\pi$, respectively.

3°. We now establish equations that connect the various solutions of the Hermite equation, $H_\nu(\pm z)$ and $e^{-z^2}H_{-\nu-1}(\pm iz)$. These depend on (9) and (10). The Wronskians that appear in (10) are easily evaluated at $z = 0$ by using the values of $H_\nu(0)$ and $H'_\nu(0)$. These values are easily calculated for $\operatorname{Re}\nu < 0$ by using (24), §20, and the duplication formula for the gamma function:

$$H_\nu(0) = \frac{2^\nu\sqrt{\pi}}{\Gamma((1-\nu)/2)}, \quad H'_\nu(0) = -\frac{2^{\nu+1}\sqrt{\pi}}{\Gamma(-\nu/2)}. \tag{19}$$

We are led to the following equations for the Hermite functions:

$$H_\nu(z) = \frac{2^\nu\Gamma(\nu+1)}{\sqrt{\pi}}e^{z^2}[e^{i\pi\nu/2}H_{-\nu-1}(iz) + e^{-i\pi\nu/2}H_{-\nu-1}(-iz)],$$

$$H_\nu(z) = e^{i\pi\nu}H_\nu(-z) + \frac{2^{\nu+1}\sqrt{\pi}}{\Gamma(-\nu)}e^{z^2+\pi i(\nu+1)/2}H_{-\nu-1}(-iz),$$

$$H_\nu(z) = e^{-i\pi\nu}H_\nu(-z) + \frac{2^{\nu+1}\sqrt{\pi}}{\Gamma(-\nu)}e^{z^2-\pi i(\nu+1)/2}H_{-\nu-1}(iz).$$

By the principle of analytic continuation, these formulas are still valid for arbitrary ν.

4°. There is also a class of functional equations connected with the symmetry of the differential equation under replacement of z by $-z$. Suppose that an equation

$$\sigma(z)u'' + \tau(z)u' + \lambda u = 0 \tag{20}$$

of hypergeometric type is invariant under replacement of z by $-z$, i.e.

$$\sigma(-z) = \sigma(z), \quad \tau(-z) = -\tau(z).$$

In this case

$$\sigma(z) = \sigma_1(z^2), \quad \tau(z) = \mu z.$$

Here $\sigma_1(s)$ is a polynomial of degree at most 1 in s, and μ is a constant. Under the substitutions $s = z^2, u(z) = v(s)$, equation (20) becomes

$$4s\sigma_1(s)v'' + 2[\sigma_1(s) + \mu s]v' + \lambda v = 0. \tag{21}$$

Equation (21) is again an equation of hypergeometric type. Therefore every solution $u(z)$ of (20) can be represented as a linear combination of any two linearly independent solutions $v_1(s)$ and $v_2(s)$ of (21). Hence we arrive at functional equations that connect functions of hypergeometric type in z with functions of hypergeometric type in $s = z^2$.

Let us consider some typical examples.

Example 1. Let $u(s)$ satisfy

$$(1-z^2)u'' - (\alpha+\beta+1)zu' - \alpha\beta u = 0,$$

which is not changed by replacing z by $-z$. If we carry this equation into canonical form by the substitution $t = (1+z)/2$, one solution will be

$$u_1(z) = F\left(\alpha, \beta, \frac{1}{2}(\alpha+\beta+1), \frac{1}{2}(1+z)\right).$$

On the other hand, the substitutions $s = z^2, u(z) = v(s)$ lead to

$$s(1-s)v'' + \left(\frac{1}{2} - \frac{1}{2}(\alpha+\beta+2)s\right)v' - \frac{1}{4}\alpha\beta v = 0,$$

whose solutions can be expressed in terms of hypergeometric functions of $s, 1-s, 1/s$, etc. Since the hypergeometric function $F(\alpha, \beta, \gamma, z)$ has simple behavior as $z \to 0$, it is natural to represent $u_1(z)$ as a linear combination of functions of $1 - s = 1 - z^2$ and determine the coefficients by using the behavior of $u_1(z)$ as $z \to -1$. Using (20), §20, we have

$$\begin{aligned}
&F\left(\alpha, \beta, \frac{1}{2}(\alpha+\beta+1), \frac{1}{2}(1+z)\right) \\
&= C_1 F\left(\frac{1}{2}\alpha, \frac{1}{2}\beta, \frac{1}{2}(\alpha+\beta+1), 1-z^2\right) \\
&+ C_2(1-z^2)^{(1-\alpha-\beta)/2} F\left(\frac{1}{2}(1-\alpha), \frac{1}{2}(1-\beta), \frac{1}{2}(3-\alpha-\beta), 1-z^2\right).
\end{aligned}$$

If $\alpha + \beta > 1$, letting $z \to -1$ yields $C_2 = 0$, $C_1 = 1$, i.e.

$$F\left(\alpha, \beta, \frac{1}{2}(\alpha+\beta+1), \frac{1}{2}(1+z)\right) = F\left(\frac{1}{2}\alpha, \frac{1}{2}\beta, \frac{1}{2}(\alpha+\beta+1), 1-z^2\right),$$

which is equivalent to

$$F\left(2\alpha, 2\beta, \alpha+\beta+\frac{1}{2}, t\right) = F\left(\alpha, \beta, \alpha+\beta+\frac{1}{2}, 4t(1-t)\right).$$

By the principle of analytic continuation, this equation remains valid for arbitrary α and β.

Example 2. Consider the equation for the Hermite function $u = H_\nu(z)$,

$$u'' - 2zu' + 2\nu u = 0.$$

This equation is not changed by replacing z by $-z$. Putting $s = z^2, u(z) = v(s)$, we obtain

$$sv'' + \left(\frac{1}{2} - s\right)v' + \frac{1}{2}\nu v = 0,$$

whose solutions can be expressed in terms of confluent hypergeometric functions:

$$v(s) = C_1 F\left(-\frac{1}{2}\nu, \frac{1}{2}, s\right) + C_2 s^{1/2} F\left(\frac{1}{2}(1-\nu), \frac{3}{2}, s\right).$$

Hence

$$H_\nu(z) = C_1 F\left(-\frac{1}{2}\nu, \frac{1}{2}, z^2\right) + C_2 z F\left(\frac{1}{2}(1-\nu), \frac{3}{2}, z^2\right).$$

It is easily seen that

$$C_1 = H_\nu(0), \quad C_2 = H_\nu'(0).$$

Therefore, using (19), we arrive at the equation

$$H_\nu(z) = \frac{2^\nu\sqrt{\pi}}{\Gamma((1-\nu)/2)} F\left(-\frac{1}{2}\nu, \frac{1}{2}, z^2\right) - \frac{2^{\nu+1}\sqrt{\pi}}{\Gamma(-\nu/2)} zF\left(\frac{1}{2} - \frac{1}{2}\nu, \frac{3}{2}, z^2\right). \tag{22}$$

By using (22) and (6), we can obtain the power series expansion of the Hermite function $H_\nu(z)$.

If $-\pi/2 < \arg z \leq \pi/2$, we can use (16) to put (22) into the form

$$H_\nu(z) = 2^\nu G\left(-\frac{1}{2}\nu, \frac{1}{2}, z^2\right). \tag{23}$$

Equations (22) and (23) make it possible to obtain an asymptotic formula for the Hermite functions by using the asymptotic formulas (derived above) for the confluent hypergeometric functions $F(\alpha, \gamma, z)$ and $G(\alpha, \gamma, z)$. In particular, for $-\pi/2 < \arg z \leq \pi/2$, we have

$$H_\nu(z) = (2z)^\nu[1 + O(1/z^2)]. \tag{24}$$

All of our functional equations may become meaningless for parameter values that lead to $F(\alpha, \beta, \gamma, z)$ or $F(\alpha, \gamma, z)$ with $\gamma = -n$ $(n = 0, 1, \ldots)$. In such singular cases we must replace $F(\alpha,\beta,\gamma,z)$ and $F(\alpha,\gamma,z)$ by

$$\phi(\alpha,\beta,\gamma,z) = F(\alpha,\beta,\gamma,z)/\Gamma(\gamma)$$

and

$$\phi(\alpha,\gamma,z) = F(\alpha,\gamma,z)/\Gamma(\gamma)$$

and remove the indetermination by L'Hospital's rule in the same way as for the Bessel functions $H_\nu^{(1,2)}(z)$ when $\nu = n$ (see §15, part 4).

Example 3. As an example of a singular case consider (16) with $\gamma = n$ ($n = 1, 2, \ldots$). Here we replace $F(\alpha,\gamma,z)$ by $\phi(\alpha,\gamma,z)$ and use the addition formula for the gamma function (see Appendix A):

$$G(\alpha,\gamma,z) = \frac{\pi}{\sin\pi\gamma}\left[\frac{\phi(\alpha,\gamma,z)}{\Gamma(\alpha-\gamma+1)} - \frac{1}{\Gamma(\alpha)}z^{1-\gamma}\phi(\alpha-\gamma+1, 2-\gamma, z)\right]. \tag{25}$$

The point $\gamma = n$ on the right-hand side of (25) is a removable singular point. We take the limit in (25) as $\gamma \to n$, calculating it by L'Hospital's rule, and obtain

$$G(\alpha,n,z) = (-1)^n \left\{\frac{\partial}{\partial\gamma}\left[\frac{\phi(\alpha,\gamma,z)}{\Gamma(\alpha-\gamma+1)}\right] - \frac{1}{\Gamma(\alpha)}\frac{\partial}{\partial\gamma}[z^{1-\gamma}\phi(\alpha-\gamma+1, 2-\gamma, z)]\right\}\bigg|_{\gamma=n}.$$

If we use the expansion of $\phi(\alpha,\gamma,z)$ in powers of z, we can now deduce the corresponding expansion for $G(\alpha,n,z)$:

$$\begin{aligned} G(\alpha,n,z) &= (-1)^n \sum_{k=0}^{\infty} \frac{(\alpha)_k z^k}{k!\Gamma(n+k)\Gamma(\alpha-n+1)}[\psi(\alpha-n+1) - \psi(n+k)] \\ &+ (-1)^n \sum_{k=0}^{\infty} \frac{(\alpha-n+1)_k z^{k-n+1}}{k!\Gamma(2-n+k)\Gamma(\alpha)}[\ln z + \psi(\alpha-n+1+k) \\ &- \psi(\alpha-n+1) - \psi(2-n+k)]. \end{aligned} \tag{26}$$

Here $\psi(z)$ is the logarithmic derivative of $\Gamma(z)$ (see Appendix A). If $2-n+k = -s$ ($s = 0, 1, \ldots$) in the last sum in (26), we must use the formulas

$$\frac{1}{\Gamma(-s)} = 0, \quad \frac{\psi(-s)}{\Gamma(-s)} = (-1)^{s+1}s!.$$

Then the last sum in (26) will fall into two parts, the first containing the terms with $0 \le k \le n-2$, and the second, the terms with $k \ge n-1$. If we replace k in the first part by $n-1-k$, and by $n-1+k$ in the second part, we finally obtain

$$G(\alpha, n, z) = \frac{(-1)^n}{(n-1)!\Gamma(\alpha-n+1)} \left\{ \sum_{k=1}^{n-1} \frac{(-1)^{k-1}(k-1)!}{(\alpha-k)_k (n-k)_k} z^{-k} \right.$$
$$\left. + \sum_{k=0}^{\infty} \frac{(\alpha)_k z^k}{k!(n)_k} [\ln z + \psi(\alpha+k) - \psi(n+k) - \psi(k+1)] \right\} \tag{27}$$

(when $n=1$ the first sum has to be taken to be zero).

4. Special cases. Let us consider the problem of finding linearly independent solutions of the hypergeometric equation for arbitrary α, β and γ. If $\gamma \neq n$ $(n = 0, \pm 1, \pm 2, \dots)$ the functions

$$u_1(z) = F(\alpha, \beta, \gamma, z)$$

and

$$u_2(z) = z^{1-\gamma} F(\alpha-\gamma+1, \beta-\gamma+1, 2-\gamma, z)$$

are linearly independent solutions. However, when $\gamma = n$, one of these functions becomes undefined and we have the problem of finding linearly independent solutions.

Let us consider this problem first when $\gamma = n$ $(n = 1, 2, 3, \dots)$ and neither α, β nor $\alpha+\beta$ is an integer. In this case the functions $F(\alpha, \beta, n, z)$ and $F(\alpha, \beta, \alpha+\beta-n+1, 1-z)$ are a pair of linearly independent solutions. To prove this we consider their behavior as $z \to 0$. We know that $F(\alpha, \beta, n, 0) = 1$. To study $F(\alpha, \beta, \alpha+\beta-n+1, 1-z)$ as $z \to 0$ we use the functional equation obtained by replacing γ by $\alpha+\beta-\gamma+1$ and z by $1-z$ in (13):

$$F(\alpha, \beta, \alpha+\beta-\gamma+1, 1-z)$$
$$= \Gamma(\alpha+\beta-\gamma+1) \left[\frac{\Gamma(1-\gamma)}{\Gamma(\alpha-\gamma+1)\Gamma(\beta-\gamma+1)} F(\alpha, \beta, \gamma, z) \right. \tag{28}$$
$$\left. + \frac{\Gamma(\gamma-1)}{\Gamma(\alpha)\Gamma(\beta)} z^{1-\gamma} F(\alpha-\gamma+1, \beta-\gamma+1, 2-\gamma, z) \right].$$

To remove the indeterminacy when $\gamma = n$ in (28) we need to replace the hypergeometric functions by

$$\phi(\alpha, \beta, \gamma, z) = \frac{1}{\Gamma(\gamma)} F(\alpha, \beta, \gamma, z).$$

By using the formula $\Gamma(z)\Gamma(1-z) = \pi/\sin\pi z$, we can rewrite (28) in the form

$$\begin{aligned} &F(\alpha,\beta,\alpha+\beta-\gamma+1,1-z) \\ &= \Gamma(\alpha+\beta-\gamma+1)\frac{\pi}{\sin\pi\gamma}\left[\frac{\phi(\alpha,\beta,\gamma,z)}{\Gamma(\alpha-\gamma+1)\Gamma(\beta-\gamma+1)}\right. \\ &\left.-\frac{1}{\Gamma(\alpha)\Gamma(\beta)}z^{1-\gamma}\phi(\alpha-\gamma+1,\beta-\gamma+1,2-\gamma,z)\right]. \end{aligned} \tag{29}$$

On the right-hand side, $\gamma = n$ is a removable singular point. Taking limits in (29) as $\gamma \to n$, by L'Hospital's rule, we find

$$\begin{aligned} &F(\alpha,\beta,\alpha+\beta-n+1,1-z) \\ &= (-1)^n\Gamma(\alpha+\beta-n+1)\left\{\frac{\partial}{\partial\gamma}\left[\frac{\phi(\alpha,\beta,\gamma,z)}{\Gamma(\alpha-\gamma+1)\Gamma(\beta-\gamma+1)}\right]\right. \\ &\left.\left.-\frac{1}{\Gamma(\alpha)\Gamma(\beta)}\frac{\partial}{\partial\gamma}[z^{1-\gamma}\phi(\alpha-\gamma+1,\beta-\gamma+1,2-\gamma,z)]\right\}\right|_{\gamma=n} \\ &= (-1)^n\Gamma(\alpha+\beta-n+1)\left\{\frac{\partial}{\partial\gamma}\left[\frac{F(\alpha,\beta,\gamma,z)}{\Gamma(\alpha-\gamma+1)\Gamma(\beta-\gamma+1)\Gamma(\gamma)}\right]\right. \\ &\left.\left.-\frac{1}{\Gamma(\alpha)\Gamma(\beta)}\frac{\partial}{\partial\gamma}\left[\frac{z^{1-\gamma}F(\alpha-\gamma+1,\beta-\gamma+1,2-\gamma,z)}{\Gamma(2-\gamma)}\right]\right\}\right|_{\gamma=n}. \end{aligned}$$

Hence

$$\begin{aligned} &F(\alpha,\beta,\alpha+\beta-n+1,1-z) \\ &= \frac{(-1)^n\Gamma(\alpha+\beta-n+1)}{\Gamma(\alpha-n+1)\Gamma(\beta-n+1)(n-1)!}\{[\psi(\alpha-n+1) \\ &\quad+\psi(\beta-n+1)-\psi(n)]F(\alpha,\beta,n,z)+\Phi(\alpha,\beta,n,z)\}, \end{aligned} \tag{30}$$

where

$$\begin{aligned} \Phi(\alpha,\beta,\gamma,z) &= \frac{\partial}{\partial\gamma}F(\alpha,\beta,\gamma,z) - \frac{\Gamma(\alpha-\gamma+1)\Gamma(\beta-\gamma+1)\Gamma(\gamma)}{\Gamma(\alpha)\Gamma(\beta)} \\ &\times\frac{\partial}{\partial\gamma}\left[z^{1-\gamma}\frac{F(\alpha-\gamma+1,\beta-\gamma+1,2-\gamma,z)}{\Gamma(2-\gamma)}\right]. \end{aligned} \tag{31}$$

We can find the power series of $\Phi(\alpha, \beta, n, z)$ by using the corresponding expansions for the hypergeometric functions. For $|z| < 1$ we have

$$\Phi(\alpha, \beta, \gamma, z) = \sum_{k=0}^{\infty} \frac{(\alpha)_k (\beta)_k}{k!(\gamma)_k} z^k [\psi(\gamma) - \psi(\gamma + k)]$$

$$+ \frac{\Gamma(\gamma)}{\Gamma(\alpha)\Gamma(\beta)} \sum_{k=0}^{\infty} \frac{\Gamma(\alpha - \gamma + 1 + k)\Gamma(\beta - \gamma + 1 + k)}{k!\Gamma(2 - \gamma + k)} z^{k+1-\gamma}$$

$$\times [\ln z + \psi(\alpha - \gamma + 1 + k) - \psi(\alpha - \gamma + 1) + \psi(\beta - \gamma + 1 + k)$$

$$- \psi(\beta - \gamma + 1) - \psi(2 - \gamma + k)].$$

Making the same transformation for $\gamma = n$ as we did in going from (26) to (27), we obtain

$$\Phi(\alpha, \beta, n, z) = \sum_{k=1}^{n-1} \frac{(-1)^{k-1}(k-1)!}{(n-k)_k (\alpha - k)_k (\beta - k)_k} z^{-k}$$

$$+ \sum_{k=0}^{\infty} \frac{(\alpha)_k (\beta)_k}{k!(n)_k} z^k [\ln z + \psi(\alpha + k) - \psi(\alpha - n + 1) + \psi(\beta + k) \tag{32}$$

$$- \psi(\beta - n + 1) + \psi(n) - \psi(n + k) - \psi(k + 1)].$$

This is still valid for $n = 1$ if we take the first sum in (32) to be zero.

We see from (30) and (32) that when α, β and $\alpha + \beta$ are not integers, the functions $F(\alpha, \beta, n, z)$ and $F(\alpha, \beta, \alpha+\beta-n+1, 1-z)$ are linearly independent, since they behave differently as $z \to 0$. Hence in this case we may take $F(\alpha, \beta, n, z)$ and $F(\alpha, \beta, \alpha+\beta-n+1, 1-z)$ as linearly independent solutions of the hypergeometric equation. Rather than taking $F(\alpha, \beta, \alpha+\beta-n+1, 1-z)$ as the second solution, it is convenient to use $\Phi(\alpha, \beta, n, z)$, since then we can weaken the restrictions on α and β. In fact, it follows from (30) that $\Phi(\alpha, \beta, \gamma, z)$ is a solution of the hypergeometric equation for $\gamma = n$ if α, β and $\alpha + \beta$ are not integers, since it is a linear combination of two solutions of this equation: $F(\alpha, \beta, n, z)$ and $F(\alpha, \beta, \alpha + \beta - n + 1, 1 - z)$. On the other hand, $\Phi(\alpha, \beta, n, z)$ and its derivatives with respect to z are analytic functions of each variable for $|\arg z| < \pi$, as follows from (31), for all values of the parameters except the case when the factor

$$\frac{\Gamma(\alpha - n + 1)\Gamma(\beta - n + 1)}{\Gamma(\alpha)\Gamma(\beta)}$$

$$= \frac{1}{(\alpha - 1)(\alpha - 2)\dots(\alpha - n + 1)(\beta - 1)(\beta - 2)\dots(\beta - n + 1)}$$

becomes infinite, that is, when $\alpha = 1, 2, \ldots, n-1$ or $\beta = 1, 2, \ldots, n-1$. Therefore for $\gamma = n$ $(n = 1, 2, \ldots)$ and

$$\frac{\Gamma(\alpha - n + 1)\Gamma(\beta - n + 1)}{\Gamma(\alpha)\Gamma(\beta)} \neq \infty$$

the functions $F(\alpha, \beta, n, z)$ and $\Phi(\alpha, \beta, n, z)$ are linearly independent solutions of the hypergeometric equation. If $\gamma = n$ $(n = 1, 2, \ldots)$ and

$$\frac{\Gamma(\alpha - n + 1)\Gamma(\beta - n + 1)}{\Gamma(\alpha)\Gamma(\beta)} = \infty,$$

then two linearly independent solutions, just as for $\gamma \neq 0, \pm 1, \ldots$, are given by $F(\alpha, \beta, \gamma, z)$ and $z^{1-\gamma}F(\alpha - \gamma + 1, \beta - \gamma + 1, 2 - \gamma, z)$, since the latter function is well defined for $\gamma = n$ and integral values of $\alpha = 1, 2, \ldots, n-1$ or $\beta = 1, 2, \ldots, n-1$. For these values of α, β, and γ this function is a polynomial since $\alpha - \gamma + 1$ or $\beta - \gamma + 1$ will take negative integral values larger than $2 - \gamma$ (see part 2).

Formula (32) for $\Phi(\alpha, \beta, n, z)$ becomes indeterminate when α or β takes one of the values $0, -1, -2, \ldots$. If we use the addition formulas for $\Gamma(z)$ and $\psi(z)$ for $z \leq 0$, then when $\alpha = -m$ $(m = 0, 1, \ldots)$ we can eliminate the indeterminacy in the following way:

$$\begin{aligned} &(\alpha)_k[\psi(\alpha + k) - \psi(\alpha - n + 1)]|_{\alpha = -m} \\ &= \begin{cases} (-1)^m (k - m - 1)! & (\alpha + k > 0), \\ (-1)^k \dfrac{m!}{(m-k)!}[\psi(m + 1 - k) - \psi(m + n)] & (\alpha + k \leq 0). \end{cases} \end{aligned}$$

Similarly we can eliminate the indeterminacy in the product

$$(\beta)_k[\psi(\beta + k) - \psi(\beta - n + 1)] \quad \text{for} \quad \beta = 0, -1, -2, \ldots.$$

It remains to consider the case when $\gamma = -n$ $(n = 0, 1, 2, \ldots)$ in the hypergeometric equation. This case can be reduced to the preceding one if we recall that the substitution $u = z^{1-\gamma}y$ leads to a hypergeometric equation for $y(z)$ with parameters $\alpha' = \alpha - \gamma + 1$, $\beta' = \beta - \gamma + 1$, $\gamma' = 2 - \gamma$ (see §20, part 2). Hence when $\gamma = -n$, linearly independent solutions of the hypergeometric equation are

$$\text{(a)} \qquad \begin{aligned} u_1(z) &= z^{n+1}F(\alpha + n + 1, \beta + n + 1, n + 2, z), \\ u_2(z) &= z^{n+1}\Phi(\alpha + n + 1, \beta + n + 1, n + 2, z), \end{aligned}$$

if

$$\frac{\Gamma(\alpha)\Gamma(\beta)}{\Gamma(\alpha+n+1)\Gamma(\beta+n+1)} \neq \infty;$$

(b)

$$u_1(z) = F(\alpha, \beta, -n, z),$$
$$u_2(z) = z^{n+1} F(\alpha+n+1, \beta+n+1, n+2, z),$$

if

$$\frac{\Gamma(\alpha)\Gamma(\beta)}{\Gamma(\alpha+n+1)\Gamma(\beta+n+1)} = \infty.$$

Thus we have constructed a complete set of the linearly independent solutions of the hypergeometric and confluent hypergeometric equations, for all possible values of the parameters.

In conclusion, we present a table of the two linearly independent solutions $u_1(z)$ and $u_2(z)$ as functions of α, β, and γ. Here $(\alpha)_k = \alpha(\alpha+1)\dots(\alpha+k-1)$, $(\alpha)_0 = 1$, $\alpha' = \alpha - \gamma + 1$, $\beta' = \beta - \gamma + 1$, $\gamma' = 2 - \gamma$.

Table 13. Linearly independent solutions of the hypergeometric equation

γ	α, β	$u_1(z)$	$u_2(z)$
$\gamma \neq 0, \pm 1, \dots$	α, β arbitrary	$F(\alpha, \beta, \gamma, z)$	$z^{1-\gamma} F(\alpha', \beta', \gamma', z)$
$\gamma = 1 + m,$	$(\alpha')_m (\beta')_m = 0,$	"	"
$m = 0, 1, \dots$	$(\alpha')_m (\beta')_m \neq 0$	"	$\Phi(\alpha, \beta, \gamma, z)$
$\gamma = 1 - m$	$(\alpha)_m (\beta)_m = 0$	"	$z^{1-\gamma} F(\alpha', \beta', \gamma', z)$
$m = 1, \dots$	$(\alpha)_m (\beta)_m \neq 0$	$z^{1-\gamma} \Phi(\alpha', \beta', \gamma', z)$	"

§ 22 Representation of various functions in terms of functions of hypergeometric type

Many special functions that arise in the solution of problems of mathematical and theoretical physics can be expressed in terms of functions of hypergeometric type: the hypergeometric functions $F(\alpha, \beta, \gamma, z)$, the confluent hypergeometric functions $F(\alpha, \gamma, z)$ and $G(\alpha, \gamma, z)$, and the Hermite functions $H_\nu(z)$. Such representations let us read off the properties of functions from the results previously obtained for functions of hypergeometric type: power series expansions, asymptotic formulas, recursion relations, and differentiation formulas. Let us consider some representative examples.

1. Some elementary functions. The functions $F(\alpha, 0, \gamma, z)$, $F(0, \gamma, z)$ and $G(0, \gamma, z)$ are particularly simple. By using power series and (21.16), we obtain

$$F(\alpha, 0, \gamma, z) = F(0, \gamma, z) = G(0, \gamma, z) = 1.$$

Applying the functional equations (20.14), (20.17), and (20.23), we now obtain

$$F(\alpha, \beta, \beta, z) = (1 - z)^{-\alpha} F(\beta - \alpha, 0, \beta, z) = (1 - z)^{-\alpha},$$
$$F(\alpha, \alpha, z) = e^z F(0, \alpha, -z) = e^z,$$
$$G(\alpha, \alpha + 1, z) = z^{-\alpha} G(0, 1 - \alpha, z) = z^{-\alpha}.$$

2. Jacobi, Laguerre, and Hermite polynomials. As we showed in §2, the polynomial solutions of the differential equation

$$\sigma(z)y'' + \tau(z)y' + \lambda y = 0 \tag{1}$$

of hypergeometric type are uniquely determined up to constant factors. Hence to find polynomials of hypergeometric type, we have only to find the polynomial solutions of (1). On the other hand, the solutions of (1) can be expressed in terms of hypergeometric, confluent hypergeometric, or Hermite functions according to the degree of $\sigma(z)$. This lets us connect the Jacobi, Laguerre and Hermite polynomials with the functions $F(\alpha, \beta, \gamma, z)$, $F(\alpha, \gamma, z)$, $G(\alpha, \gamma, z)$, and $H_\nu(z)$.

1) *Jacobi polynomials.* The differential equation (1) for $P_n^{(\alpha,\beta)}(z)$ is

$$(1 - z^2)y'' + [\beta - \alpha - (\alpha + \beta + 2)z]y' + n(n + \alpha + \beta + 1)y = 0.$$

After the substitution $z = 1 - 2s$ this becomes the hypergeometric equation

$$s(1-s)y'' + [\gamma_1 - (\alpha_1 + \beta_1 + 1)s]y' - \alpha_1\beta_1 y = 0,$$

with $\alpha_1 = -n$, $\beta_1 = n + \alpha + \beta + 1$, $\gamma_1 = \beta + 1$. The polynomial solutions of this equation are

$$y(z) = F(\alpha_1, \beta_1, \gamma_1, s) = F(-n, n + \alpha + \beta + 1, \alpha + 1, (1-z)/2).$$

Therefore

$$P_n^{(\alpha,\beta)}(z) = C_n F(-n, n + \alpha + \beta + 1, \alpha + 1, (1-z)/2).$$

The constants C_n are easily found, for example by taking $z = 1$ (see §5, part 2). We find

$$P_n^{(\alpha,\beta)}(z) = \frac{\Gamma(n+\alpha+1)}{n!\Gamma(\alpha+1)} F\left(-n, n+\alpha+\beta+1, \alpha+1, \frac{1-z}{2}\right). \quad (2)$$

By using the equation (see §6, part 6)

$$P_n^{(\alpha,\beta)}(z) = (-1)^n P_n^{(\beta,\alpha)}(-z)$$

we can obtain an equivalent form of (2):

$$P_n^{(\alpha,\beta)}(z) = \frac{(-1)^n\Gamma(n+\beta+1)}{n!\Gamma(\beta+1)} F\left(-n, n+\alpha+\beta+1, \beta+1, \frac{1+z}{2}\right). \quad (3)$$

Putting $\alpha = \beta = 0$ in (2) and (3), we obtain two equivalent representations of the Legendre polynomials as hypergeometric functions:

$$P_n(z) = F\left(-n, n+1, 1, \frac{1-z}{2}\right) = (-1)^n F\left(-n, n+1, 1, \frac{1+z}{2}\right).$$

2) *Laguerre polynomials.* The differential equation

$$zy'' + (1 + \alpha - z)y' + ny = 0$$

for the Laguerre polynomials $L_n^\alpha(z)$ has the particular solution

$$y(z) = F(-n, 1+\alpha, z),$$

which is a polynomial. Therefore

$$L_n^\alpha(z) = C_n F(-n, 1+\alpha, z).$$

The constant C_n can be found by putting $z = 0$ (§5, part 2). We obtain

$$L_n^\alpha(z) = \frac{\Gamma(n+\alpha+1)}{n!\Gamma(\alpha+1)} F(-n, 1+\alpha, z).$$

By using (21.17), $L_n^\alpha(z)$ can also be expressed in terms of the confluent hypergeometric function $G(\alpha, \gamma, z)$ of the second kind:

$$L_n^\alpha(z) = \frac{(-1)^n}{n!} G(-n, 1+\alpha, z).$$

3) *Hermite polynomials.* The differential equation

$$y'' - 2zy' + 2ny = 0$$

for the Hermite polynomials has, as particular solutions, the Hermite functions $H_n(z)$, which are polynomials of degree n. Indeed, the functional equation (21.22) yields

$$H_{2n}(z) = \frac{2^{2n}\sqrt{\pi}}{\Gamma(1/2-n)} F\left(-n, \frac{1}{2}, z^2\right),$$

$$H_{2n+1}(z) = -\frac{2^{2n+2}\sqrt{\pi}}{\Gamma(-1/2-n)} zF\left(-n, \frac{3}{2}, z^2\right).$$

Comparing the coefficients of the leading terms of the Hermite functions $H_\nu(z)$ for $\nu = n$ and for the Hermite polynomials shows that when $\nu = n$ the $H_\nu(z)$ are the same as $H_n(z)$.

3. Classical orthogonal polynomials of a discrete variable. Now let us consider the connection between the classical orthogonal polynomials of a discrete variable, and hypergeometric functions. We can obtain it from the Rodrigues formula (12.22):

$$y_n(x) = \frac{B_n}{\rho(x)} \nabla^n \rho_n(x).$$

We can show by induction that for any $f(x)$

$$\nabla^n f(x) = \sum_{k=0}^{n} (-1)^k \frac{n!}{k!(n-k)!} f(x-k) = \sum_{k=0}^{n} \frac{(-n)_k}{k!} f(x-k).$$

Therefore

$$y_n(x) = B_n \sum_{k=0}^{n} \frac{(-n)_k}{k!} \frac{\rho_n(x-k)}{\rho(x)}.$$

In particular, for the Meixner polynomials $m_n^{(\gamma,\mu)}(x)$,

$$\rho_n(x) = \rho(x+n)\prod_{k=1}^{n}\sigma(x+k) = \mu^{x+n}\frac{\Gamma(\gamma+x+n)}{\Gamma(x+1)\Gamma(\gamma)}.$$

Hence

$$\begin{aligned}\frac{\rho_n(x-k)}{\rho(x)} &= \mu^{n-k}\frac{\Gamma(\gamma+x+n-k)\Gamma(x+1)}{\Gamma(x-k+1)\Gamma(\gamma+x)}\\ &= \mu^{n-k}\frac{\Gamma(\gamma+x+n)}{\Gamma(\gamma+x)}\frac{\Gamma(x+1)}{\Gamma(x+1-k)}\frac{\Gamma(\gamma+x+n-k)}{\Gamma(\gamma+x+n)}\\ &= \mu^{n-k}(\gamma+x)_n\frac{x(x-1)\cdots(x-k+1)}{(\gamma+x+n-1)(\gamma+x+n-2)\cdots(\gamma+x+n-k)}\\ &= \mu^{n-k}(\gamma+x)_n\frac{(-x)_k}{(-\gamma-x-n+1)_k}.\end{aligned}$$

Therefore

$$m_n^{(\gamma,\mu)}(x) = (\gamma+x)_n F(-n,-x,-\gamma-x-n+1,1/\mu).$$

Similarly we can obtain, for the Kravchuk polynomials $k_n^{(p)}(x)$ and the Charlier polynomials $c_n^{(\mu)}(x)$,

$$\begin{aligned}k_n^{(p)}(x) &= (-N+x)_n\frac{p^n}{n!}F(-n,-x,N-n-x+1,-q/p),\\ c_n^{(\mu)}(x) &= (-x)_n\mu^{-n}F(-n,x-n+1,\mu).\end{aligned}$$

The Hahn polynomials $h_n^{(\alpha,\beta)}(x)$ can conveniently be expressed in terms of generalized hypergeometric functions (see §21, part 2):

$$\begin{aligned}h_n^{(\alpha,\beta)}(x) = {}&\frac{(\beta+1+x)_n(-N+1+n)_n}{n!}\\ &\times {}_3F_2(-n,-x,N+\alpha-x;N-x-n,-\beta-x-n;1).\end{aligned}$$

Remark. If we compare the formulas that express the Jacobi and Meixner polynomials in terms of hypergeometric functions, we can find a connection between these polynomials:

$$m_n^{(\gamma,\mu)}(x) = n!P_n^{(\gamma-1,-\gamma-n-x)}\left(\frac{2-\mu}{\mu}\right).$$

Similarly, we can find a connection between the Laguerre and Charlier polynomials:

$$c_n^{(\mu)}(x) = \frac{n!}{(-\mu)^n} L_n^{x-n}(\mu).$$

4. Functions of the second kind. The connection between the functions $Q_n(z)$ for the classical orthogonal polynomials and functions of hypergeometric type is most easily found if we start directly from the integral representations for $Q_n(z)$ (see §11, part 1).

1) *Jacobi functions of the second kind.* The integral representation for the Jacobi functions $Q_n^{(\alpha,\beta)}(z)$ of the second kind has the form

$$Q_n^{(\alpha,\beta)}(z) = \frac{(-1)^n}{2^n(1-z)^\alpha(1+z)^\beta} \int_{-1}^{1} \frac{(1-s)^{n+\alpha}(1+s)^{n+\beta}}{(s-z)^{n+1}} ds. \tag{4}$$

Putting $s = 2t - 1$, we obtain

$$Q_n^{(\alpha,\beta)}(z) = -\frac{2^{n+\alpha+\beta+1}}{(1-z)^\alpha(1+z)^{n+\beta+1}} \int_0^1 t^{n+\beta}(1-t)^{n+\alpha} \left(1 - \frac{2}{1+z}t\right)^{-n-1} dt.$$

If we compare this formula with the integral representation (20.18) for the hypergeometric functions, we find the following representation for $Q_n^{(\alpha,\beta)}(z)$:

$$\begin{aligned} Q_n^{(\alpha,\beta)}(z) = &-\frac{2^{n+\alpha+\beta+1}}{(1-z)^\alpha(1+z)^{n+\beta+1}} \frac{\Gamma(n+\alpha+1)\Gamma(n+\beta+1)}{\Gamma(2n+\alpha+\beta+2)} \\ &\times F(n+1, n+\beta+1, 2n+\alpha+\beta+2, 2/(1+z)). \end{aligned}$$

Similarly, if we take $s = 1 - 2t$ in (4), we obtain

$$\begin{aligned} Q_n^{(\alpha,\beta)}(z) = &\frac{(-1)^n 2^{n+\alpha+\beta+1}}{(1-z)^{n+\alpha+1}(1+z)^\beta} \frac{\Gamma(n+\alpha+1)\Gamma(n+\beta+1)}{\Gamma(2n+\alpha+\beta+2)} \\ &\times F(n+1, n+\alpha+1, 2n+\alpha+\beta+2, 2/(1-z)). \end{aligned}$$

2) *Laguerre functions of the second kind.* The integral representation for the Laguerre functions $Q_n^\alpha(z)$ of the second kind has the form

$$Q_n^\alpha(z) = \frac{1}{e^{-z} z^\alpha} \int_0^\infty \frac{e^{-s} s^{n+\alpha}}{(s-z)^{n+1}} ds.$$

Let $z < 0$. Putting $s = -zt$, we obtain

$$Q_n^\alpha(z) = e^z z^{-\alpha}(-z)^\alpha \int_0^\infty e^{zt} t^{n+\alpha}(1+t)^{-n-1}\,dt.$$

Comparing this with the integral representation (20.22) for the confluent hypergeometric function of the second kind, we obtain

$$Q_n^\alpha(z) = e^z z^{-\alpha}(-z)^\alpha \Gamma(n+\alpha+1) G(n+\alpha+1, \alpha+1, -z). \tag{5}$$

Since the definition of $Q_n^\alpha(z)$ involves the factor z^α, we must introduce a cut along the real axis for $z > 0$ in order to have $Q_n^\alpha(z)$ single-valued, i.e. we must suppose $0 < \arg z < 2\pi$. Therefore when $z < 0$ we must take $z^{-\alpha} = e^{-i\pi\alpha}(-z)^{-\alpha}$. We obtain

$$Q_n^\alpha(z) = e^{-i\pi\alpha}\Gamma(n+\alpha+1)e^z G(n+\alpha+1, \alpha+1, -z).$$

This equation was obtained for $z < 0$, but by the principle of analytic continuation it remains valid for all z.

3) *Hermite functions of the second kind.* The integral representation for the Hermite functions of the second kind is

$$Q_n(z) = (-1)^n n! e^{z^2} \int_{-\infty}^{\infty} e^{-\xi^2}(\xi - z)^{-n-1}\,d\xi.$$

In order to represent $Q_n(z)$ in terms of Hermite functions we use the fact that $Q_n(z)$ satisfies the same equation as the Hermite polynomials. Hence it can be represented as a linear combination of two linearly independent solutions of this equation:

$$Q_n(z) = A_n H_n(z) + B_n e^{z^2} H_{-n-1}(-iz), \tag{6}$$

or

$$Q_n(z) = C_n H_n(z) + D_n e^{z^2} H_{-n-1}(iz). \tag{7}$$

To determine the coefficients in these expressions we use the asymptotic formulas for $Q_n(z)$ and the Hermite functions as $z \to \infty$. Let $z = x + iy$, $y \to \infty$. Then by (11.7),

$$Q_n(z) = -\frac{e^{-y^2}}{(iy)^{n+1}} n! \sqrt{\pi}\left[1 + O\left(\frac{1}{y}\right)\right].$$

On the other hand, by (21.24) we have

$$H_n(z) = (2iy)^n[1 + O(1/y^2)],$$
$$H_{-n-1}(-iz) = (2y)^{-n-1}[1 + O(1/y^2)].$$

Hence $A_n = 0$, $B_n = 2^{n+1}n!\sqrt{\pi}e^{-i\pi(n-1)/2}$. We finally obtain

$$Q_n(z) = 2^{n+1}n!\sqrt{\pi}e^{z^2-i\pi(n-1)/2}H_{-n-1}(-iz) \quad (\text{Im}\, z > 0).$$

Similarly, we find from the equation $Q_n(\bar{z}) = \bar{Q}_n(z)$ (the bar denotes the complex conjugate) that

$$Q_n(z) = 2^{n+1}n!\sqrt{\pi}e^{z^2+i\pi(n-1)/2}H_{-n-1}(iz)$$

when $\text{Im}\, z < 0$.

5. Bessel functions. The connection between Bessel functions and the confluent hypergeometric functions of the first and second kinds is easily deduced from the integral representations of these functions. For example (see §17, part 3) we have

$$I_\nu(z) = \frac{(z/2)^\nu}{\sqrt{\pi}\Gamma(\nu+\frac{1}{2})}\int_{-1}^{1}(1-t^2)^{\nu-1/2}\cosh zt\, dt,$$

$$K_\nu(z) = \sqrt{\frac{\pi}{2z}}e^{-z}\frac{1}{\Gamma(\nu+\frac{1}{2})}\int_0^\infty e^{-t}t^{\nu-1/2}(1+t/(2z))^{\nu-1/2}dt.$$

If we replace t by zt the integral representation of $K_\nu(z)$ turns out to involve the integral representation for $G(\nu+1/2, 2\nu+1, 2z)$:

$$K_\nu(z) = \sqrt{\pi}(2z)^\nu e^{-z}G(\nu+1/2, 2\nu+1, 2z).$$

To establish the connection of $I_\nu(z)$ with a confluent hypergeometric function, we observe that we may replace $\cosh zt$ by e^{zt} in the integrand for $I_\nu(z)$, since

$$e^{zt} = \cosh zt + \sinh zt,$$

and the integral of an odd function is zero over a symmetric interval. After replacing t by $2t-1$ in the resulting integral representation we arrive at the following integral:

$$I_\nu(z) = \frac{(2z)^\nu e^{-z}}{\sqrt{\pi}\Gamma(\nu+1/2)} \int_0^1 e^{2zt}[t(1-t)]^{\nu-1/2}\,dt.$$

Hence

$$I_\nu(z) = \frac{(2z)^\nu e^{-z}}{\sqrt{\pi}} \frac{\Gamma(\nu+\frac{1}{2})}{\Gamma(2\nu+1)} F\left(\nu+\frac{1}{2}, 2\nu+1, 2z\right).$$

By the duplication formula for the gamma function we have

$$\sqrt{\pi}\Gamma(2\nu+1) = 2^{2\nu}\Gamma\left(\nu+\frac{1}{2}\right)\Gamma(\nu+1).$$

Hence we finally have

$$I_\nu(z) = \frac{(z/2)^\nu}{\Gamma(\nu+1)} e^{-z} F\left(\nu+\frac{1}{2}, 2\nu+1, 2z\right).$$

6. Elliptic integrals. The complete *elliptic integrals of the first and the second kinds* are

$$K(z) = \int_0^{\pi/2} (1 - z^2\sin^2\phi)^{-1/2}\,d\phi,$$

$$E(z) = \int_0^{\pi/2} (1 - z^2\sin^2\phi)^{1/2}\,d\phi.$$

Putting $\sin^2\phi = t$, we have the following integral representations:

$$K(z) = \frac{1}{2}\int_0^1 t^{-1/2}(1-t)^{-1/2}(1-z^2t)^{-1/2}\,dt,$$

$$E(z) = \frac{1}{2}\int_0^1 t^{-1/2}(1-t)^{-1/2}(1-z^2t)^{1/2}\,dt.$$

Comparing these integrals with the integral representations of the hypergeometric functions, we obtain

$$K(z) = \frac{\pi}{2} F\left(\frac{1}{2}, \frac{1}{2}, 1, z^2\right),$$

$$E(z) = \frac{\pi}{2} F\left(\frac{1}{2}, -\frac{1}{2}, 1, z^2\right).$$

The connection between the elliptic integrals and the hypergeometric functions makes it possible to study $K(z)$ and $E(z)$ for complex values of z.

7. Whittaker functions. One of the special cases of the generalized equation of hypergeometric type is Whittaker's equation

$$u'' + \left(-\frac{1}{4} + \frac{k}{z} + \frac{\frac{1}{4} - \mu^2}{z^2}\right) u = 0, \tag{8}$$

where k and μ are constants. Equation (8) is transformed into

$$zy'' + (2\mu + 1 - z)y' + (k - \mu - 1/2)y = 0,$$

by the substitution $u = z^{\mu+1/2}e^{-z/2}y$; the solutions of that equation are

$$y_1(z) = F\left(\frac{1}{2} - k + \mu, 2\mu + 1, z\right),$$

$$y_2(z) = G\left(\frac{1}{2} - k + \mu, 2\mu + 1, z\right).$$

Hence the Whittaker equation has solutions

$$u_1(z) = M_{k\mu}(z) = z^{\mu+1/2}e^{-z/2}F\left(\frac{1}{2} - k + \mu, 2\mu + 1, z\right),$$

$$u_2(z) = W_{k\mu}(z) = z^{\mu+1/2}e^{-z/2}G\left(\frac{1}{2} - k + \mu, 2\mu + 1, z\right),$$

which are known as *Whittaker functions.*

The function $M_{k\mu}(z)$ has simple behavior as $z \to 0$, and $W_{k\mu}(z)$ behaves simply as $z \to \infty$.

Since the Whittaker equation is unchanged if μ is replaced by $-\mu$ or if k is replaced by $-k$ and z by $-z$, it is also satisfied by $M_{k,-\mu}(z)$ and $M_{-k,\pm\mu}(-z)$, $W_{k,-\mu}(z)$ and $W_{-k,\pm\mu}(-z)$. These functions are connected by many functional equations for confluent hypergeometric functions. We have, for example,

$$M_{-k,\mu}(-z) = (-z)^{\mu+1/2} e^{z/2} F\left(\frac{1}{2} + k + \mu, 2\mu+1, -z\right)$$
$$= (-z)^{\mu+1/2} e^{-z/2} F\left(\frac{1}{2} - k + \mu, 2\mu+1, z\right),$$

i.e., $M_{k\mu}(z)$ and $M_{-k\mu}(-z)$ are linearly dependent. It follows from (20.23) that

$$W_{k,-\mu}(z) = W_{k\mu}(z).$$

§ 23 Definite integrals containing functions of hypergeometric type

The definite integrals that arise in applications and contain functions of hypergeometric type can usually be evaluated either by using integral representations for those functions or by using their expansions in series. We confine ourselves to a few examples.

1) To evaluate the integral

$$\int_0^\infty e^{-\lambda x} x^\nu F(\alpha, \gamma, kx)dx \qquad (\mathrm{Re}\,\lambda > \mathrm{Re}\,k,\ \mathrm{Re}\,\nu > -1)$$

it is convenient to use (20.19), supposing temporarily that $\mathrm{Re}\,\gamma > \mathrm{Re}\,\alpha > 0$ and $\lambda > k > 0$:

$$\int_0^\infty e^{-\lambda x} x^\nu F(\alpha, \gamma, kx)dx$$
$$= \frac{\Gamma(\gamma)}{\Gamma(\alpha)\Gamma(\gamma-\alpha)} \int_0^1 t^{\alpha-1}(1-t)^{\nu-\alpha-1} dt \int_0^\infty e^{-\lambda x + kxt} x^\nu dx$$
$$= \frac{\Gamma(\nu+1)}{\lambda^{\nu+1}} \frac{\Gamma(\gamma)}{\Gamma(\alpha)\Gamma(\gamma-\alpha)} \int_0^1 t^{\alpha-1}(1-t)^{\nu-\alpha-1} \left(1 - \frac{k}{\lambda} t\right)^{-\nu-1} dt.$$

The integral representation (20.18) yields

$$\int_0^\infty e^{-\lambda x} x^\nu F(\alpha, \gamma, kx)dx = \frac{\Gamma(\nu+1)}{\lambda^{\nu+1}} F(\alpha, \nu+1, \gamma, k/\lambda).$$

The resulting formula can be extended to arbitrary values of α, γ, λ, and k by means of the principle of analytic continuation.

2) The integral

$$\int_0^\infty e^{-a^2x^2} J_\nu(bx) x^\rho dx$$

is easily evaluated by using the series expansion of $J_\nu(bx)$. We have

$$\int_0^\infty e^{-a^2x^2} J_\nu(bx) x^\rho dx = \int_0^\infty e^{-a^2x^2} \left[\sum_{k=0}^\infty \frac{(-1)^k (bx/2)^{\nu+2k}}{k!\Gamma(k+\nu+1)}\right] x^\rho dx$$

$$= \sum_{k=0}^\infty \frac{(-1)^k (b/2)^{\nu+2k}}{k!\Gamma(k+\nu+1)} \int_0^\infty e^{-a^2x^2} x^{\nu+\rho+2k} dx.$$

On the other hand

$$\int_0^\infty e^{-a^2x^2} x^{\nu+\rho+2k} dx = \frac{1}{2a^{\nu+\rho+2k+1}} \int_0^\infty e^{-t} t^{(\nu+\rho-1)/2+k} dt$$

$$= \frac{\Gamma((\nu+\rho+1)/2+k)}{2a^{\nu+\rho+2k+1}}.$$

Hence

$$\int_0^\infty e^{-a^2x^2} J_\nu(bx) x^\rho dx = \frac{1}{2a^{\rho+1}} \sum_{k=0}^\infty (-1)^k \left(\frac{b}{2a}\right)^{\nu+2k} \frac{\Gamma((\nu+\rho+1)/2+k)}{k!\Gamma(\nu+1+k)}.$$

If we use (21.6) for the confluent hypergeometric function and (20.17), our integral is expressed as a confluent hypergeometric function:

$$\int_0^\infty e^{-a^2x^2} J_\nu(bx) x^\rho dx$$

$$= \frac{\Gamma((\nu+\rho+1)/2)}{\Gamma(\nu+1)} \frac{(b/2a)^\nu}{2a^{\rho+1}} F((\nu+\rho+1)/2, \nu+1, -b^2/(4a^2))$$

$$= \frac{\Gamma((\nu+\rho+1)/2)}{\Gamma(\nu+1)} \frac{(b/2a)^\nu}{2a^{\rho+1}} e^{-b^2/(4a^2)} F((\nu+1-\rho)/2, \nu+1, b^2/(4a^2)). \tag{1}$$

3) Let us consider the *Sonine-Gegenbauer integral*

$$\int_0^\infty \frac{K_\mu(a\sqrt{x^2+y^2})}{(x^2+y^2)^{\mu/2}} J_\nu(bx) x^{\nu+1} dx \quad (a>0, b>0, y>0, \operatorname{Re}\nu > -1).$$

To evaluate this integral we use Sommerfeld's integral for Macdonald's function and formula (1) with $\rho = \nu + 1$. We have

$$\int_0^\infty \frac{K_\mu(a\sqrt{x^2+y^2})}{(x^2+y^2)^{\mu/2}} J_\nu(bx) x^{\nu+1}\,dx$$

$$= \frac{a^\mu}{2^{\mu+1}} \int_0^\infty J_\nu(bx)x^{\nu+1}\,dx \int_0^\infty e^{-t-a^2(x^2+y^2)/(4t)} \frac{dt}{t^{\mu+1}}$$

$$= \frac{a^\mu}{2^{\mu+1}} \int_0^\infty e^{-t-a^2y^2/(4t)} \frac{dt}{t^{\mu+1}} \int_0^\infty e^{-a^2x^2/(4t)} J_\nu(bx)x^{\nu+1}\,dx$$

$$= 2^{\nu-\mu} a^{\mu-2\nu-2} b^\nu \int_0^\infty e^{-t(1+b^2/a^2)-a^2y^2/(4t)} \frac{dt}{t^{\mu-\nu}}$$

$$= \frac{2^{\nu-\mu}b^\nu}{a^\mu}(a^2+b^2)^{\mu-\nu-1} \int_0^\infty e^{-u-y^2(a^2+b^2)/(4u)} \frac{du}{u^{\mu-\nu}}$$

$$= \frac{b^\nu}{a^\mu}\left(\frac{\sqrt{a^2+b^2}}{y}\right)^{\mu-\nu-1} K_{\mu-\nu-1}(y\sqrt{a^2+b^2}).$$

Therefore

$$\int_0^\infty \frac{K_\mu(a\sqrt{x^2+y^2})}{(x^2+y^2)^{\mu/2}} J_\nu(bx)x^{\nu+1}\,dx$$
$$= \frac{b^\nu}{a^\mu}\left(\frac{\sqrt{a^2+b^2}}{y}\right)^{\mu-\nu-1} K_{\mu-\nu-1}(y\sqrt{a^2+b^2}). \tag{2}$$

Let us mention some corollaries of (2).

a) Let $\mu = 1/2$. Since

$$K_{1/2}(z) = \left(\frac{1}{2}\pi/z\right)^{1/2} e^{-z}$$

we have

$$\int_0^\infty \frac{e^{-a\sqrt{x^2+y^2}}}{\sqrt{x^2+y^2}} J_\nu(bx)x^{\nu+1}\,dx$$
$$= (2/(\pi b))^{1/2}\left(\frac{by}{\sqrt{a^2+b^2}}\right)^{\nu+1/2} K_{\nu+1/2}(y\sqrt{a^2+b^2}). \tag{3}$$

In particular, when $\nu = 0$ we obtain

$$\int_0^\infty \frac{e^{-a\sqrt{x^2+y^2}}}{\sqrt{x^2+y^2}} J_0(bx)x\,dx = \frac{e^{-y\sqrt{a^2+b^2}}}{\sqrt{a^2+b^2}}. \tag{4}$$

When $y \to 0$ and $\nu + 1/2 > 0$, formula (3) yields

$$\int_0^\infty e^{-ax} J_\nu(bx)x^\nu\,dx = \frac{1}{\sqrt{\pi(a^2+b^2)}} \left(\frac{2b}{a^2+b^2}\right)^\nu \Gamma\left(\nu + \frac{1}{2}\right). \tag{5}$$

In obtaining (5) we used the fact that

$$K_\nu(z) \approx \frac{\pi}{2\sin\pi\nu} \frac{(z/2)^{-\nu}}{\Gamma(-\nu+1)} = \frac{\Gamma(\nu)}{2}\left(\frac{z}{2}\right)^{-\nu}.$$

for $\nu > 0$ and $z \to 0$.

b) Let $\nu < 2\mu - 3/2$ and $a \to 0$ in (2). Since then

$$K_\mu(az) \approx \frac{\Gamma(\mu)}{2}\left(\frac{az}{2}\right)^{-\mu},$$

we obtain

$$\int_0^\infty \frac{J_\nu(bx)x^{\nu+1}}{(x^2+y^2)^\mu}dx = \left(\frac{b}{2y}\right)^{\mu-1} \frac{y^\mu}{\Gamma(\mu)} K_{\mu-\nu-1}(by). \tag{6}$$

Taking $\mu = 3/2$ here, and $\nu = 0$, we obtain

$$\int_0^\infty \frac{J_0(bx)x}{(x^2+y^2)^{3/2}}dx = \frac{e^{-by}}{y}. \tag{7}$$

Formulas (2)–(7) were derived under various restrictions on the parameters. The principle of analytic continuation lets us extend them to wider domains of parameter values. In particular, (6) is valid for

$$-1 < \text{Re}\,\nu < 2\text{Re}\,\mu - \frac{1}{2}.$$

Taking $\mu = 1/2$, $\nu = 0$, we obtain

$$\int_0^\infty \frac{xJ_0(bx)}{\sqrt{x^2+y^2}}dx = \frac{e^{-by}}{b}.$$

Chapter V

Solution of some problems of Mathematical Physics, Quantum Mechanics, and Numerical Analysis

The theory of special functions is a branch of mathematics whose roots penetrate deeply into analysis, the theory of functions of a complex variable, the theory of group representations, and theoretical and mathematical physics. Thanks to these connections, special functions have a wide domain of applicability. In the present chapter we discuss a number of examples of the application of special functions to the solution of some significant problems in mathematical physics, quantum mechanics, and numerical analysis.

§ 24 Reduction of partial differential equations to ordinary differential equations by the method of separation of variables

1. General outline of the method of separation of variables. Generalized equations of hypergeometric type arise, as a rule, when the equations of mathematical physics and quantum mechanics are solved by the method of separation of variables. This is a method of obtaining particular solutions of equations of the form

$$Lu = 0, \tag{1}$$

where L is representable in the form

$$L = L_1 L_2 + M_1 M_2. \tag{2}$$

Here the operators L_1 and M_1 act only on a subset of the variables, and L_2 and M_2 act on the others; a product of operators means the result of applying them successively. It is assumed that the operators L_i and M_i $(i = 1, 2)$ are linear, i.e. that

$$L_i(C_1u + C_2v) = C_1L_iu + C_2L_iv,$$
$$M_i(C_1u + C_2v) = C_1M_iu + C_2M_iv$$

(C_1, C_2 constants).

Example. Let $Lu = u_{xx} + u_{yy}$. Then

$$L_1 = \partial^2/\partial x^2, \quad L_2 = E, \quad M_1 = E, \quad M_2 = \partial^2/\partial y^2,$$

where E is the identity.

For operators of the form (2) we can look for solutions of (1) of the form $u = u_1u_2$, where u_1 involves only the first set of variables and u_2 involves the others. Since

$$L_1L_2(u_1u_2) = L_1u_1 \cdot L_2u_2,$$
$$M_1M_2(u_1u_2) = M_1u_1 \cdot M_2u_2,$$

the equation $Lu = 0$ can be rewritten in the form

$$\frac{L_1u_1}{M_1u_1} = -\frac{M_2u_2}{L_2u_2}.$$

Since $L_1u_1/(M_1u_1)$ is independent of the second group of variables and $M_2u_2/(L_2u_2)$ is independent of the first group, we must have

$$\frac{L_1u_1}{M_1u_1} = -\frac{M_2u_2}{L_2u_1} = \lambda,$$

where λ is a constant. Hence we obtain equations each containing functions of only some of the variables:

$$L_1u_1 = \lambda M_1u_1, \quad M_2u_2 = -\lambda L_2u_2. \tag{3}$$

Since L is linear, a linear combination of solutions,

$$u = \sum_i C_iu_{1i}u_{2i} \qquad (C_i,\ \text{constants}),$$

corresponding to the various admissible values $\lambda = \lambda_i$, will be a solution of (1). Under certain conditions having to do with the completeness of the set of particular solutions, every solution of $Lu = 0$ can be represented in the form $\sum_i C_iu_{1i}u_{2i}$.

We have reduced the solution of the original equation to the solution of equations with fewer variables. In the most interesting case the resulting equations can be reduced, by the same method, to a collection of ordinary differential equations.

2. Application of curvilinear coordinate systems. We have given a general description of the method of separation of variables for equations $Lu = 0$, where L is a linear operator of a certain structure. In specific problems, when one is to solve an equation $Lu = 0$ subject to boundary conditions, the method is applicable only when the variables can be separated in the boundary conditions as well as in the equations. Here one often introduces new independent variables that reflect the symmetry of the problem. A system of curvilinear coordinates should be chosen so that:

1) the boundaries of the domain where the problem is to be solved are coordinate surfaces;

2) transformation to curvilinear coordinates leads to an equation that can be separated.

Example. Let us consider the Helmholtz equation $\Delta u + k^2 u = 0$ ($\Delta = \partial^2/\partial x^2 + \partial^2/\partial y^2 + \partial^2/\partial z^2$ is the Laplacian). There are eleven systems of curvilinear coordinates in which the variables separate. These lead, as a rule, to generalized equations of hypergeometric type.

As examples we shall discuss the Helmholtz equation in parabolic cylinder coordinates and in rotational paraboloidal coordinates.* Parabolic cylinder coordinates ξ, η, ζ are related to Cartesian coordinates by the formulas

$$x = \xi\eta, \quad y = \frac{1}{2}(\xi^2 - \eta^2), \quad z = \zeta;$$

and rotational paraboloidal coordinates ξ, η, ϕ, by the formulas

$$x = \xi\eta\cos\phi, \quad y = \xi\eta\sin\phi, \quad z = \frac{1}{2}(\xi^2 - \eta^2).$$

In the first case the Helmholtz equation $\Delta u + k^2 u = 0$ becomes

$$\frac{1}{\xi^2+\eta^2}\left(\frac{\partial^2 u}{\partial \xi^2} + \frac{\partial^2 u}{\partial \eta^2}\right) + \frac{\partial^2 u}{\partial \zeta^2} + k^2 u = 0, \tag{4}$$

and in the second,

$$\frac{1}{\xi^2+\eta^2}\left[\frac{1}{\xi}\frac{\partial}{\partial \xi}\left(\xi\frac{\partial u}{\partial \xi}\right) + \frac{1}{\eta}\frac{\partial}{\partial \eta}\left(\eta\frac{\partial u}{\partial \eta}\right)\right] + \frac{1}{(\xi\eta)^2}\frac{\partial^2 u}{\partial \phi^2} + k^2 u = 0. \tag{5}$$

Let us find particular solutions of (4) by separation of variables, by taking

$$u = U(\xi)V(\eta)W(\zeta). \tag{6}$$

* In quantum mechanics textbooks rotational paraboloidal coordinates are often called parabolic coordinates.

Substituting (6) into equation (4), we obtain

$$\frac{1}{\xi^2+\eta^2}\left[\frac{U''(\xi)}{U(\xi)}+\frac{V''(\eta)}{V(\eta)}\right]=-\left[\frac{W''(\zeta)}{W(\zeta)}+k^2\right].$$

The left-hand side of this equation is independent of ζ, and the right-hand side is independent of ξ and η. Hence

$$\frac{1}{\xi^2+\eta^2}\left[\frac{U''(\xi)}{U(\xi)}+\frac{V''(\eta)}{V(\eta)}\right]=\lambda, \tag{7}$$

$$\frac{W''(\zeta)}{W(\zeta)}+k^2=-\lambda, \tag{8}$$

where λ is a constant.

Writing (7) in the form

$$\frac{U''(\xi)}{U(\xi)}-\lambda\xi^2=-\left[\frac{V''(\eta)}{V(\eta)}-\lambda\eta^2\right],$$

we find similarly that

$$\frac{U''(\xi)}{U(\xi)}-\lambda\xi^2=\mu, \quad \frac{V''(\eta)}{V(\eta)}-\lambda\eta^2=-\mu,$$

where μ is a constant.

Hence we obtain the following equations for $U(\xi), V(\eta)$, and $W(\zeta)$:

$$U''-(\lambda\xi^2+\mu)U=0, \tag{9}$$

$$V''-(\lambda\eta^2-\mu)V=0, \tag{10}$$

$$W''+(k^2+\lambda)W=0. \tag{11}$$

In a similar way, if we seek a solution of (5) in the form

$$u=U(\xi)V(\eta)W(\phi),$$

we obtain the equations

$$U''+\frac{1}{\xi}U'+(k^2\xi^2-\lambda\xi^{-2}+\mu)U=0, \tag{12}$$

$$V''+\frac{1}{\eta}V'+(k^2\eta^2-\lambda\eta^{-2}-\mu)V=0, \tag{13}$$

$$W''+\lambda W=0, \tag{14}$$

for $U(\xi), V(\eta)$ and $W(\phi)$. Equations (11) and (14) can be solved in terms of elementary functions. Equations (10) and (13) lead to (9) and (12) if we substitute $-\mu$ for μ. Hence we need only solve (9) and (12).

Equation (9) is a generalized equation of hypergeometric type. For (12), it is natural to make the preliminary substitution $\xi^2 = t$, which transforms (12) into a generalized equation of hypergeometric type,

$$\frac{d^2U}{dt^2} + \frac{1}{t}\frac{dU}{dt} + \frac{1}{4t^2}(k^2t^2 + \mu t - \lambda)U = 0. \tag{15}$$

Equations (9) and (15) can be carried respectively into the Hermite equation and the confluent hypergeometric equation (see §20) by the method described in §1.

§ 25 Boundary value problems of mathematical physics

When partial differential equations are solved by the method of separation of variables, as described in §24, the problem reduces to the solution of ordinary differential equations. The solutions of these equations can, in many interesting problems of mathematical physics, be expressed in terms of special functions. In order to obtain such solutions of the partial differential equations in specific cases, we have to impose additional conditions on the solutions so that the problems will have unique solutions. These conditions in turn lead to conditions on the solutions of the ordinary differential equations and thus to boundary value problems. Finally, the study of the properties of the solutions of boundary value problems as applied to the differential equations of special functions allows one to obtain interesting properties of special functions. Let us consider, in greater detail, the solution of boundary value problems by the method of separation of variables.

1. Sturm-Liouville problems. The method of separation of variables, as discussed in §24, is extensively used in mathematical physics for solving partial differential equations of the form

$$\rho(x, y, z)\left[A(t)\frac{\partial^2 u}{\partial t^2} + B(t)\frac{\partial u}{\partial t}\right] = Lu, \tag{1}$$

where

$$Lu = \operatorname{div}[k(x, y, z)\operatorname{grad} u] - q(x, y, z)u.$$

If $A(t) = 1, B(t) = 0$, equation (1) describes the propagation of a vibration, for example of electromagnetic or acoustic waves; when $A(t) = 0$,

$B(t) = 1$, it describes transfer processes, for example heat transfer or the diffusion of particles in a medium; when $A(t) = 0$ and $B(t) = 0$, it describes the corresponding time-independent processes.

The solutions of partial differential equations usually involve arbitrary functions. For example, the general solution of $\partial^2 u/\partial x \partial y = 0$ is $u(x, y) = f(x) + g(y)$, where f and g are arbitrary differentiable functions. Consequently if we are to have a unique solution of a partial differential equation corresponding to an actual physical problem, we must impose some supplementary conditions. The most typical conditions are initial or boundary conditions. For equation (1), initial conditions are the values of $u(x, y, z, t)$ and $\partial u(x, y, z, t)/dt$ (if $A(t) = 0$, it is enough to give only $u(x, y, z, t)|_{t=0}$). The simplest boundary conditions have the form

$$\left[\alpha(x, y, z)u + \beta(x, y, z)\frac{\partial u}{\partial n}\right]\Big|_S = 0. \tag{2}$$

Here $\alpha(x, y, z)$ and $\beta(x, y, z)$ are given functions; S is the surface bounding the domain where (1) is to be solved; $\partial u/\partial n$ is the derivative in the direction of the outward normal to S.

Let us outline the solution of equation (1), with initial and boundary conditions of the kind discussed above, by the method of separation of variables. Particular solutions of (1) under the boundary conditions (2) can be found by the method of separation of variables if we look for a solution of the form

$$u(x, y, z, t) = T(t)v(x, y, z).$$

We obtain the equations

$$A(t)T'' + B(t)T' + \lambda T = 0, \tag{3}$$

$$Lv + \lambda \rho v = 0, \tag{4}$$

where λ is a constant. Equation (3) is an ordinary differential equation which, for typical problems in mathematical physics, can easily be solved explicitly. To solve (4) we are to use the boundary condition that follows from (2), namely

$$\left[\alpha(x, y, z)v + \beta(x, y, z)\frac{\partial v}{\partial n}\right]\Big|_S = 0. \tag{5}$$

Consequently we have arrived at the following *boundary value problem*: to find a nontrivial solution of (4) under condition (5). A value of λ for which this problem has a nontrivial solution ($v(x, y, z) \not\equiv 0$) is an *eigenvalue*, and the corresponding $v(x, y, z)$ is an *eigenfunction*.

In typical problems of mathematical physics, the eigenfunctions and the eigenvalues λ can be enumerated. Let $v_n(x, y, z)$ be the eigenfunction corresponding to the eigenvalue $\lambda = \lambda_n (n = 0, 1, \dots)$. We then look for a solution of (1) under the boundary condition (2) and the corresponding initial conditions in the form

$$u(x, y, z, t) = \sum_{n=0}^{\infty} T_n(t) v_n(x, y, z),$$

where $T_n(t)$ is the solution of (3) for $\lambda = \lambda_n$. For the initial conditions to be satisfied, the values of $T_n(t)$ and $T_n'(t)$ at $t = 0$ must be chosen so that

$$u(x, y, z, t)|_{t=0} = \sum_{n=0}^{\infty} T_n(0) v_n(x, y, z),$$

$$\frac{\partial}{\partial t} u(x, y, z, t)\Big|_{t=0} = \sum_{n=0}^{\infty} T_n'(0) v_n(x, y, z).$$

Consequently in order to solve the problem for equation (1) we need to require that arbitrary functions of x, y, and z (in the present case, $u|_{t=0}$ and $\partial u/\partial t|_{t=0}$) can be expanded in series of the eigenfunctions $v_n(x, y, z)$, i.e. that the system of eigenfunctions $v_n(x, y, z)$ is *complete.**

The problem is much simpler if the boundary value problem (4)–(5) that we obtain by separation of variables reduces to a one-dimensional problem, i.e. to an equation of the form

$$Ly + \lambda \rho y = 0, \tag{6}$$

with

$$Ly = \frac{d}{dx}\left[k(x)\frac{dy}{dx}\right] - q(x)y, \qquad k(x) > 0, \quad \rho(x) > 0.$$

---

* The representation of a solution in the form $\sum_{n=0}^{\infty} T_n(t) v_n(x,y,z)$ is useful not only for equations of the form (1), but also for the more general nonhomogeneous equations

$$\rho(x, y, z)\left[A(t)\frac{\partial^2 u}{\partial t^2} + B(t)\frac{\partial u}{\partial t}\right] = Lu + F(x, y, z, t)$$

(see, for example, [V3]).

Equation (6) is considered on an interval (a, b) with boundary conditions of the form

$$\begin{aligned} \alpha_1 y(a) + \beta_1 y'(a) = 0, \\ \alpha_2 y(b) + \beta_2 y'(b) = 0 \end{aligned} \tag{7}$$

(α_j, β_j, given constants).

This is a *Sturm-Liouville problem.* The functions $k(x), k'(x), q(x)$ and $\rho(x)$ will be assumed continuous for $x \in [a, b]$.

2. Basic properties of the eigenvalues and eigenfunctions. Let us consider the basic properties of the solutions of a Sturm-Liouville problem. The simplest properties are obtained from the identity

$$\int_{x_1}^{x_2} (fLg - gLf)dx = k(x)W(f,g)\Big|_{x_1}^{x_2}, \tag{8}$$

where

$$W(f,g) = \begin{vmatrix} f & g \\ f' & g' \end{vmatrix}$$

is the Wronskian. Let $y_1(x)$ and $y_2(x)$ be two solutions of the Sturm-Liouville problem corresponding to different eigenvalues λ_1 and λ_2. By (7), we have

$$\begin{aligned} \alpha_1 y_1(a) + \beta_1 y_1'(a) = 0, \\ \alpha_1 y_2(a) + \beta_1 y_2'(a) = 0. \end{aligned}$$

If we consider these equations as homogeneous linear equations in the constants α_1 and β_1, we see that there can be a nontrivial solution only when the determinant of the system, namely the Wronskian $W(y_1, y_2)$, is zero for $x = a$. Similarly, $W(y_1, y_2)|_{x=b} = 0$. With $x_1 = a$, $x_2 = b$, $f(x) = y_1(x)$, $g(x) = y_2(x)$, we find from (8) that

$$\int_a^b (y_1 Ly_2 - y_2 Ly_1)dx = 0.$$

Because of (6), and because $\lambda_1 \neq \lambda_2$, we can rewrite this in the form

$$\int_a^b y_1(x)y_2(x)\rho(x)dx = 0 \qquad (\lambda_1 \neq \lambda_2). \tag{9}$$

Consequently *two eigenfunctions of the Sturm-Liouville problem* (6)–(7), *corresponding to different eigenvalues, are orthogonal on* (a, b) *with weight function* $\rho(x)$.

It is easily shown by using this property that *the eigenvalues of a Sturm-Liouville problem are real* if the coefficients in (6) and the constants α_i and β_i in (7) are real. In fact, suppose that $y(x)$ is an eigenfunction corresponding to a complex eigenvalue λ. Taking complex conjugates in (6) and (7), we see that the conjugate function $y^*(x)$ is an eigenfunction corresponding to the eigenvalue λ^*. If we suppose that $\lambda \neq \lambda^*$, equation (9) yields

$$\int_a^b |y(x)|^2 \rho(x) dx = 0,$$

which is impossible since $\rho(x) > 0$ and $y(x) \not\equiv 0$.

In problems of physical interest we are often required to determine eigenfunctions and eigenvalues of boundary value problems in cases when the coefficients in (6) have singularities as $x \to a$ (for example, when $k(x) \to 0$ or $q(x) \to \infty$, etc.). All the properties that we have discussed for the eigenfunctions and eigenvalues of Sturm-Liouville problems remain valid under rather general conditions on the coefficients of (6) as $x \to a$. In such cases the first boundary condition (7) is often replaced by the requirement that the solution of the problem should be bounded as $x \to a$.

For the Sturm-Liouville problem for equations without singular points, the eigenfunctions are determined by homogeneous boundary conditions of the form (7) at both $x = a$ and $x = b$. Here the properties of orthogonality of the eigenfunctions and the reality of the eigenvalues depend on the property of self-adjointness of L for the class of functions that have continuous second derivatives on (a, b):

$$\int_a^b (fLg - gLf) dx = 0.$$

By (8), we have

$$\int_a^b (fLg - gLf) dx = k(x) W(f, g) \Big|_a^b .$$

If f and g satisfy homogeneous boundary conditions at both $x = a$ and $x = b$, then L is self-adjoint since

$$W(f, g) = (fg' - gf') \Big|_{x=a,b} = 0.$$

Now let $x = a$ be a singular point of the equation. Then the properties of the eigenfunctions and eigenvalues of the Sturm-Liouville problem for equations without singularities are evidently preserved for equations with singularities provided that the boundedness condition at $x = a$ implies

$$k(x)(fg' - gf')\Big|_{x=a} = 0.$$

A classification of the eigenfunctions and eigenvalues of Sturm-Liouville problems can be based on the oscillatory properties of the solutions of the equation.

3. Oscillation properties of the solutions of a Sturm-Liouville problem. Oscillation properties of the solutions of

$$[k(x)y']' + g(x)y = 0 \tag{10}$$

with $k(x) > 0$ can be studied by means of the substitutions

$$y = r(x)\sin\phi(x), \quad ky' = r(x)\cos\phi(x). \tag{11}$$

We obtain the following equations for the unknown functions $r(x)$ and $\phi(x)$:

$$k(x)y' = k(x)(r'\sin\phi + r\phi'\cos\phi) = r\cos\phi,$$
$$g(x)y = -[k(x)y']' = -r'\cos\phi + r\phi'\sin\phi = gr\sin\phi.$$

Hence

$$r'\sin\phi + r\phi'\cos\phi = \frac{r}{k}\cos\phi,$$
$$-r'\cos\phi + r\phi'\sin\phi = gr\sin\phi.$$

Solving for ϕ' and r', we obtain the following differential equations:

$$\begin{aligned} \phi' &= \frac{1}{k(x)}\cos^2\phi + g(x)\sin^2\phi, \\ r' &= \frac{1}{2}r\left(\frac{1}{k} - g\right)\sin 2\phi. \end{aligned} \tag{12}$$

From the second equation we obtain

$$r(x) = r(x_0)\exp\left\{\frac{1}{2}\int_{x_0}^{x}\left[\frac{1}{k(t)} - g(t)\right]\sin 2\phi(t)dt\right\},$$

from which it follows that $r(x)$ has a fixed sign. Therefore any variations of sign of $y(x)$ and $y'(x)$ must result from variations of the signs of $\sin\phi(x)$ and $\cos\phi(x)$, i.e. we can study the oscillations of the solutions of (10) just by studying the behavior of the solution of the first equation in (12).

Theorem 1 (Comparison theorem). *Let $\phi(x)$ and $\bar{\phi}(x)$ be solutions of*

$$\phi' = \frac{1}{k(x)}\cos^2\phi + g(x)\sin^2\phi,$$

$$\bar{\phi}' = \frac{1}{\bar{k}(x)}\cos^2\bar{\phi} + \bar{g}(x)\sin^2\bar{\phi},$$

with $\bar{\phi}(x_0) = \phi(x_0)$. If $1/\bar{k}(x) \geq 1/k(x)$ and $\bar{g}(x) \geq g(x)$, then

$$\bar{\phi}(x) \geq \phi(x) \text{ for } x > x_0,$$

$$\bar{\phi}(x) \leq \phi(x) \text{ for } x < x_0.$$

Proof. Put

$$\frac{1}{k_\nu(x)} = \frac{1}{k(x)} + \nu\left[\frac{1}{\bar{k}(x)} - \frac{1}{k(x)}\right],$$

$$g_\nu(x) = g(x) + \nu[\bar{g}(x) - g(x)],$$

where the parameter ν takes values between 0 and 1: $0 \leq \nu < 1$. Consider the equation

$$\phi_\nu' = \frac{1}{k_\nu(x)}\cos^2\phi_\nu + g_\nu(x)\sin^2\phi_\nu \tag{13}$$

with the initial condition $\phi_\nu(x_0) = \phi(x_0)$. Since $k_0(x) = k(x)$, $k_1(x) = \bar{k}(x)$, $g_0(x) = g(x)$, and $g_1(x) = \bar{g}(x)$, we have $\phi_0(x) = \phi(x)$ and $\phi_1(x) = \bar{\phi}(x)$. Let $\psi_\nu(x) = \partial\phi_\nu(x)/\partial\nu$. By (13), we have

$$\psi_\nu' = a_\nu(x)\psi_\nu + b_\nu(x), \quad \psi_\nu(x_0) = 0,$$

where

$$a_\nu = (g_\nu - 1/k_\nu)\sin 2\phi_\nu,$$

$$b_\nu = (1/\bar{k} - 1/k)\cos^2\phi_\nu + (\bar{g} - g)\sin^2\phi_\nu.$$

Evidently $b_\nu(x) \geq 0$. The solution of the linear inhomogeneous equation for $\psi_\nu(x)$ has the form

$$\psi_\nu(x) = \int_{x_0}^{x} b_\nu(t)\exp\left[\int_t^x a_\nu(s)ds\right]dt.$$

It is clear from this expression that $\psi_\nu(x) \geq 0$ for $x > x_0$ and $\psi_\nu(x) \leq 0$ for $x < x_0$. The conclusion of the theorem follows from the evident equation

$$\bar{\phi}(x) - \phi(x) = \phi_1(x) - \phi_0(x) = \int_0^1 \frac{\partial \phi_\nu(x)}{\partial \nu} d\nu = \int_0^1 \psi_\nu(x) d\nu.$$

This completes the proof of the theorem.

Remark. If we have strict inequality in one of the hypotheses of the theorem, on part of the interval (x_0, x), then we also have strict inequality in the corresponding conclusion. This happens because $b_\nu(t) > 0$ on the subinterval under consideration.

We also notice the following property of $\phi(x)$. Since $\phi'(x) > 0$ at the points where $\phi(x) = \pi n (n = 0, \pm 1, \ldots)$ it follows that if $\phi(x_0) \geq \pi n$ we have $\phi(x) > \pi n$ for $x > x_0$. For, in the contrary case there would be a point $x_1 > x_0$ such that $\phi(x_1) = \pi n$, $\phi'(x_1) \leq 0$, which is impossible. In particular, if $\phi(x_0) \geq 0$ we have $\phi(x) > 0$ for $x > x_0$.

The number of zeros of $y(x)$ on (a, b) is (because $y = r \sin \phi$), equal to the number of points in this interval where $\phi(x) = \pi n$. It therefore follows that the number of zeros of $y(x)$ is equal to the number of integers between $\phi(a)/\pi$ and $\phi(b)/\pi$.

We next consider the oscillations of the solutions of the Sturm-Liouville problem

$$\begin{gathered} [k(x)y']' + g(x, \lambda)y = 0, \quad g(x, \lambda) = \lambda \rho(x) - q(x), \\ \alpha_1 y(a) + \beta_1 y'(a) = 0, \\ \alpha_2 y(b) + \beta_2 y'(b) = 0, \\ k(x) > 0 \text{ and } \rho(x) > 0 \text{ for } x \in [a, b]. \end{gathered} \tag{14}$$

After the substitutions (11) we obtained the following equation for $\phi(x)$:

$$\phi' = \frac{1}{k(x)} \cos^2 \phi + g(x, \lambda) \sin^2 \phi.$$

The boundary conditions (14) can be rewritten, by means of (11), in the form

$$\cot \phi(a) = -\alpha_1 k(a)/\beta_1,$$
$$\cot \phi(b) = -\alpha_2 k(b)/\beta_2.$$

The first condition will be satisfied if we put

$$\phi(a) = \cot^{-1}(-\alpha_1 k(a)/\beta_1)$$

(when $\beta_1 = 0$ we take $\phi(a) = 0$). Then $0 \leq \phi(a) < \pi$ and consequently $\phi(b) > 0$.

The second boundary condition serves to determine the eigenvalues λ:

$$\phi(b) \equiv \phi(b, \lambda) = \cot^{-1}(-\alpha_2 k(b)/\beta_2) + \pi n,$$

where n is a nonnegative integer (if $\beta_2 = 0$ we take $\cot^{-1}(-\alpha_2 k(b)/\beta_2) = \pi$).

Theorem 2 (Oscillation theorem). *The Sturm-Liouville problem has an infinite set of eigenvalues $\lambda_0 < \lambda_1 < \lambda_2 < \ldots$. The eigenfunction corresponding to $\lambda = \lambda_n$ has n zeros on (a, b).*

Proof. Let $\lambda = \lambda_n$ be the roots of the equation

$$\phi(b, \lambda) = \cot^{-1}\left(-\frac{\alpha_2 k(b)}{\beta_2}\right) + \pi n, \tag{15}$$

where n is a nonnegative integer. Since $0 \leq \phi(a) < \pi$, and $\pi n < \phi(b, \lambda) \leq \pi(n + 1)$, there are n integers between $\phi(a)/\pi$ and $\phi(b, \lambda_n)/\pi$. As we saw above, this means that the function $y(x) = r(x) \sin \phi(x)$ corresponding to the eigenvalue $\lambda = \lambda_n$ has n zeros on (a, b).

Let us now show that (15) has exactly one root for each given $n = 0, 1, \ldots$. Since the function $g(x, \lambda) = \lambda\rho(x) - q(x)$ increases with λ, and $\phi(a)$ is independent of λ, it follows from the comparison theorem (proved above) that $\phi(x) = \phi(x, \lambda)$ is a monotonic increasing function of λ for a given $x > a$. Therefore, for a given n, equation (15) can have only one root $\lambda = \lambda_n$, and $\lambda_{n+1} > \lambda_n$. Consequently the eigenvalues can be indexed by the integers $n = 0, 1, 2, \ldots$.

To show that (15) has a zero for each $n = 0, 1, \ldots$, it is enough to show that

$$\lim_{\lambda \to -\infty} \phi(b, \lambda) = 0, \qquad \lim_{\lambda \to +\infty} \phi(b, \lambda) = +\infty. \tag{16}$$

We use the comparison theorem. In the Sturm-Liouville problem, let us replace $k(x)$ and $g(x, \lambda)$ by constants $\bar{k}, \bar{g}(\lambda)$ or by $\tilde{k}, \tilde{g}(\lambda)$, where

$$\frac{1}{\bar{k}} \leq \frac{1}{k(x)} \leq \frac{1}{\tilde{k}}, \quad \bar{g}(\lambda) \leq g(x, \lambda) \leq \tilde{g}(\lambda).$$

Then the corresponding functions $\phi(x), \bar{\phi}(x)$ and $\tilde{\phi}(x)$ will satisfy

$$\phi' = \frac{1}{k(x)}\cos^2\phi + g(x,\lambda)\sin^2\phi,$$

$$\bar{\phi}' = \frac{1}{\bar{k}}\cos^2\bar{\phi} + \bar{g}(\lambda)\sin^2\bar{\phi},$$

$$\tilde{\phi}' = \frac{1}{\tilde{k}}\cos^2\tilde{\phi} + \tilde{g}(\lambda)\sin^2\tilde{\phi}.$$

If we replace the constants α_1, β_1 in the boundary conditions (14) by $\bar{\alpha}_1, \bar{\beta}_1$ or $\tilde{\alpha}_1, \tilde{\beta}_1$, respectively, chosen* so that $\phi(a) = \bar{\phi}(a) = \tilde{\phi}(a)$, we shall then have by the comparison theorem

$$\bar{\phi}(x,\lambda) \leq \phi(x,\lambda) \leq \tilde{\phi}(x,\lambda)$$

for $x > a$, and, in particular,

$$\bar{\phi}(b,\lambda) \leq \phi(b,\lambda) \leq \tilde{\phi}(b,\lambda).$$

Therefore the limit relations (16) follow from the corresponding relations for $\bar{\phi}(b,\lambda)$ and $\tilde{\phi}(b,\lambda)$.

To determine $\bar{\phi}(x,\lambda)$ and $\tilde{\phi}(x,\lambda)$, we solve the following equations for $\bar{y}(x,\lambda)$ and $\tilde{y}(x,\lambda)$:

$$\bar{y}'' + \frac{\bar{g}(\lambda)}{\bar{k}}\bar{y} = 0, \tag{17}$$

$$\tilde{y}'' + \frac{\tilde{g}(\lambda)}{\tilde{k}}\tilde{y} = 0. \tag{18}$$

Since $g(x,\lambda) = \lambda\rho(x) - q(x)$, we therefore see that $\lim_{\lambda\to-\infty} g(x,\lambda) = -\infty$ and $\lim_{\lambda\to+\infty} g(x,\lambda) = +\infty$; we may then suppose that the corresponding conditions are also satisfied by $\bar{g}(\lambda)$ and $\tilde{g}(\lambda)$. Let us show that $\lim_{\lambda\to-\infty}\bar{\phi}(b,\lambda) = 0$ and $\lim_{\lambda\to-\infty}\tilde{\phi}(b,\lambda) = 0$. The solution of (17) that satisfies $\bar{\alpha}_1\bar{y}(a) + \bar{\beta}_1\bar{y}'(a) = 0$ has the form

$$\bar{y}(x,\lambda) = \begin{cases} A\left[(\bar{\alpha}_1/\omega)\sinh\omega(x-a) - \bar{\beta}_1\cosh\omega(x-a)\right] & \text{for } \bar{g}(\lambda) < 0, \\ A\sin[\omega(x-a) + \phi_0] & \text{for } \bar{g}(\lambda) > 0, \end{cases}$$

---

* This condition will be satisfied if

$$\frac{\alpha_1}{\beta_1}k(a) = \frac{\bar{\alpha}_1}{\bar{\beta}_1}\bar{k} = \frac{\tilde{\alpha}_1}{\tilde{\beta}_1}\tilde{k}.$$

where

$$\omega = |\bar{g}/\bar{k}|^{1/2}, \quad \bar{\alpha}_1 \sin\phi_0 + \bar{\beta}_1 \omega \cos\phi_0 = 0.$$

When $x > a$ the function $\bar{y}(x,\lambda)$ has no zeros if $\bar{g}(\lambda) \to -\infty$, since in that case

$$\bar{y}(x,\lambda) \approx \begin{cases} -A\bar{\beta}_1 \cosh\omega(x-a) & \text{for } \bar{\beta}_1 \neq 0, \\ A(\bar{\alpha}_1/\omega)\sinh\omega(x-a) & \text{for } \bar{\beta}_1 = 0. \end{cases}$$

This means that $0 < \bar{\phi}(x,\lambda) < \pi$ for $x > a$ if $\lambda \to -\infty$. Moreover, it is easy to see that

$$\cot\bar{\phi}(x,\lambda) = \bar{k}\frac{\bar{y}'(x,\lambda)}{\bar{y}(x,\lambda)} \to +\infty, \quad \lambda \to -\infty.$$

This means that $\lim_{\lambda\to-\infty} \bar{\phi}(x,\lambda) = 0$.

Now let $\lambda \to +\infty$. Then it is clear from the explicit form of $\bar{y}(x,\lambda)$ that this function can have arbitrarily many zeros on (a,b), i.e. $\bar{\phi}(b,\lambda) \geq \pi n$ for $n > 0$ provided that λ is sufficiently large. Consequently $\lim_{\lambda\to+\infty} \bar{\phi}(b,\lambda) = +\infty$.

We have established the asymptotic behavior of $\bar{\phi}(x,\lambda)$, and corresponding results for $\tilde{\phi}(x,\lambda)$ can be established in a similar way. Since

$$\bar{\phi}(x,\lambda) \leq \phi(x,\lambda) \leq \tilde{\phi}(x,\lambda),$$

we have $\lim_{\lambda\to-\infty} \phi(b,\lambda) = 0$, $\lim_{\lambda\to+\infty} \phi(b,\lambda) = +\infty$. This completes the proof of the theorem.

By a similar discussion we can obtain simple inequalities for the eigenvalues. Let $\bar{\lambda}_n$ and $\tilde{\lambda}_n$ correspond to $\bar{k}(x), \bar{g}(x,\lambda)$ and to $\tilde{k}(x), \tilde{g}(x,\lambda)$, respectively, where

$$\frac{1}{\bar{k}(x)} \leq \frac{1}{k(x)} \leq \frac{1}{\tilde{k}(x)}, \quad \bar{g}(x,\lambda) \leq g(x,\lambda) \leq \tilde{g}(x,\lambda),$$

$$\alpha_1 k(a)/\beta_1 = \bar{\alpha}_1 \bar{k}(a)/\bar{\beta}_1 = \tilde{\alpha}_1 \tilde{k}(a)/\tilde{\beta}_1,$$

$$\alpha_2 k(b)/\beta_2 = \bar{\alpha}_2 \bar{k}(b)/\bar{\beta}_2 = \tilde{\alpha}_2 \tilde{k}(b)/\tilde{\beta}_2,$$

and $\alpha_i, \beta_i, \bar{\alpha}_i, \bar{\beta}_i, \tilde{\alpha}_i, \tilde{\beta}_i$ are the corresponding constants in boundary conditions of the form (14).

Since $\bar{\phi}(a) = \phi(a) = \tilde{\phi}(a)$, we have $\bar{\phi}(b,\lambda) \leq \phi(b,\lambda) \leq \tilde{\phi}(b,\lambda)$, by the comparison theorem. On the other hand,

$$\phi(b,\lambda_n) = \tan^{-1}(-\alpha_2 k(b)/\beta_2) + \pi n,$$

$$\bar{\phi}(b,\bar{\lambda}_n) = \tan^{-1}(-\bar{\alpha}_2 \bar{k}(b)/\bar{\beta}_2) + \pi n,$$

$$\tilde{\phi}(b,\tilde{\lambda}_n) = \tan^{-1}(-\tilde{\alpha}_2 \tilde{k}(b)/\tilde{\beta}_2) + \pi n,$$

whence $\phi(b,\lambda_n) = \bar{\phi}(b,\bar{\lambda}_n) = \tilde{\phi}(b,\tilde{\lambda}_n)$. Consequently $\tilde{\lambda}_n \le \lambda_n \le \bar{\lambda}_n$ because $\phi(b,\lambda), \bar{\phi}(b,\lambda)$ and $\tilde{\phi}(b,\lambda)$ increase with λ.

We also discuss a useful method of obtaining lower bounds for the eigenvalues when $\alpha_1\beta_1 \le 0$ and $\alpha_2\beta_2 \ge 0$, a case that is important in applications. We multiply the equation

$$[k(x)y']' + [\lambda\rho(x) - q(x)]y = 0$$

by $y(x)$ and integrate from $x = a$ to b. We have

$$\lambda = -\frac{\int_a^b yLy\,dx}{\int_a^b y^2\rho\,dx} = \frac{\int_a^b qy^2dx - \int_a^b y\frac{d}{dx}\left(k\frac{dy}{dx}\right)dx}{\int_a^b y^2\rho\,dx}.$$

On the other hand,

$$-\int_a^b y\frac{d}{dx}\left(k\frac{dy}{dx}\right)dx = -kyy'\Big|_a^b + \int_a^b k\left(\frac{dy}{dx}\right)^2 dx.$$

To estimate the integrated terms, we use the boundary conditions (14), multiplying the first by $(\alpha_1 y' + \beta_1 y)|_{x=a}$ and the second by $(\alpha_2 y' + \beta_2 y)|_{x=b}$. We obtain

$$yy'|_{x=a} = -\frac{\alpha_1\beta_1}{\alpha_1^2+\beta_1^2}(y^2+y'^2)|_{x=a} \ge 0,$$

$$yy'|_{x=b} = -\frac{\alpha_2\beta_2}{\alpha_2^2+\beta_2^2}(y^2+y'^2)|_{x=b} \le 0.$$

Therefore

$$-\int_a^b y\frac{d}{dx}\left(k\frac{dy}{dx}\right)dx \ge 0,$$

from which it follows that

$$\lambda \ge \int_a^b qy^2dx \Big/ \int_a^b y^2\rho\,dx.$$

Since $\rho(x) > 0$, we have, by the mean value theorem,

$$\int_a^b qy^2dx = \left(\frac{q}{\rho}\right)\Big|_{x=x^*} \int_a^b y^2\rho\,dx, \quad x^* \in (a,b).$$

Therefore

$$\lambda \geq \min_{x\in(a,b)} \frac{q(x)}{\rho(x)}. \tag{19}$$

There will be strict inequality when $y(x) \neq$ const., since

$$\int_a^b k\left(\frac{dy}{dx}\right)^2 dx > 0.$$

4. Expansion of functions in eigenfunctions of a Sturm-Liouville problem. Many boundary value problems of mathematical physics can be solved by using the *expansions of functions in terms of the eigenfunctions of a Sturm-Liouville problem*,

$$f(x) = \sum_{n=0}^{\infty} a_n y_n(x). \tag{20}$$

Here $y_n(x)$ is the eigenfunction corresponding to the eigenvalue $\lambda = \lambda_n$; the coefficients a_n are found by using the orthogonality of the eigenfunctions:

$$a_n = \int_a^b f(x)y_n(x)\rho(x)dx \bigg/ \int_a^b y_n^2(x)\rho(x)dx. \tag{21}$$

In the special case when $k(x) = 1, \rho(x) = 1, q(x) = 0$ and $\beta_1 = \beta_2 = 0$, the eigenfunctions $y_n(x)$ have the form

$$y_n(x) = A_n \sin\frac{\pi n}{l}(x-a), \quad \lambda = \sqrt{\frac{\pi n}{l}}$$

and when $\alpha_1 = \alpha_2 = 0$,

$$y_n(x) = B_n \cos\frac{\pi n}{l}(x-a), \quad \lambda_n = \sqrt{\frac{\pi n}{l}} \quad (l = b-a).$$

In these special cases (20) becomes the Fourier sine or cosine expansion.

In the general case, conditions for the validity of (20) can be obtained from conditions for expanding a function in Fourier series, as was done in §8 for the classical orthogonal polynomials (see the equiconvergence theorem).

5. Boundary value problems for Bessel's equation. As an example of the solution of a physical problem by separation of variables we consider the solution of the heat flow equation

$$\partial u/\partial t = a^2 \Delta u$$

in the infinite cylinder $r < r_0$ with boundary condition

$$(\alpha u + \beta \partial u/\partial r)|_{r=r_0} = 0 \tag{22}$$

and with arbitrary initial conditions, independent of distance along the cylinder (α and β are constants).

Using cylindrical coordinates, we naturally suppose that $u = u(r, \phi, t)$. We look for particular solutions by separation of variables, taking

$$u = T(t)R(r)\Phi(\phi).$$

Substituting this into the heat flow equation,

$$\frac{1}{a^2}\frac{\partial u}{\partial t} = \frac{1}{r}\left(r\frac{\partial u}{\partial r}\right) + \frac{1}{r^2}\frac{\partial^2 u}{\partial \phi^2},$$

we obtain

$$\frac{1}{a^2}\frac{T'}{T} = \frac{1}{rR}(rR')' + \frac{1}{r^2}\frac{\Phi''}{\Phi} = -\lambda.$$

Here λ is a constant since the left-hand side of the equation is independent of r and ϕ, and the right-hand side is independent of t. The equation for $T(t)$ yields

$$T(t) = e^{-\lambda a^2 t}.$$

We also have

$$\frac{r}{R}(rR')' + \lambda r^2 = -\frac{\Phi''}{\phi} = \mu \qquad (\mu = \text{ const }).$$

Solving for $\Phi(\phi)$, we obtain

$$\Phi(\phi) = A\cos\sqrt{\mu}\phi + B\sin\sqrt{\mu}\phi.$$

Since, for physical reasons, $u(r, \phi, t)$ must be single-valued, Φ must be periodic,

$$\Phi(\phi + 2\pi) = \Phi(\phi),$$

whence $\mu = n^2, n = 0, 1, \ldots$. Therefore $R(r)$ must satisfy

$$R'' + \frac{1}{r}R' + \left(\lambda - \frac{n^2}{r^2}\right) R = 0, \tag{23}$$

which is a special case of the Lommel equation (14.4). For physical reasons, $u(r, \phi, t)$ should be bounded for $r \le r_0$, and in particular as $r \to 0$. Hence, up to a constant factor,

$$R(r) = J_n(\sqrt{\lambda} r).$$

By (22), $R(r)$ must satisfy the boundary condition

$$[\alpha R(r) + \beta R'(r)]|_{r=r_0} = 0, \tag{24}$$

whence we obtain an equation for determining the possible values of λ:

$$\alpha J_n(z) + \gamma z J_n'(z) = 0. \tag{25}$$

Here

$$z = \sqrt{\lambda} r_0, \quad \gamma = \beta / r_0.$$

The general solution of our problem can be represented as a superposition of the particular solutions:

$$u(r, \phi, t) = \sum_{n=0}^{\infty} \sum_{m=0}^{\infty} e^{-\lambda_{nm} a^2 t} (A_{nm} \cos n\phi + B_{nm} \sin n\phi) J_n(\sqrt{\lambda_{nm}} r)$$

(summation over all different eigenvalues λ).

The constants A_{nm} and B_{nm} are determined from the initial conditions by using the orthogonality of the eigenfunctions.

A natural generalization of (23)–(24) is the problem of finding the eigenvalues and eigenfunctions for

$$\frac{d}{dx}\left(x \frac{dy}{dx}\right) + \left(\lambda x - \frac{\nu^2}{x}\right) y = 0 \qquad (\nu \ge 0) \tag{26}$$

with the boundary condition $[\alpha y(x) + \beta y'(x)]|_{x=l} = 0$ and a boundedness condition as $x \to 0$. For a given λ and $\nu \ge 0$ the condition at $x = 0$ is satisfied by only one of the two linearly independent solutions of (26):

$$y_\lambda(x) = J_\nu(kx) \qquad (k = \sqrt{\lambda}).$$

Equation (26) has a singular point at $x = 0$. Consequently we cannot know that the basic properties of the eigenvalues and eigenfunctions of a Sturm-Liouville system are still available until we know that

$$k(x)W[y_{\lambda_1}(x), y_{\lambda_2}(x)] \underset{x\to 0}{\longrightarrow} 0.$$

Using the power series of $J_\nu(kx)$ and setting $k(x) = x$, we obtain

$$\begin{aligned}
&x[y_{\lambda_1}(x)y'_{\lambda_2}(x) - y_{\lambda_2}(x)y'_{\lambda_1}(x)] \\
&\quad = x\left[J_\nu(k_1x)\frac{d}{dx}J_\nu(k_2x) - J_\nu(k_2x)\frac{d}{dx}J_\nu(k_1x)\right] \\
&= x\left\{\frac{1}{2^{2\nu}\Gamma^2(\nu+1)}[(k_1x)^\nu \nu k_2(k_2x)^{\nu-1} - (k_2x)^\nu \nu k_1(k_1x)^{\nu-1}]\right. \\
&\quad \left. + O(x^{2\nu+1})\right\} = O(x^{2\nu+2}) \to 0, \qquad x \to 0.
\end{aligned}$$

We obtain the following propositions.

1) *The eigenfunctions for our problem are*

$$y_{\nu n}(x) = J_\nu(k_nx) \qquad (k_n = \sqrt{\lambda_n},\ n = 0, 1, \ldots),$$

and the eigenvalues $\lambda = \lambda_n$ *are determined by the equation*

$$\alpha J_\nu(z) + \gamma z J'_\nu(z) = 0, \tag{27}$$

where $z = kl$, $k = \sqrt{\lambda}$, $\gamma = \beta/l$.

If $\alpha/\gamma + \nu < 0$, equation (27) will have a single root corresponding to an eigenvalue $\lambda < 0$. For this value of λ we have to replace $\sqrt{\lambda}$ and $J_\nu(\sqrt{\lambda}x)$ by $i\sqrt{-\lambda}$ and $e^{i\pi\nu/2}I_\nu(\sqrt{-\lambda}x)$, respectively.

2) *The eigenfunctions* $J_\nu(k_nx)$ *are orthogonal with weight function* $\rho(x) = x$ *on* $(0, l)$, *i.e.*

$$\int_0^l J_\nu(k_nx)J_\nu(k_mx)x\,dx = 0 \quad (m \neq n).$$

The squared norms $N_{\nu n}^2$ of the eigenfunctions $J_\nu(k_n x)$ are easily calculated by integration by parts:

$$\begin{aligned} N_{\nu n}^2 &= \int_0^l J_\nu^2(k_n x)x\,dx = \frac{1}{k_n^2}\int_0^{k_n l} J_\nu^2(z)z\,dz \\ &= \frac{1}{k_n^2}\int_0^{k_n l} J_\nu^2(z)d\frac{z^2}{2} = \frac{1}{k_n^2}\left[\frac{z^2}{2}J_\nu^2(z)\bigg|_{z=k_n l} - \int_0^{k_n l} z^2 J_\nu(z)J_\nu'(z)dz\right]. \end{aligned} \tag{28}$$

Since according to the Bessel equation

$$\begin{aligned} \int_0^{k_n l} z^2 J_\nu(z)J_\nu'(z)dz &= \int_0^{k_n l} J_\nu'(z)[\nu^2 J_\nu(z) - zJ_\nu'(z) - z^2 J_\nu''(z)]dz \\ &= \frac{1}{2}\left\{\nu^2 J_\nu^2(z) - [zJ_\nu'(z)]^2\right\}\bigg|_{z=k_n l} \end{aligned}$$

we finally obtain

$$\begin{aligned} N_{\nu n}^2 &= \left\{\frac{l^2}{2}J_\nu^2(z) + \frac{1}{2k_n^2}[(zJ_\nu'(z))^2 - \nu^2 J_\nu^2(z)]\right\}\bigg|_{z=k_n l} \\ &= \frac{l^2}{2}\left\{[J_\nu'(z)]^2 + \left(1 - \frac{\nu^2}{z^2}\right)J_\nu^2(z)\right\}\bigg|_{z=k_n l}. \end{aligned} \tag{29}$$

Equation (26) has a singular point at $x = 0$. However, it can be shown that the oscillation theorem still holds, so that (27) has an infinite number of roots $\lambda_0 < \lambda_1 < \lambda_2 < \ldots$, and the eigenfunctions $y_\lambda(x)$ corresponding to the eigenvalues $\lambda = \lambda_n$ have n zeros in $(0, l)$. By the comparison theorem, $\lambda = \lambda_n$ increases with ν.

6. Dini and Fourier-Bessel expansions. Fourier-Bessel integral. An expansion

$$f(x) = \sum_{n=0}^{\infty} a_n J_\nu(k_n x), \tag{30}$$

where

$$a_n = \frac{1}{N_{\nu n}^2}\int_0^l x f(x) J_\nu(k_n x)dx, \tag{31}$$

is a *Dini expansion* of $f(x)$. Here λ_n are the roots of (27) and the square of the norm is given by (29). If (27) reduces to $J_\nu(z) = 0$, that is, if $\gamma = 0$, then (30) is a *Fourier-Bessel expansion*. We have the following theorem.

Theorem 3. *Let $\sqrt{x}f(x)$ be absolutely integrable on $[0, l]$ and let $\nu \geq -1/2$. Then, for $0 < x < l$, the expansion (30) is equiconvergent with the ordinary Fourier series.*

The theory of Fourier-Bessel and Dini expansions is described in detail in [W2].

In mathematical physics one often needs a limiting form of the Fourier-Bessel expansion that is obtained from (30) by letting $l \to \infty$. We shall give only a heuristic derivation. By (29)–(31),

$$f(x) = \sum_{n=0}^{\infty} \frac{\int_0^l x f(x) J_\nu(k_n x)\,dx}{\frac{1}{2} l^2 [J'_\nu(k_n l)]^2} J_\nu(k_n x), \tag{32}$$

where k_n are determined from

$$J_\nu(k_n l) = 0. \tag{33}$$

The first few terms of (32) do not contribute very much because of the factor l^2 in the denominator. Hence we may use the asymptotic value of k_n for large n. Using (33), we obtain

$$\cos\left(k_n l - \frac{\pi\nu}{2} - \frac{\pi}{4}\right) \approx 0,$$

whence

$$k_n l \approx \pi n + \text{ const.}$$

Calculating $J'_\nu(k_n l)$ by the differentiation formula, we have

$$[J'_\nu(k_n l)]^2 = [J_{\nu+1}(k_n l)]^2 \approx \frac{2}{\pi(k_n l)} \sin^2\left(k_n l - \frac{\pi\nu}{2} - \frac{\pi}{4}\right) \approx \frac{2}{\pi k_n l}$$

(we took $\sin^2(k_n l - \pi\nu/2 - \pi/4) \approx 1$, since $\cos(k_n l - \pi\nu/2 - \pi/4) \approx 0$). Since $\Delta k_n = k_{n+1} - k_n \approx \pi/l$, the expansion (32) can be put in the form

$$f(x) \approx \sum_{k_n=0}^{\infty} k_n J_\nu(k_n x) \Delta k_n \int_0^l x f(x) J_\nu(k_n x)\,dx.$$

Since $\Delta k_n \to 0$ as $l \to \infty$, if we replace summation over k_n by integration, we obtain

$$f(x) = \int_0^\infty kF(k)J_\nu(kx)dk, \tag{34}$$

$$F(k) = \int_0^\infty xf(x)J_\nu(kx)dx. \tag{35}$$

The expansion (34) is a *Fourier-Bessel integral.*

Conditions for a function $f(x)$ to be expandable in a Fourier-Bessel integral are discussed in [W2]. We have the following theorem.

Theorem 4. *Let $\sqrt{x}f(x)$ be absolutely integrable on $(0,\infty)$ and let $\nu \geq -1/2$. Then, for $x > 0$, the expansions* (34) *and* (35) *are equiconvergent with the corresponding Fourier integral expansions.*

We may observe that when $\nu = \pm 1/2$ the expansions (30) and (34) reduce to the series and integral expansions of $\sqrt{x}f(x)$ in terms of cosines ($\nu = -1/2$) or sines ($\nu = 1/2$).

§ 26 Solution of some basic problems in quantum mechanics

In §9 we discussed a general method of solving quantum mechanics problems for a state with a discrete energy spectrum, when these problems can be reduced by the method of separation of variables to differential equations of the form

$$u'' + \frac{\tilde{\tau}(x)}{\sigma(x)}u' + \frac{\tilde{\sigma}(x)}{\sigma^2(x)}u = 0, \tag{1}$$

Here $\sigma(x)$ and $\tilde{\sigma}(x)$ are polynomials of degree at most 2, and $\tilde{\tau}(x)$ is of degree at most 1. In the present section we shall discuss the solution of some basic quantum mechanics problems by the method of §9. Observe that the differential equation (1) governs such fundamental problems as the motion of a particle in a central field, the harmonic oscillator, the solution of the Schrödinger, Dirac and Klein-Gordon equations for the Coulomb field, and the motion of a charged particle in a homogeneous electric or magnetic field. In addition, (1) governs models of many problems in atomic, molecular and nuclear physics connected with the study of scattering, interactions of neutrons with attracting nuclei, and analysis of the rotation-vibration spectra of molecules (for example, the solution of the Schrödinger equation with the Morse, Kratzer, Wood-Saxon, and Pöschl-Teller potentials) (see [F2]).

To find the energy eigenvalues and the eigenfunctions for the Schrödinger, Dirac or Klein-Gordon equation, we use the method of separation of variables on some interval (a, b). The energy E appears as a parameter in the coefficients of (1). Supplementary requirements have to be imposed, depending on the problem. These are usually equivalent to the following conditions on the solutions of (1): $u(x)\sqrt{\tilde{\rho}(x)}$ is to be bounded and of integrable square on (a, b). Here $\tilde{\rho}(x)$ is a solution of $(\sigma\tilde{\rho})' = \tilde{\tau}\tilde{\rho}$, where $\tilde{\rho}(x)$ appears when (1) is put into self-adjoint form:

$$(\sigma\tilde{\rho}u')' + \tilde{\rho}(x)\frac{\tilde{\sigma}(x)}{\sigma(x)}u = 0.$$

As we showed in §9, this problem can be solved in the following way. First we transform (1) into the equation

$$\sigma(x)y'' + \tau(x)y' + \lambda y = 0$$

of hypergeometric type by the substitution $u = \phi(x)y$. We choose the method of reduction so that the function

$$\tau(x) = \tilde{\tau}(x) + 2\pi(x)$$

has a negative derivative and a zero on (a, b). We also suppose that $\sigma(x) > 0$ for $x \in (a, b)$. The eigenvalues are determined by

$$\lambda + n\tau' + \frac{1}{2}n(n-1)\sigma'' = 0 \qquad (n = 0, 1, \ldots),$$

and the eigenfunctions $y_n(x)$ are the polynomials

$$y_n(x) = \frac{B_n}{\rho(x)}\frac{d^n}{dx^n}[\sigma^n(x)\rho(x)],$$

of degree n; they are orthogonal on (a, b) with weight $\rho(x)$ satisfying the equation $(\sigma\rho)' = \tau\rho$ (B_n is a normalizing constant).

Let us consider some typical quantum mechanics problems that can be studied by this method.

1. Solution of the Schrödinger equation for a central field. A basic problem in the quantum theory of the atom is the problem of the motion of an electron in a central attractive force field. This is partly because a description of the motion of electrons in an atom by means of the central force approximation is the basis for calculating various aspects of atomic structure (see [H1]). Such a description makes it easier to understand the behavior of atoms and to find their energy states without having to solve the extremely difficult quantum mechanical many-body problem.

To find the wave function $\psi(\mathbf{r})$ of a particle in a central force field with potential $U(r)$, we have to solve the Schrödinger equation

$$\Delta\psi + \frac{2M}{\hbar^2}[E - U(r)]\psi = 0 \tag{2}$$

($\hbar$ is Planck's constant, M is the mass of the particle, and $U(r)$ is the potential energy).

We shall try to find a solution of (2) by separating variables in spherical coordinates, putting

$$\psi(\mathbf{r}) = F(r)Y(\theta, \phi).$$

Proceeding as for Laplace's equation (see §10), we obtain the equations

$$\Delta_{\theta,\phi}Y + \lambda Y = 0, \tag{3}$$

$$\frac{1}{r^2}\frac{d}{dr}\left(r^2\frac{dF}{dr}\right) + \left[\frac{2M}{\hbar^2}(E - U(r)) - \frac{\lambda}{r^2}\right]F(r) = 0, \tag{4}$$

for $Y(\theta, \phi)$ and $F(r)$. As we showed before, equation (3) has solutions that are bounded and single-valued for $0 \le \theta \le \pi$, $0 \le \phi \le 2\pi$, only when $\lambda = l(l+1)$; in this case $Y(\theta, \phi) = Y_{lm}(\theta, \phi)$ is a spherical harmonic. Since

$$\frac{1}{r^2}\frac{d}{dr}\left(r^2\frac{dF}{dr}\right) = \frac{1}{r}\frac{d^2}{dr^2}(rF),$$

the substitution $R(r) = rF(r)$ takes (4) into the equation

$$R'' + \left[\frac{2M}{\hbar^2}(E - U(r)) - \frac{l(l+1)}{r^2}\right]R = 0. \tag{5}$$

For a state with a discrete spectrum the wave function $\psi(\mathbf{r})$ should satisfy the normalization condition

$$\int |\psi(\mathbf{r})|^2 r^2 dr\, d\Omega = 1.$$

Since

$$\int |Y_{lm}(\theta, \phi)|^2 d\Omega = 1,$$

the normalization condition for $R(r)$ is

$$\int_0^\infty R^2(r)dr = 1. \tag{6}$$

Here we assume that $F(r) = (1/r)R(r)$ is bounded as $r \to 0$.

2. Solution of the Schrödinger equation for the Coulomb field. The only atom for which the Schrödinger equation can be solved exactly is the simplest atom — that of hydrogen. This, however, does not diminish, but rather increases, the importance of the exact solution for hydrogen, since an explicit analytical solution can often be useful as the starting point in approximate calculations for more complicated quantum-mechanical systems.

For a quantum-mechanical description of the hydrogen atom we need to discuss the relative motion of the electron (mass m, charge $-e$) and the nucleus (mass M, charge e). We can, in fact, solve the more general problem in which the charge of the nucleus is Ze. This problem is of direct physical interest, since except for relativistic effects the calculated energy eigenvalues correspond to the observable energy levels of the hydrogen atom ($Z = 1$), or of a singly ionized helium atom ($Z = 2$), etc. In addition, this model of a hydrogen-like atom is useful, for example, in investigating the spectra of the alkali elements, as well as the x-ray spectra of atoms with large Z.

Our problem of the motion of an electron is easily reduced to the problem of the motion of a single object, namely a particle with the reduced mass (see, for example, [J1])

$$\mu = \frac{mM}{m+M} \approx m$$

in the Coulomb field $U(r) = -Ze^2/r$, i.e. to the solution of the Schrödinger equation

$$\Delta\psi + \frac{2\mu}{\hbar^2}\left(E + \frac{Ze^2}{r}\right)\psi = 0.$$

Since the potential energy $U(r)$ is negative and tends to zero at infinity, it follows from physical considerations that there is a discrete spectrum only when $E < 0$ [S1].

Changing to spherical coordinates, we obtain the equation

$$R'' + \left[\frac{2\mu}{\hbar^2}\left(E + \frac{Ze^2}{r}\right) - \frac{l(l+1)}{r^2}\right]R = 0 \tag{7}$$

for the radial function $R(r)$. It is convenient to transform (7) to dimensionless form, using the atomic system of units in which the units of charge, length, and energy are the charge e of the electron ($e > 0$), and

$$a_0 = \hbar^2/(\mu e^2), \qquad E_0 = e^2/a_0.$$

Equation (7) now becomes

$$R'' + \left[2\left(E + \frac{Z}{r}\right) - \frac{l(l+1)}{r^2}\right] R = 0. \tag{8}$$

The requirement that $\psi(\mathbf{r})$ is bounded and of integrable square now reduces to the boundedness of $R(r)/r$ as $r \to 0$ and the normalization condition (6).

Equation (8) is a generalized equation of hypergeometric type, with

$$\sigma(r) = r, \quad \tilde{\tau}(r) = 0, \quad \tilde{\sigma}(r) = 2Er^2 + 2Zr - l(l+1).$$

We have the same kind of problem for equation (8) that was discussed in §9. In fact, $\tilde{\rho}(r) = 1/r$ in the present case. Therefore the requirements that $\sqrt{\tilde{\rho}(r)}R(r)$ is of integrable square on $(0, \infty)$ and is bounded as $r \to 0$ follow from (6) and the boundedness of $R(r)/r$ as $r \to 0$. Hence we can follow the method used above. We reduce (8) to the equation

$$\sigma(r)y'' + \tau(r)y' + \lambda y = 0$$

of hypergeometric type by setting $R(r) = \phi(r)y(r)$, where $\phi(r)$ is a solution of

$$\phi'/\phi = \pi(r)/\sigma(r).$$

In the present case the polynomial $\pi(r)$ is

$$\pi(r) = \frac{1}{2} \pm \sqrt{\frac{1}{4} - 2Er^2 - 2Zr + l(l+1) + kr}.$$

The constant k is to be determined by making the expression under the square root sign have a double zero. We thus obtain the possible forms of $\pi(r)$:

$$\pi(r) = \frac{1}{2} \pm \begin{cases} \sqrt{-2E}\,r + l + \frac{1}{2} & \text{for } k = 2Z + (2l+1)\sqrt{-2E}, \\ \sqrt{-2E}\,r - l - \frac{1}{2} & \text{for } k = 2Z - (2l+1)\sqrt{-2E}. \end{cases}$$

We must now choose $\pi(r)$ so that the function

$$\tau(r) = \tilde{\tau}(r) + 2\pi(r)$$

will have a negative derivative, and a zero on $(0, +\infty)$. This condition is satisfied by

$$\tau(r) = 2(l + 1 - \sqrt{-2E}r),$$

which corresponds to

$$\pi(r) = l + 1 - \sqrt{-2E}\,r, \quad \phi(r) = r^{l+1}\exp\{-\sqrt{-2E}\,r\},$$
$$\lambda = 2[Z - (l+1)\sqrt{-2E}, \quad \rho(r) = r^{2l+1}\exp\{\sqrt{-2E}\,r\}.$$

The energy eigenvalues E are determined by the equation

$$\lambda + n\tau' + \frac{1}{2}n(n-1)\sigma'' = 0 \qquad (n = 0, 1, \ldots),$$

which yields

$$E = -\frac{Z^2}{2(n + l + 1)^2}. \tag{9}$$

The values of E are determined by the number $n + l + 1$, which is known as the principal quantum number.

In the present case the eigenfunctions $y(r) = y_{nl}(r)$ are

$$y_{nl}(r) = \frac{B_{nl}}{r^{2l+1}\exp(-2Zr/(n+l+1))}\frac{d^n}{dr^n}\left[r^{n+2l+1}\exp\left(-\frac{2Zr}{n+l+1}\right)\right]$$

and are, except for a numerical factor, the Laguerre polynomials $L_n^{2l+1}(x)$ with* $x = 2Zr/(n+l+1)$. Hence the radial function $R(r) = R_{nl}(r)$ is

$$R_{nl}(r) = C_{nl}e^{-x/2}x^{l+1}L_n^{2l+1}(x). \tag{10}$$

It is easily verified that $R_{nl}(r)$ satisfies the original requirement

$$\int_0^\infty R_{nl}^2(r)dr < \infty.$$

The constant C_{nl} is determined by making this integral equal 1, i.e.

$$\frac{n+l+1}{2Z}C_{nl}^2\int_0^\infty e^{-x}x^{2l+2}[L_n^{2l+1}(x)]^2\,dx = 1. \tag{11}$$

* In quantum mechanics textbooks it is customary to denote the number of zeros of $R(r)$ by n_r, and the principal quantum number by n. With this notation, n should be replaced by $n-l-1$ in all formulas.

The integral in (11) can be evaluated by using the recursion relation for Laguerre polynomials (see §6). We have

$$xL_n^{2l+1} = 2(n+l+1)L_n^{2l+1} - (n+1)L_{n+1}^{2l+1} - (n+2l+1)L_{n-1}^{2l+1}. \qquad (12)$$

Hence, by the orthogonality of the Laguerre polynomials, we obtain

$$\int_0^\infty e^{-x}x^{2l+2}[L_n^{2l+1}(x)]^2\,dx$$

$$= \int_0^\infty e^{-x}x^{2l+1}L_n^{2l+1}(x)[2(n+l+1)L_n^{2l+1}(x)+\ldots]\,dx$$

$$= 2(n+l+1)\int_0^\infty e^{-x}x^{2l+1}[L_n^{2l+1}(x)]^2\,dx = 2(n+l+1)d_n^2,$$

where d_n^2 is the square of the norm of $L_n^{2l+1}(x)$. Therefore

$$C_{nl}^2 = \frac{Z}{(n+l+1)^2 d_n^2} = \frac{Zn!}{(n+l+1)^2(n+2l+1)!}. \qquad (13)$$

The simplest radial function occurs for $n=0$:

$$R_{0l}(r) = \frac{1}{l+1}\left(\frac{Z}{(2l+1)!}\right)^{1/2} e^{-x/2}x^{l+1}.$$

The most complicated radial function occurs for $l=0$: it has the largest possible number of zeros for a given energy. However, the dependence of the wave function on θ and ϕ is very simple in this case: for $l=0$ the wave function has spherical symmetry, since

$$Y_{00}(\theta,\phi) = 1/\sqrt{4\pi}.$$

Example 1. Knowing the radial functions $R_{nl}(r)$, we can calculate various properties of hydrogen-like atoms, in particular, the average potential energy $\bar{U}_{nl}$ for the electrostatic interaction between the electron and the nucleus, and the average distance $\bar{r}_{nl}$ between the electron and the nucleus.

Using (10) and (13), we obtain

$$\bar{U}_{nl} = -\int_0^\infty \frac{Z}{r} R_{nl}^2(r)dr$$

$$= -ZC_{nl}^2 \int_0^\infty e^{-x} x^{2l+1} [L_n^{2l+1}(x)]^2 dx = -ZC_{nl}^2 d_n^2 = -\frac{Z^2}{(n+l+1)^2}.$$

Therefore the total energy E of the electron (see (9)) is equal to half the average potential energy.

Moreover,

$$\bar{r}_{nl} = \int_0^\infty r R_{nl}^2(r)dr = C_{nl}^2 \left(\frac{n+l+1}{2Z}\right)^2 \int_0^\infty e^{-x} x^{2l+1} [x L_n^{2l+1}(x)]^2 dx.$$

To evaluate the integral it is enough to apply the recursion relation (12) and use the orthogonality of the Laguerre polynomials:

$$\bar{r}_{nl} = C_{nl}^2 \left(\frac{n+l+1}{2Z}\right)^2 [(n+1)^2 d_{n+1}^2 + 4(n+l+1)^2 d_n^2 + (2l+1)^2 d_{n-1}^2]$$

$$= C_{nl}^2 \left(\frac{n+l+1}{2Z}\right)^2 \frac{(2l+1)!}{n!} 2[3(n+l+1)^2 - l(l+1)]$$

$$= \frac{1}{2Z}[3(n+l+1)^2 - l(l+1)].$$

Example 2. To find the electrostatic potential at a given point in a hydrogen-like atom, by using the hydrogen-like wave functions.

Let the electron, which is in the Coulomb field of the nucleus of charge Ze, be in the stationary state specified by the quantum numbers n, l, m. Since the mass of the electron is small in comparison with the mass of the nucleus, it will be sufficiently accurate to suppose that the nucleus is at rest at the origin, $r = 0$. Let us find the average potential $V(\mathbf{r})$ at the point $\mathbf{r}$, due to the electron and the nucleus.

Since the potential of the nucleus, in the units we are using, is Z/r, we have

$$V(\mathbf{r}) = \frac{Z}{r} - \int \frac{|\psi_{nlm}(\mathbf{r}')|^2}{|\mathbf{r} - \mathbf{r}'|} (r')^2 dr' d\Omega'.$$

In calculating the integral, it is convenient to use the generating function for the Legendre polynomials and the addition theorem for spherical harmonics (see §10, part 5):

$$\frac{1}{|\mathbf{r}-\mathbf{r}'|} = \sum_{s=0}^{\infty} \frac{r_<^s}{r_>^{s+1}} \left[\frac{4\pi}{2s+1} \sum_{m'=-s}^{s} Y_{sm'}^*(\theta', \phi') Y_{sm'}(\theta, \phi) \right].$$

Since

$$\psi_{nlm}(\mathbf{r}') = \frac{1}{r'} R_{nl}(r') Y_{lm}(\theta', \phi'),$$

we have

$$\begin{aligned} &\int \frac{|\psi_{nlm}(\mathbf{r}')|^2}{|\mathbf{r}-\mathbf{r}'|} (r')^2 dr' d\Omega \\ &= \sum_{s=0}^{\infty} \frac{4\pi}{2s+1} \sum_{m'} Y_{sm'}(\theta, \phi) \int \frac{r_<^s}{r_>^{s+1}} R_{nl}^2(r') dr' \\ &\times \int Y_{lm}(\theta', \phi') Y_{lm}^*(\theta' \phi') Y_{sm'}^*(\theta', \phi') d\Omega'. \end{aligned} \tag{14}$$

If we introduce the explicit forms of the spherical harmonics, the ϕ' integration shows that the sum over m' reduces to the single term corresponding to $m' = 0$. We finally obtain

$$\begin{aligned} V(\mathbf{r}) = \frac{Z}{r} - \sum_{s=0}^{\infty} \frac{4\pi}{2s+1} Y_{s0}(\theta, \phi) \int_0^{\infty} \frac{r_<^s}{r_>^{s+1}} R_{nl}^2(r') dr' \\ \times \int Y_{lm}(\theta', \phi') Y_{lm}^*(\theta', \phi') Y_{s0}^*(\theta', \phi') d\Omega. \end{aligned}$$

The integral

$$\int_0^{\infty} \frac{r_<^{s+1}}{r_>^{s+1}} R_{nl}^2(r') dr'$$

can be written in the form

$$\int_0^{\infty} \frac{r_<^s}{r_>^{s+1}} R_{nl}^2(r') dr' = \frac{1}{r^{s+1}} \int_0^{r} (r')^s R_{nl}^2(r') dr' + r^s \int_r^{\infty} \frac{R_{nl}^2(r')}{(r')^{s+1}} dr'.$$

The integral of the product of three spherical harmonics reduces to the integral of a product of three functions $\Theta_{lm}(\cos\theta)$:

$$\int Y_{lm}(\theta',\phi')Y^*_{lm}(\theta',\phi')Y^*_{s0}(\theta',\phi')d\Omega = \frac{1}{\sqrt{2\pi}}\int_{-1}^{1}\Theta^2_{lm}(x)\Theta_{s0}(x)dx.$$

The last integral can be expressed in terms of Clebsch-Gordan coefficients or Wigner coefficients; tables of these are available. See, for example, [D1], [E1], [R2] and [V1]. Because of the orthogonality of the $\Theta_{lm}(x)$, the integral is different from zero only for $s = 0, 2\dots, 2l$, i.e. the sum in (14) has only finitely many terms.

When the electron is in the ground state ($n = 0, l = 0$), all the integrals are easily evaluated. We obtain

$$V(r) = \frac{Z-1}{r} + \left(Z + \frac{1}{r}\right)e^{-2Zr}.$$

For small r, $V(r) \approx Z/r$ (as one would expect), and as $r \to \infty$ we have $V(r) \approx (Z-1)/r$ (potential of the nucleus screened by an electron).

3. Solution of the Klein-Gordon equation for the Coulomb field. We have considered the solution of the Schrödinger equation for a charged particle in a Coulomb field. If the energy of the particle is comparable with its rest mass energy Mc^2 (M is the mass of the particle, c the speed of light), the Schrödinger equation is no longer applicable and we have to use a relativistic generalization of the Schrödinger equation, i.e. depending on the value of the internal angular momentum of the particle (spin) we must use either the Klein-Gordon or the Dirac equation.(See [D1], [S1].)

Let us first consider the Klein-Gordon equation, which describes the motion of a particle with charge $-e$ ($e > 0$) with integral spin and mass M in a Coulomb field of potential energy $u(r) = -Ze^2/r$. This problem arises, for example, in the study of pions in the field of an atomic nucleus. If we use a system of units in which the mass M, Planck's constant $\hbar$, and c (the speed of light) are all 1, the Klein-Gordon equation becomes

$$\Delta\psi + \left[\left(E + \frac{\mu}{r}\right)^2 - 1\right]\psi = 0 \quad \left(\mu = \frac{Ze^2}{\hbar c} \approx \frac{Z}{137}\right). \tag{15}$$

For bound states, $0 < E < 1$.

We look for particular solutions of (15) by separating variables in spherical coordinates, setting $\psi(\mathbf{r}) = F(r)Y(\theta, \phi)$. Carrying out the same operations as for Laplace's equation (see §10), we obtain the equations

$$\Delta_{\theta\phi} Y + \lambda Y = 0, \tag{16}$$

$$\frac{1}{r^2}\frac{d}{dr}\left(r^2\frac{dF}{dr}\right) + \left[\left(E + \frac{\mu}{r}\right)^2 - 1 - \frac{\lambda}{r^2}\right] F(r) = 0. \tag{17}$$

As we showed above, (16) has solutions that are bounded and single-valued for $0 \le \theta \le \pi$ and $0 \le \phi \le 2\pi$, only when $\lambda = l(l+1)$; in that case $Y(\theta, \phi) = Y_{lm}(\theta, \phi)$, the spherical harmonics. The substitution $R(r) = rF(r)$ transforms (17) into the equation

$$R'' + \left[\left(E + \frac{\mu}{r}\right)^2 - 1 - \frac{l(l+1)}{r^2}\right] R = 0. \tag{18}$$

This is a generalized equation of hypergeometric type with

$$\sigma(r) = r, \quad \tilde{\tau}(r) = 0, \quad \tilde{\sigma}(r) = (Er + \mu)^2 - r^2 - l(l+1).$$

The function $R(r)$ is to satisfy the normalization condition

$$\int_0^\infty R^2(r)dr = 1 \tag{19}$$

and is to be bounded as $r \to 0$. We note that for the solution of the corresponding Schrödinger equation we used the stronger requirement that $R(r)/r$ was bounded as $r \to 0$.

Equation (18) has a singular point at $r = 0$. Let us investigate the behavior of $R(r)$ for $r \sim 0$. Since

$$\left(E + \frac{\mu}{r}\right)^2 - 1 - \frac{l(l+l)}{r^2} \approx \frac{\mu^2 - l(l+1)}{r^2},$$

as $r \to 0$, the behavior of $R(r)$ near $r = 0$ is approximately described by the Euler equation

$$R'' + \frac{\mu^2 - l(l+1)}{r^2} R = 0,$$

whose solutions are

$$R(r) = C_1 r^{\nu+1} + C_2 r^{-\nu},$$

where

$$\nu = -1/2 + \sqrt{(l+1/2)^2 - \mu^2}$$

(we shall suppose that $\mu < l + 1/2$). The requirement that $R(r)$ is bounded as $r \to 0$ requires $C_2 = 0$, i.e. $R(r) \approx C_1 r^{\nu+1}$ as $r \to 0$.

Our problem for (18) is a problem of the type that we considered in §9. In fact, in this case $\tilde{\rho}(r) = 1/r$ and the boundedness of $\sqrt{\tilde{\rho}(r)}R(r)$ as $r \to 0$ and the integrability of the square of this function on $(0, \infty)$ follow from the behavior of $R(r)$ as $r \to 0$ and from (19). Hence we can apply the method that we used in §9.

We transform (18) into an equation of hypergeometric type,

$$\sigma(r)y'' + \tau(r)y' + \lambda y = 0,$$

by putting $R(r) = \phi(r)y(r)$, where $\phi(r)$ satisfies

$$\phi'/\phi = \pi(r)/\sigma(r).$$

In the present case

$$\pi(r) = \frac{1}{2} \pm \sqrt{(l+1/2)^2 - \mu^2 - 2\mu E r + (1-E^2)r^2 + kr}.$$

The constant k is to be chosen so that the expression under the square root sign has a double zero. We obtain the following possible forms for $\pi(r)$:

$$\pi(r) = \frac{1}{2} \pm \begin{cases} \sqrt{1-E^2}\, r + \nu + \frac{1}{2} & \text{for } k = 2\mu E + (2\nu+1)\sqrt{1-E^2}, \\ \sqrt{1-E^2}\, r - \nu - \frac{1}{2} & \text{for } k = 2\mu E - (2\nu+1)\sqrt{1-E^2}. \end{cases}$$

We must select $\pi(r)$ so that the function

$$\tau(r) = \tilde{\tau}(r) + 2\pi(r)$$

has a negative derivative, and a zero on $(0, +\infty)$. The function

$$\tau(r) = 2(\nu + 1 - ar), \qquad \text{where } a = \sqrt{1-E^2},$$

satisfies these requirements, and correspondingly we have

$$\pi(r) = \nu + 1 - ar, \qquad \phi(r) = r^{\nu+1}e^{-ar},$$

$$\lambda = 2\left[\mu E - (\nu+1)a\right], \qquad \rho(r) = r^{2\nu+1}e^{-2ar},$$

$$\nu = -\frac{1}{2} + \left[\left(l + \frac{1}{2}\right)^2 - \mu^2\right]^{1/2}.$$

The energy eigenvalues are determined from the equation

$$\lambda + n\tau' + \frac{1}{2}n(n-1)\sigma'' = 0,$$

which yields

$$E = E_n = \left[1 + \left(\frac{\mu}{n+\nu+1}\right)^2\right]^{-1/2} \quad (n = 0, 1, \ldots). \tag{20}$$

The corresponding eigenfunctions are

$$y_n(r) = \frac{B_{nl}}{r^{2\nu+1}e^{-2ar}} \frac{d^n}{dr^n}(r^{n+2\nu+1}e^{-2ar});$$

up to a numerical factor they are the Laguerre polynomials $L_n^{2\nu+1}(x)$, where $x = 2ar$. The eigenfunctions $R(r) = R_{nl}(r)$ are

$$R_{nl}(r) = C_{nl}x^{\nu+1}e^{-x/2}L_n^{2\nu+1}(x).$$

It is easily verified that $R_{nl}(r)$ satisfy the restriction $\int_0^\infty R_{nl}^2(r)dr < \infty$. The constants C_{nl} are determined from (19) by the method used in solving the corresponding Schrödinger equation.

Let us consider passing to the nonrelativistic limit. In this case, μ is small. We estimate everything else as $\mu \to 0$:

$$\nu \approx l, \quad E \approx 1 - \frac{\mu^2}{2(n+l+1)^2}, \quad a = \sqrt{1-E^2} \approx \frac{\mu}{n+l+1},$$

$$R_{nl}(r) \approx C_{nl}x^{l+1}e^{-x/2}L_n^{2l+1}(x), \quad x = \frac{2\mu}{n+l+1}r.$$

These agree with the formulas obtained in part 2 for the solution of the Schrödinger equation, if we remember that in our system of units μr goes over into Zr for the atomic system of units, and the energy

$$E = 1 - \frac{\mu^2}{2(n+l+1)^2}$$

contains the rest mass energy $E_0 = 1$ of the particle.

4. Solution of the Dirac equation for the Coulomb field. Now let us consider the Dirac equation for a charged particle with half-integral spin in the field

$$U(r) = -\frac{Ze^2}{r}.$$

1°. In the present case the wave function of the particle has four components $\psi_k(r)$ $(k = 1, \ldots, 4)$. If we use the system of units in which the mass M of the particle, Planck's constant $\hbar$, and the speed of light are all equal to 1, the Dirac equation will have the form (see [B3])

$$\left.\begin{aligned}
i\left(E + \frac{\mu}{r} + 1\right)\psi_1 + \frac{\partial \psi_3}{\partial z} + \frac{\partial \psi_4}{\partial x} - i\frac{\partial \psi_4}{\partial y} = 0,\\
i\left(E + \frac{\mu}{r} + 1\right)\psi_2 - \frac{\partial \psi_4}{\partial z} + \frac{\partial \psi_3}{\partial x} + i\frac{\partial \psi_3}{\partial y} = 0,\\
i\left(E + \frac{\mu}{r} - 1\right)\psi_3 + \frac{\partial \psi_1}{\partial z} + \frac{\partial \psi_2}{\partial x} - i\frac{\partial \psi_2}{\partial y} = 0,\\
i\left(E + \frac{\mu}{r} - 1\right)\psi_4 - \frac{\partial \psi_2}{\partial z} + \frac{\partial \psi_1}{\partial x} + i\frac{\partial \psi_1}{\partial y} = 0.
\end{aligned}\right\} \tag{21}$$

Here E and μ are the same as for the Klein-Gordon equation, and $0 < E < 1$.

In spherical coordinates (r, θ, ϕ) the variables in (21) can be separated by looking for a solution that has the form

$$\begin{aligned}
\begin{pmatrix}\psi_1(r)\\ \psi_2(r)\end{pmatrix} &= f(r)\Omega_{jlm}(\theta, \phi),\\
\begin{pmatrix}\psi_3(r)\\ \psi_4(r)\end{pmatrix} &= (-1)^{(l-l'+1)/2} g(r)\Omega_{jl'm}(\theta, \phi).
\end{aligned} \tag{22}$$

Here j is a quantum number specifiying the total angular momentum of the particle $(j = 1/2, 3/2, \ldots)$; l and l' are orbital angular momentum quantum numbers, which, for a given j, can have the values $j - 1/2$ and $j + 1/2$, with $l' = 2j - l$; the quantum number m takes half-integral values between the numbers $-j$ and j.

The functions $\Omega_{jlm}(\theta, \phi)$ and $\Omega_{jl'm}(\theta, \phi)$ contain the dependence of the wave function on the angular variables and are called *spherical spinors*. They can be expressed in terms of spherical harmonics and Clebsch-Gordan coefficients, which arise in the addition of orbital and spin moments of electrons. Spherical spinors are connected with the spherical harmonics $Y_{lm}(\theta, \phi)$ as follows ([A2], [B3]):

$$\Omega_{jlm} = \begin{pmatrix} \sqrt{\dfrac{j+m}{2l+1}} Y_{l,m-1/2}(\theta,\phi) \\ \sqrt{\dfrac{j-m}{2l+l}} Y_{l,m+1/2}(\theta,\phi) \end{pmatrix} \qquad \text{for } l = j-1/2,$$

$$\Omega_{jlm} = \begin{pmatrix} -\sqrt{\dfrac{j-m+1}{2l+1}} Y_{l,m-1/2}(\theta,\phi) \\ \sqrt{\dfrac{j+m+1}{2l+1}} Y_{l,m+1/2}(\theta,\phi) \end{pmatrix} \qquad \text{for } l = j+1/2.$$

By substituting (22) into (21) we obtain a system of equations for $f(r)$ and $g(r)$:

$$\begin{aligned} f' + \frac{1+\kappa}{r} f - \left(E + 1 + \frac{\mu}{r}\right) g = 0, \\ g' + \frac{1-\kappa}{r} g + \left(E - 1 + \frac{\mu}{r}\right) f = 0, \end{aligned} \tag{23}$$

where

$$\kappa = \begin{cases} -(l+1) & \text{for } l = j - 1/2, \\ l & \text{for } l = j + 1/2. \end{cases}$$

We remark that, as will be shown later, in the nonrelativistic approximation $| f(r) | \gg | g(r) |$.

The conditions that determine $f(r)$ and $g(r)$ for states with discrete spectra lead to the following requirements: $rf(r)$ and $rg(r)$ must be bounded as $r \to 0$ and satisfy

$$\int_0^\infty r^2 \left[f^2(r) + g^2(r)\right] dr = 1. \tag{24}$$

2°. Let us write the system (23) in matrix form. Let

$$u = \begin{pmatrix} u_1 \\ u_2 \end{pmatrix} \equiv \begin{pmatrix} f(r) \\ g(r) \end{pmatrix}, \quad u' = \begin{pmatrix} u_1' \\ u_2' \end{pmatrix}.$$

Then

$$u' = Au, \tag{25}$$

where

$$A = \begin{pmatrix} a_{11} & a_{12} \\ a_{21} & a_{22} \end{pmatrix} = \begin{pmatrix} -\dfrac{1+\kappa}{r} & 1 + E + \dfrac{\mu}{r} \\ 1 - E - \dfrac{\mu}{r} & -\dfrac{1-\kappa}{r} \end{pmatrix}.$$

To find $u_1(r)$, we eliminate $u_2(r)$ from (25), obtaining a second order differential equation

$$u_1''-\left(a_{11}+a_{22}+\frac{a_{12}'}{a_{12}}\right)u_1'+\left(a_{11}a_{22}-a_{12}a_{21}-a_{11}'+\frac{a_{12}'}{a_{12}}a_{11}\right)u_1=0. \quad (26)$$

Similarly, eliminating $u_1(r)$, we obtain an equation for $u_2(r)$:

$$u_2''-\left(a_{11}+a_{22}+\frac{a_{21}'}{a_{21}}\right)u_2'+\left(a_{11}a_{22}-a_{12}a_{21}-a_{22}'+\frac{a_{21}'}{a_{21}}a_{22}\right)u_2=0. \quad (27)$$

The components of A have the form

$$a_{ik}=b_{ik}+c_{ik}/r,$$

where b_{ik} and c_{ik} are constants. Equations (26) and (27) are not generalized equations of hypergeometric type. This is connected with the fact that

$$\frac{a_{12}'}{a_{12}}=-\frac{c_{12}}{c_{12}r+b_{12}r^2},$$

from which we can see that the coefficients of $u_1'(r)$ and $u_1(r)$ in (26) have the form

$$a_{11}+a_{22}+\frac{a_{12}'}{a_{12}}=\frac{p_1(r)}{r}-\frac{c_{12}}{c_{12}r+b_{12}r^2},$$

$$a_{11}a_{22}-a_{12}a_{21}-a_{11}'+\frac{a_{12}'}{a_{12}}a_{11}=\frac{p_2(r)}{r^2}-\frac{c_{12}}{c_{12}r+b_{12}r^2}\,\frac{c_{11}+b_{11}r}{r}$$

($p_1(r)$ and $p_2(r)$ are polynomials of degrees at most 1 and 2, respectively). Equation (26) will become a generalized equation of hypergeometric type with $\sigma(r)=r$ if either b_{12} or c_{12} is zero. The following argument is useful. By a linear transformation

$$\begin{pmatrix} v_1 \\ v_2 \end{pmatrix}=C\begin{pmatrix} u_1 \\ u_2 \end{pmatrix}$$

with a nonsingular matrix C that is independent of r we obtain equations for $v_1(r)$ and $v_2(r)$ of the same form as the original ones. In fact, (25) is replaced by

$$v'=\tilde{A}v, \quad (28)$$

where

$$v=\begin{pmatrix} v_1 \\ v_2 \end{pmatrix}, \quad \tilde{A}=CAC^{-1}=\begin{pmatrix} \tilde{a}_{11} & \tilde{a}_{12} \\ \tilde{a}_{21} & \tilde{a}_{22} \end{pmatrix}.$$

The coefficients $\tilde{a}_{ik}$ are evidently linear combinations of the a_{ik}. Hence they have the form

$$\tilde{a}_{ik} = \tilde{b}_{ik} + \tilde{c}_{ik}/r$$

($\tilde{b}_{ik}$ and $\tilde{c}_{ik}$ are constants).

The equations for $v_1(r)$ and $v_2(r)$ are similar to (26) and (27):

$$v_1'' - \left(\tilde{a}_{11} + \tilde{a}_{22} + \frac{\tilde{a}'_{12}}{\tilde{a}_{12}}\right) v_1' + \left(\tilde{a}_{11}\tilde{a}_{22} - \tilde{a}_{12}\tilde{a}_{21} - \tilde{a}'_{11} + \frac{\tilde{a}'_{12}}{\tilde{a}_{12}}\tilde{a}_{11}\right) v_1 = 0, \tag{29}$$

$$v_2'' - \left(\tilde{a}_{11} + \tilde{a}_{22} + \frac{\tilde{a}'_{21}}{\tilde{a}_{21}}\right) v_2' + \left(\tilde{a}_{11}\tilde{a}_{22} - \tilde{a}_{12}\tilde{a}_{21} - \tilde{a}'_{22} + \frac{\tilde{a}'_{21}}{\tilde{a}_{21}}\tilde{a}_{22}\right) v_2 = 0. \tag{30}$$

Notice that the calculation of the coefficients in (29) and (30) is facilitated by the similarity of A and $\tilde{A}$:

$$\tilde{a}_{11} + \tilde{a}_{22} = a_{11} + a_{22}, \quad \tilde{a}_{11}\tilde{a}_{22} - \tilde{a}_{12}\tilde{a}_{21} = a_{11}a_{22} - a_{12}a_{21}.$$

For (29) to be an equation of generalized hypergeometric type, it is sufficient that either $\tilde{b}_{12} = 0$ or $\tilde{c}_{12} = 0$. Similarly for (30): either $\tilde{b}_{21} = 0$ or $\tilde{c}_{21} = 0$. These conditions impose restrictions on the choice of C. Let

$$C = \begin{pmatrix} \alpha & \beta \\ \gamma & \delta \end{pmatrix}.$$

Then

$$C^{-1} = \frac{1}{\Delta}\begin{pmatrix} \delta & -\beta \\ -\gamma & \alpha \end{pmatrix}, \quad \Delta = \alpha\delta - \beta\gamma,$$

$$\tilde{A} = \frac{1}{\Delta}\begin{pmatrix} a_{11}\alpha\delta - a_{12}\alpha\gamma + a_{21}\beta\delta - a_{22}\beta\gamma & a_{12}\alpha^2 - a_{21}\beta^2 + (a_{22} - a_{11})\alpha\beta \\ a_{21}\delta^2 - a_{12}\gamma^2 + (a_{11} - a_{22})\gamma\delta & -a_{11}\beta\gamma + a_{12}\alpha\gamma - a_{21}\beta\delta + a_{22}\alpha\delta \end{pmatrix}.$$

The condition	$\tilde{b}_{12} = 0$	yields	$(1+E)\,\alpha^2 - (1-E)\,\beta^2 = 0,$
" "	$\tilde{c}_{12} = 0$	"	$2\kappa\alpha\beta + \mu\left(\alpha^2 + \beta^2\right) = 0,$
" "	$\tilde{b}_{21} = 0$	"	$(1+E)\,\gamma^2 - (1-E)\,\delta^2 = 0,$
" "	$\tilde{c}_{21} = 0$	"	$2\kappa\gamma\delta + \mu\left(\gamma^2 + \delta^2\right) = 0.$

We see that there are several possibilities for α, β, γ, δ. All quantum mechanics textbooks use the same one, namely $\tilde{b}_{12} = 0$, $\tilde{b}_{21} = 0$. We shall instead determine α, β, γ, δ so that $\tilde{c}_{12} = 0, \tilde{c}_{21} = 0$ (we will show later that this is preferable to taking $\tilde{b}_{12} = 0, \tilde{b}_{21} = 0$). This condition will be satisfied if

$$C = \begin{pmatrix} \mu & \nu - \kappa \\ \nu - \kappa & \mu \end{pmatrix},$$

where $\nu = \sqrt{\kappa^2 - \mu^2}$. We obtain the following system of equations for $v_1(r)$ and $v_2(r)$:

$$v_1' = \left(-\frac{\nu + 1}{r} + \frac{E\mu}{\nu}\right) v_1 + \left(1 + \frac{E\kappa}{\nu}\right) v_2, \tag{31}$$

$$v_2' = \left(1 - \frac{E\kappa}{\nu}\right) v_1 + \left(\frac{\nu - 1}{r} - \frac{E\mu}{\nu}\right) v_2. \tag{32}$$

If $1 + E\kappa/\nu \neq 0$, we can eliminate $v_2(r)$ from (31) and (32) and obtain a differential equation for $v_1(r)$:

$$v_1'' + \frac{2}{r} v_1' + \frac{(E^2 - 1) r^2 + 2E\mu r - \nu(\nu + 1)}{r^2} v_1 = 0. \tag{33}$$

Now let $1 + E\kappa/\nu = 0$, i.e. $E = -\nu/\kappa$, which is possible only if $\kappa < 0$, since $\nu > 0$ and $E > 0$. In this case a solution of (31) has the form

$$v_1(r) = C_1 r^{-\nu-1} \exp\left(\frac{E\mu}{\nu} r\right).$$

This satisfies the conditions of the problem only if $C_1 = 0$. Then we find from (32) that

$$v_2(r) = C_2 r^{\nu-1} \exp(-E\mu r/\nu).$$

It is clear that this $v_2(r)$, with $C_2 \neq 0$, does satisfy the conditions of the problem.

3°. Now let us consider the solution of (33). We need to know the behavior of $v_1(r)$ as $r \to 0$. Since

$$|\,(E^2 - 1) r^2 + 2E\mu r\,| \ll \nu(\nu + 1)$$

as $r \to 0$, the behavior of $v_1(r)$ in the neighborhood of $r = 0$ will be approximately described by the Euler equation

$$r^2 v_1'' + 2r v_1' - \nu(\nu + 1) v_1 = 0,$$

whose solution has the form

$$v_1(r) = C_1 r^{\nu} + C_2 r^{-\nu-1}.$$

From the requirements on $v_1(r)$ it follows that $C_2 = 0$. Hence $v_1(r) \approx C_1 r^\nu$ for $r \sim 0$.

Equation (33) is a generalized equation of hypergeometric type for which $\sigma(r) = r$, $\tilde{\tau}(r) = 2$, $\tilde{\sigma}(r) = \left(E^2 - 1\right) r^2 + 2E\mu r - \nu(\nu + 1)$. Our problem for (33) belongs to the class of problems that were studied in §9. In fact, in the present case $\tilde{\rho}(r) = r$. The required integrability of the square of $\sqrt{\tilde{\rho}(r)}v_1(r)$ on $(0, \infty)$ and boundedness as $r \to 0$ follow from (24) and the behavior of $v_1(r)$ as $r \to 0$. Hence we can use the method discussed in §9. We transform (33) into an equation of hypergeometric type

$$\sigma(r)y'' + \tau(r)y' + \lambda y = 0$$

by the substitution $v_1 = \phi(r)y$, where $\phi(r)$ satisfies

$$\phi'/\phi = \pi(r)/\sigma(r)$$

($\pi(r)$ is a polynomial of degree 1 at most). From the four possible forms of $\pi(r)$ we select the one for which the function $\tau(r) = \tilde{\tau}(r) + 2\pi(r)$ has a negative derivative, and a zero on $(0, +\infty)$. The function $\tau(r) = 2(\nu + 1 - ar)$ satisfies these requirements, with $a = \sqrt{1 - E^2}$, $\nu = \sqrt{\kappa^2 - \mu^2}$, and we have

$$\begin{aligned} \pi(r) &= -ar + \nu, & \phi(r) &= r^\nu e^{-ar}, \\ \lambda &= 2\left[E\mu - (\nu + 1)a\right], & \rho(r) &= r^{2\nu+1}e^{-2ar}. \end{aligned}$$

The values $E = E_n$ of the energy are determined by

$$\lambda + n\tau' + \frac{1}{2}n(n - 1)\sigma'' = 0 \qquad (n = 0, 1, \ldots),$$

whence

$$\mu E - (n + \nu + 1)a = 0, \tag{34}$$

and the eigenfunctions are found from the Rodrigues formula

$$y_n(r) = \frac{C_n}{\rho(r)}\frac{d^n}{dr^n}\left[\sigma^n(r)\rho(r)\right] = C_n r^{-2\nu-1}e^{2ar}\frac{d^n}{dr^n}\left(r^{n+2\nu+1}e^{-2ar}\right). \tag{35}$$

The functions $y_n(r)$ are, up to constant multiples, the Laguerre polynomials $L_n^{2\nu+1}(x)$, with $x = 2ar$.

The eigenvalue $E=-\nu/\kappa$ that we found above satisfies (34) with $n=-1$. Consequently it is natural to replace n by $n-1$ in (34) and (35) and define the eigenvalues by

$$\mu E-(n+\nu)a=0 \qquad (n=0,1\ldots); \qquad a=\sqrt{1-E^2}. \tag{36}$$

The eigenfunctions $v_1(r)$ will have the form

$$v_1(r)=\begin{cases} A_n x^\nu e^{-x/2} L_{n-1}^{2\nu+1}(x) & (n=1,2\ldots), \\ 0 \quad (n=0). \end{cases} \tag{37}$$

It is easily verified that $rv_1(r)$ satisfies the original requirement of being of integrable square.

4°. From (31) with $E=E_n$ $(n=1,2,\ldots)$, we have

$$v_2(r)=\frac{1}{1+E\kappa/\nu}\left[v_1'(r)+\left(\frac{\nu+1}{r}-\frac{E\mu}{\nu}\right)v_1(r)\right].$$

If we again replace $v_1(r)$ by its value, we find

$$v_2(r)=x^{\nu-1}e^{-x/2}y(x),$$

where $y(x)$ is a polynomial of degree n. To determine $y(x)$ we eliminate $v_1(r)$ between (31) and (32):

$$v_2''+\frac{2}{r}v_2'+\frac{(E^2-1)r^2+2E\mu r+\nu(1-\nu)}{r^2}v_2=0. \tag{38}$$

We obtain the following differential equation for $y(x)$:

$$xy''+(2\nu-x)y'+ny=0. \tag{39}$$

Equation (39) is an equation of hypergeometric type. Its only polynomial solutions are Laguerre polynomials $y(x)=B_nL_n^{2\nu-1}(x)$, whence

$$v_2(r)=B_nx^{\nu-1}e^{-x/2}L_n^{2\nu-1}(x). \tag{40}$$

It is easily seen that our previous solution for $E=-\nu/\kappa$ is included for $n=0$.

To obtain the connection between the constants A_n and B_n in (37) and (40) we compare the two sides of (31) as $r \to 0$, using the formula (see §5)

$$L_n^\alpha(0) = \frac{\Gamma(n+\alpha+1)}{n!\Gamma(\alpha+1)}.$$

We have

$$2a\nu A_n L_{n-1}^{2\nu+1}(0) = -2a(\nu+1)A_n L_{n-1}^{2\nu+1}(0) + \left(1+\frac{E\kappa}{\nu}\right)B_n L_n^{2\nu-1}(0),$$

whence

$$A_n = \frac{\nu+E\kappa}{an(n+2\nu)}B_n \qquad (n=1,2,\ldots).$$

Since

$$n(n+2\nu) = (n+\nu)^2 - \nu^2 = \frac{E^2\mu^2}{a^2} - \nu^2 = \frac{E^2\kappa^2-\nu^2}{a^2},$$

we have

$$A_n = \frac{a}{E\kappa-\nu}B_n.$$

Since we know $v_1(r)$ and $v_2(r)$, we can find $f(r)$ and $g(r)$:

$$\begin{pmatrix} f \\ g \end{pmatrix} = C^{-1}\begin{pmatrix} v_1 \\ v_2 \end{pmatrix}, \quad C^{-1} = \frac{1}{2\nu(\kappa-\nu)}\begin{pmatrix} \mu & \kappa-\nu \\ \kappa-\nu & \mu \end{pmatrix}.$$

Therefore

$$f(r) = \frac{B_n}{2\nu(\kappa-\nu)}x^{\nu-1}e^{-x/2}\left[f_1 x L_{n-1}^{2\nu+1}(x) + f_2 L_n^{2\nu-1}(x)\right],$$

$$g(r) = \frac{B_n}{2\nu(\kappa-\nu)}x^{\nu-1}e^{-x/2}\left[g_1 x L_{n-1}^{2\nu+1}(x) + g_2 L_n^{2\nu-1}(x)\right],$$

where

$$f_1 = \frac{a\mu}{E\kappa-\nu}, \quad f_2 = \kappa-\nu, \quad g_1 = \frac{a(\kappa-\nu)}{E\kappa-\nu}, \quad g_2 = \mu,$$

$$a = \sqrt{1-E^2}, \quad E = E_n = \left\{1+\left(\frac{\mu}{n+\nu}\right)^2\right\}^{-1/2}.$$

The formulas for $f(r)$ and $g(r)$ remain valid for $n=0$; in this case the terms containing $L_{n-1}^{2\nu+1}(x)$ have to be taken to be zero.

5°. We can calculate B_n by using (24). We have

$$\int_0^\infty r^2\left[f^2(r)+g^2(r)\right]dr$$

$$= \frac{B_n^2}{4\nu^2(\kappa-\nu)^2(2a)^3}\int_0^\infty e^{-x}x^{2\nu}\left\{\left[f_1 x L_{n-1}^{2\nu+1}(x)+f_2 L_n^{2\nu-1}(x)\right]^2\right.$$

$$\left.+\left[g_1 x L_{n-1}^{2\nu+1}(x)+g_2 L_n^{2\nu-1}(x)\right]^2\right\}dx = 1.$$

Two types of integrals occur in the calculation:

$$J_1 = \int_0^\infty e^{-x}x^{\alpha+1}\left[L_n^\alpha(x)\right]^2 dx,$$

$$J_2 = \int_0^\infty e^{-x}x^\alpha L_{n-1}^\alpha(x)\,L_n^{\alpha-2}(x)\,dx.$$

The integral J_1 can be evaluated in terms of the square of a norm,

$$d_n^2 = \int_0^\infty e^{-x}x^\alpha\left[L_n^\alpha(x)\right]^2 dx = \frac{1}{n!}\Gamma(n+\alpha+1),$$

by using the recursion relation

$$xL_n^\alpha(x) = -(n+1)L_{n+1}^\alpha(x)+(2n+\alpha+1)L_n^\alpha(x)-(n+\alpha)L_{n-1}^\alpha(x)$$

and the orthogonality property

$$\int_0^\infty e^{-x}x^\alpha L_n^\alpha(x)L_m^\alpha(x)dx = 0 \quad (m\neq n).$$

We find

$$J_1 = (2n+\alpha+1)\int_0^\infty e^{-x}x^\alpha\left[L_n^\alpha(x)\right]^2 dx = \frac{1}{n!}(2n+\alpha+1)\Gamma(n+\alpha+1).$$

To calculate J_2 it is enough to expand $L_n^{\alpha-2}(x)$ in terms of the polynomials $L_n^\alpha(x)$:

$$L_n^{\alpha-2}(x) = c_1 L_n^\alpha(x) + c_2 L_{n-1}^\alpha(x) + \dots$$

The coefficients c_1 and c_2 are easily found by equating coefficients of x^n and x^{n-1} on the two sides of this equation:

$$c_1 = 1, \quad c_2 = -2.$$

This yields

$$J_2 = -2\int_0^\infty e^{-x}x^\alpha \left[L_{n-1}^\alpha(x)\right]^2 dx = -2\frac{\Gamma(n+\alpha)}{(n-1)!}.$$

Consequently

$$B_n = 2a^2\left(\frac{(\kappa-\nu)(E\kappa-\nu)\,n!}{\mu\Gamma(n+2\nu)}\right)^{1/2}.$$

Observe that this formula also applies when $n = 0$.

6°. We finally obtain

$$\begin{pmatrix} f(r) \\ g(r)\end{pmatrix} = \frac{a^2}{\nu}\left(\frac{(E\kappa-\nu)n!}{\mu(\kappa-\nu)\Gamma(n+2\nu)}\right)^{1/2} x^{\nu-1}e^{-x/2}\begin{pmatrix} f_1 & f_2 \\ g_1 & g_2\end{pmatrix}\begin{pmatrix} xL_{n-1}^{2\nu+1}(x) \\ L_n^{2\nu-1}(x)\end{pmatrix}, \tag{41}$$

where

$$\mu = \frac{Ze^2}{\hbar c}, \qquad a = \sqrt{1-E^2}, \qquad \nu = \sqrt{\kappa^2-\mu^2},$$

$$f_1 = \frac{a\mu}{E\kappa-\nu}, \quad f_2 = \kappa-\nu, \quad g_1 = \frac{a(\kappa-\nu)}{E\kappa-\nu}, \quad g_2 = \mu.$$

When $n = 0$ the term $xL_{n-1}^{2\nu+1}$ has to be taken to be zero.

Let us consider the nonrelativistic limit. In that case $\mu \approx Z/137$ is small. Let us estimate the other terms as $\mu \to 0$. We have

$$E \approx 1-\mu^2/(2N^2), \quad N = n+\nu,$$

$$a = \sqrt{1-E^2} \approx \mu/N, \quad \nu - |\kappa| \approx -\mu^2/(2|\kappa|).$$

Now we estimate the order of f_1, f_2 and g_1, g_2 as functions of μ.

1) Let $l = j - 1/2$. Then $\kappa = -(l+1)$, $\kappa - \nu \approx 2\kappa$, $E\kappa - \nu \approx 2\kappa$, and consequently

$$\begin{pmatrix} f_1 & f_2 \\ g_1 & g_2\end{pmatrix} \sim \begin{pmatrix} \mu^2 & 1 \\ \mu & \mu\end{pmatrix}.$$

2) Let $l = j + 1/2$. Then $\kappa = l$, $\kappa - \nu = \mu^2/(2l)$, $E\kappa - \nu = (E-1)\kappa + (\kappa - \nu) \approx \mu^2(N^2 - l^2)/(2lN^2)$, whence

$$\begin{pmatrix} f_1 & f_2 \\ g_1 & g_2 \end{pmatrix} \sim \begin{pmatrix} 1 & \mu^2 \\ \mu & \mu \end{pmatrix}.$$

It is then clear that $| g(r) | \ll | f(r) |$ in all cases, and

$$f(r) \approx \pm \frac{2\mu^{\frac{3}{2}}}{N^2} \sqrt{\frac{(N-l-1)!}{(N+l)!}} x^l e^{-x/2} L_{N-l-1}^{2l+1}(x). \tag{42}$$

The plus sign corresponds to $l = j + 1/2$; the minus, to $l = j - 1/2$. In the nonrelativistic case, $N = n + \nu$ is equal to $n + |\kappa| = n + |j + 1/2|$; it corresponds to the principal quantum number for the Schrödinger equation. The formula (42) for $f(r)$ agrees exactly with the corresponding solution of the Schrödinger equation.

It is interesting to observe that the representation of $f(r)$ and $g(r)$ in the form (41) is well adapted to the transition to the nonrelativistic case, since one of the coefficients f_1, f_2, g_1, g_2 is much larger than the others as $\mu \to 0$. However, in the representation of $f(r)$ and $g(r)$ which usually appears in the literature there is an overlap in the orders of the coefficients. Hence to establish the identity of the nonrelativistic limit with the solution of the Schrödinger equation one has to make another appeal to the recursion relations for the hypergeometric functions.

7°. The Laguerre polynomials that appear in the formulas for $f(r)$ and $g(r)$ can be expressed in terms of confluent hypergeometric functions, by using the formula

$$L_n^\alpha(x) = \frac{\Gamma(n+\alpha+1)}{n!\Gamma(\alpha+1)} F(-n, \alpha+1, x).$$

Then (41) becomes

$$\begin{pmatrix} f(r) \\ g(r) \end{pmatrix} = D_n x^{\nu-1} e^{-x/2} \begin{pmatrix} \overline{f}_1 & \overline{f}_2 \\ \overline{g}_1 & \overline{g}_2 \end{pmatrix} \begin{pmatrix} xF(-n+1, 2\nu+2, x) \\ F(-n, 2\nu, x) \end{pmatrix},$$

where

$$D_n = \frac{1}{\nu\Gamma(2\nu+2)} \left(\frac{(E\kappa - \nu)\Gamma(n+2\nu)}{\mu(\kappa-\nu)n!} \right)^{1/2},$$

$$\overline{f}_1 = a\mu(E\kappa + \nu), \qquad \overline{f}_2 = 2\nu a^2(2\nu+1)(\kappa - \nu),$$

$$\overline{g}_1 = a(\kappa - \nu)(E\kappa + \nu), \quad \overline{g}_2 = 2\mu\nu a^2(2\nu+1).$$

Our expressions for $f(r)$ and $g(r)$ remain valid, up to normalizing factors, for states in the continous spectrum if we replace the integral values of n by $n = n(E)$, which is connected with a given E by (36), i.e.

$$n = \frac{\mu E}{i\sqrt{E^2-1}} - \nu, \qquad E > 1.$$

We have now discussed several basic problems in quantum mechanics. Many other representative quantum mechanics problems can be solved by the same methods.

5. Clebsch-Gordan coefficients and their connection with the Hahn polynomials. We learn in courses in quantum mechanics that when the Hamiltonian of a physical system is invariant under rotation of the coordinate axes, the square of the angular momentum operator and its component in any given direction (for instance, the z direction) commute with the Hamiltonian. This means that there are states for which the wave functions are simultaneous eigenfunctions of this pair of commuting operators. We shall consider the properties of these operators in more detail.

Let us denote the angular momentum operator and its components along the coordinate axes, in units of Planck's constant $\hbar$, by $\mathbf{J}$ and J_x, J_y, J_z, respectively. The operators J_x, J_y, and J_z satisfy the following commutation relations:

$$\left.\begin{aligned} J_xJ_y - J_yJ_x &= iJ_z, \\ J_yJ_z - J_zJ_y &= iJ_x, \\ J_zJ_x - J_xJ_z &= iJ_y. \end{aligned}\right\} \tag{43}$$

It follows from these relations that the operators $J^2 = J_x^2 + J_y^2 + J_z^2$ and J_z commute and have a common system of eigenfunctions ψ_{jm} which satisfy

$$\left.\begin{aligned} J^2\psi_{jm} &= j(j+1)\psi_{jm}, \\ J_z\psi_{jm} &= m\psi_{jm}, \end{aligned}\right\} \tag{44}$$

$$J_\pm\psi_{jm} = [(j \mp m)(j \pm m + 1)]^{1/2}\, \psi_{j,m\pm 1}. \tag{45}$$

Here $J_\pm = J_x \pm iJ_y$, the system $\{\psi_{jm}\}$ is orthonormal, the quantum number j can take only nonnegative integral or half-integral values, and the quantum number m can take the values $m = -j, -j+1, \ldots, j-1, j$.

An important problem in quantum mechanics is the addition of angular momenta, which may be described as follows. Let a physical system consist of two subsystems whose angular momentum operators $\mathbf{J}_1$ and $\mathbf{J}_2$ commute, and whose wave functions are $\psi_{j_1 m_1}$ and $\psi_{j_2 m_2}$. In this case the operator $\mathbf{J} = \mathbf{J}_1 + \mathbf{J}_2$ is the total angular momentum operator of the system and satisfies the commutation relations (43). Hence there must be wave functions Φ_{jm} of the operators J^2 and J_z, satisfying (44) and (45). The problem is to express ϕ_{jm} in terms of the known functions $\psi_{j_1 m_1}$ and $\psi_{j_2 m_2}$.

1°. This problem can be solved in the following way. The eigenfunction of $J_z = J_{1z} + J_{2z}$, corresponding to the eigenvalue $m = m_1 + m_2$, is easily seen to be the product $\psi_{j_1 m_1} \psi_{j_2 m_2}$. To construct the eigenfunction ϕ_{jm} we must find a linear combination of the products $\psi_{j_1 m_1} \cdot \psi_{j_2 m_2}$ with a given $m = m_1 + m_2$, which will be an eigenfunction of J^2. Since J^2 commutes with J_1^2 and J_2^2, the quantum numbers j_1 and j_2 must be considered fixed in this linear combination, i.e.

$$\phi_{jm} = \sum_{m_1, m_2} \langle j_1 m_1 j_2 m_2 \mid jm \rangle \psi_{j_1 m_1} \psi_{j_2 m_2}. \tag{46}$$

The numbers $\langle j_1 m_1 j_2 m_2 \mid jm \rangle$ are the Clebsch-Gordan coefficients.

Since $m = m_1 + m_2$ and, in addition,

$$-j_1 \leq m_1 \leq j_1, \quad -j_2 \leq m_2 \leq j_2, \quad -j \leq m \leq j,$$

the quantum number j can have the values

$$\mid j_1 - j_2 \mid \leq j \leq j_1 + j_2. \tag{47}$$

Accordingly, we shall take Clebsch-Gordan coefficients to be zero if they do not satisfy these conditions on m_1, m_2, m, and j.

The orthonormality of the functions ϕ_{jm}, $\phi_{j_1 m_1}$ and $\phi_{j_2 m_2}$ leads to the following condition on the Clebsch-Gordan coefficients:

$$\sum_{m_1, m_2} \mid \langle j_1 m_1 j_2 m_2 \mid jm \rangle \cdot \mid^2 = 1. \tag{48}$$

The Clebsch-Gordan coefficients play a fundamental role in quantum mechanics. We can use them, for example, in constructing the wave functions for complex systems (nuclei, atoms, molecules). Their theory is fully developed in many sources (see, for example, [E1], [R2],[V1]). Without going into a complete discussion, we shall give a simple derivation of their explicit form and establish the connection between Clebsch-Gordan coefficients and classical orthogonal polynomials of a discrete variable*. We start from the relations (45) for ψ_{jm}, $\psi_{j_1m_1}$ and $\psi_{j_2m_2}$. If we apply the operator $J_\pm = J_{1\pm} + J_{2\pm}$ to both sides of (46), we obtain the following recurrence relations for the Clebsch-Gordan coefficients:

$$\begin{aligned}&\alpha^j_{m-1}\langle j_1m_1j_2m_2 \mid j,m-1\rangle \\ &= \alpha^{j_1}_{m_1}\langle j_1,m_1+1,j_2m_2 \mid jm\rangle + \alpha^{j_2}_{m_2}\langle j_1m_1j_2,m_2+1 \mid jm\rangle,\end{aligned} \tag{49}$$

$$\begin{aligned}&\alpha^j_{m}\langle j_1m_1j_2m_2 \mid j,m+1\rangle \\ &= \alpha^{j_1}_{m_1-1}\langle j_1,m_1-1,j_2m_2 \mid jm\rangle + \alpha^{j_2}_{m_2-1}\langle j_1m_1j_2,m_2-1 \mid jm\rangle.\end{aligned} \tag{50}$$

Here $\alpha^j_m = [(j-m)(j+m+1)]^{1/2}$.

2°. Let us put (49) and (50) into a simpler form. We have only to observe that in (49) the change of any of the indices m_1, m_2, or m in $\langle j_1m_1j_2m_2 \mid jm\rangle$ is accompanied by the appearance of a factor with the same index. Hence if the values of j_1, j_2 and j are fixed, the transformation

$$\langle j_1m_1j_2m_2 \mid jm\rangle = A(m_1)\,B(m_2)\,C(m)u_m(m_1,m_2) \tag{51}$$

lets us, by appropriate choice of the factors A,B, C, obtain an equation for $u_m(m_1,m_2)$ with arbitrary constant coefficients:

$$cu_{m-1}(m_1,m_2) = au_m(m_1+1,m_2) + bu_m(m_1,m_2+1)\,. \tag{52}$$

Here we need to satisfy the conditions

$$\alpha^{j_1}_{m_1}\frac{A(m_1+1)}{A(m_1)} = a, \quad \alpha^{j_2}_{m_2}\frac{B(m_2+1)}{B(m_2)} = b, \quad \alpha^{j}_{m-1}\frac{C(m+1)}{C(m)} = c. \tag{53}$$

If we introduce the notation

$$\beta^j_m = \prod_{s=-j}^{m-1}\alpha^j_s \qquad \left(\beta^j_{-j}=1\right),$$

particular solutions of (53) can be given in the form

$$A(m_1) = \frac{a^{j_1+m_1}}{\beta^{j_1}_{m_1}}, \quad B(m_2) = \frac{b^{j_2+m_2}}{\beta^{j_2}_{m_2}}, \quad C(m) = \frac{\beta^j_m}{c^{j+m}}. \tag{54}$$

---

* The analogy between the Clebsch-Gordan coefficients and the Jacobi polynomials was first noticed in [G2]. See also [K2], [N1].

Since the numbers m_1 and m_2 in $\langle j_1 m_1 j_2 m_2 \mid jm \rangle$ are connected by $m_1 + m_2 = m$, we can denote $u_m(m_1, m_2) = u_m(m_1, m - m_1)$ by $u_m(m_1)$. With this notation we can write (52) in the form

$$cu_{m-1}(m_1) = au_m(m_1 + 1) + bu_m(m_1). \tag{55}$$

It is convenient to take $a = 1$, $b = -1$, $c = 1$. Then (55) becomes

$$u_{m-1}(m_1) = \Delta u_m(m_1), \tag{56}$$

where

$$\Delta f(x) = f(x+1) - f(x),$$

$$\langle j_1 m_1 j_2 m_2 \mid jm \rangle = (-1)^{j_2 + m_2} \frac{\beta^j_m}{\beta^{j_1}_{m_1} \beta^{j_2}_{m_2}} u_m(m_1) \quad (m_2 = m - m_1). \tag{57}$$

Similarly, (50) can be written in the form

$$v_{m+1}(m_1) = \nabla v_m(m_1), \tag{58}$$

where

$$\nabla f(x) = f(x) - f(x-1),$$

$$\langle j_1 m_1 j_2 m_2 \mid jm \rangle = (-1)^{j_1 + m_1} \frac{\beta^{j_1}_{m_1} \beta^{j_2}_{m_2}}{\beta^j_m} v_m(m_1) \qquad (m_2 = m - m_1). \tag{59}$$

3°. From (56) we have

$$u_m(m_1) = \Delta u_{m+1}(m_1) = \Delta^2 u_{m+2}(m_1) = \cdots = \Delta^{j-m} u_j(m_1). \tag{60}$$

We determine $u_j(m_1)$ from (58) with $m = j$. Since

$$\langle j_1 m_1 j_2 m_2 \mid j, j+1 \rangle = 0,$$

we have $v_{j+1}(m_1) = 0$. Hence, by (58), $v_j(m_1)$ is independent of m_1, i.e. $v_j(m_1) = C$, where C depends only on j_1, j_2, and j. Hence we have, by (57) and (59),

$$\begin{aligned} u_j(m_1) &= (-1)^{j_2 + j - m_1} \frac{\beta^{j_1}_{m_1} \beta^{j_2}_{j-m_1}}{\beta^j_j} \langle j_1 m_1 j_2, j - m_1 \mid jj \rangle \\ &= (-1)^{j_1 + j_2 + j} C \left(\frac{\beta^{j_1}_{m_1} \beta^{j_2}_{j-m_1}}{\beta^j_j} \right)^2 . \end{aligned} \tag{61}$$

Consequently, using (57) and (60), we obtain

$$\langle j_1 m_1 j_2 m_2 \mid jm\rangle = (-1)^{j_1+j_2+j+j_2+m_2} C \frac{\beta^j_m}{\left(\beta^j_j\right)^2 \beta^{j_1}_{m_1}\beta^{j_2}_{m_2}} \Delta^{j-m}\left[\left(\beta^{j_1}_{m_1}\beta^{j_2}_{j-m_1}\right)^2\right]. \qquad (62)$$

Since

$$\beta^j_m = \left[\frac{(2j)!\,(j+m)!}{(j-m)!}\right]^{1/2},$$

we can transform the formula (62) for the Clebsch-Gordan coefficients into

$$\langle j_1 m_1 j_2 m_2 \mid jm\rangle = (-1)^{j_2+m_2} D \left[\frac{(j_1-m_1)!\,(j_2-m_2)!\,(j+m)!}{(j_1+m_1)!\,(j_2+m_2)!\,(j-m)!}\right]^{1/2} \times \Delta^{j-m}\left[\frac{(j_1+m_1)!\,(j_2+j-m_1)!}{(j_1-m_1)!\,(j_2-j+m_1)!}\right], \qquad (63)$$

where D is a constant depending only on j_1, j_2, and j. We can find $\mid D \mid$ from the normalization condition (48) for $m = j$:

$$\mid D \mid^2 (2j)! \sum_{m_1} \frac{(j_1+m_1)!\,(j_2+j-m_1)!}{(j_1-m_1)!\,(j_2-j+m_1)!} = 1. \qquad (64)$$

It is clear from (63) and (64) that a Clebsch-Gordan coefficient is determined only up to a phase factor $e^{i\delta}$, which is usually found from the auxiliary condition

$$\langle j_1 m_1 j_2 m_2 \mid jm\rangle \mid_{m_1=j_1, m=j} \geq 0,$$

which is equivalent to requiring $D = (-1)^{j-j_1+j_2} \mid D \mid$.

4°. The sum in (64) can be evaluated by means of the formula

$$\Gamma(x)\Gamma(y) = \Gamma(x+y) \int_0^1 t^{x-1}(1-t)^{y-1}\,dt.$$

We have

$$(j_1+m_1)!)(j_2+j-m_1)! = (j_1+j_2+j+1)! \int_0^1 t^{j_1+m_1}(1-t)^{j_2+j-m_1}\,dt.$$

Hence

$$\sum_{m_1} \frac{(j_1+m_1)!\,(j_2+j-m_1)!}{(j_1-m_1)!\,(j_2-j+m_1)!}$$

$$= \frac{(j_1+j_2+j+1)!}{(j_1+j_2-j)!} \sum_{m_1} C_{j_1+j_2-j}^{j_1-m_1} \int_0^1 t^{j_1+m_1}(1-t)^{j_2+j-m_1}\,dt,$$

where

$$C_n^k = \frac{n!}{k!(n-k)!}.$$

On the other hand

$$\sum_{m_1} C_{j_1+j_2-j}^{j_1-m_1} t^{j_1+m_1}(1-t)^{j_2+j-m_1}$$

$$= t^{j_1+j-j_2}(1-t)^{j_2+j-j_1} \sum_{m_1} C_{j_1+j_2-j}^{j_1-m_1}(1-t)^{j_1-m_1} t^{j_2-j+m_1}$$

$$= t^{j_1+j-j_2}(1-t)^{j_2+j-j_1}.$$

Consequently

$$\sum_{m_1} \frac{(j_1+m_1)!(j_2+j-m_1)!}{(j_1-m_1)!(j_2-j+m_1)!} = \frac{(j_1+j_2+j+1)!(j_1+j-j_2)!(j_2+j-j_1)!}{(j_1+j_2-j)!(2j+1)!}.$$

This yields

$$D = (-1)^{j-j_1+j_2}\left[\frac{(2j+1)\,(j_1+j_2-j)!}{(j_1+j_2+j+1)!\,(j+j_1-j_2)!\,(j-j_1+j_2)!}\right]^{1/2}. \quad (65)$$

5°. If we expand the power of the difference operator in (63) by the formula

$$\Delta^n f(m_1) = \sum_{k=0}^{n} (-1)^{n+k} \frac{n!}{k!(n-k)!} f(m_1+k),$$

we obtain a formula for the Clebsch-Gordan coefficients as a sum of a finite number of terms of the form

$$\langle j_1 m_1 j_2 m_2 \,|\, jm\rangle$$

$$= (-1)^{j_1-m_1}\left[\frac{(2j+1)(j_1+j_2-j)!(j-m)!(j+m)!(j_1-m_1)!(j_2-m_2)!}{(j_1+j_2+j+1)!(j+j_1-j_2)!(j-j_1+j_2)!(j_1+m_1)!(j_2+m_2)!}\right]^{1/2}$$

$$\times \sum_k \frac{(-1)^k (j_1+m_1+k)!(j_2+j-m_1-k)!}{k!(j-m-k)!(j_1-m_1-k)!(j_2-j+m_1+k)!} \quad (66)$$

(the derivation makes essential use of the fact that $2(j+j_2-j_1)$ is even). From (66) we can obtain the following symmetry relations for the Clebsch-Gordan coefficients:

$$\langle j_1 m_1 j_2 m_2 \mid jm\rangle = (-1)^{j_1+j_2-j}\langle j_1, -m, j_2, -m_2 \mid, -m\rangle; \tag{67}$$

$$\langle j_1 m_1 j_2 m_2 \mid jm\rangle = (-1)^{j_1+j_2-j}\langle j_2 m_2 j_1 m_1 \mid jm\rangle; \tag{68}$$

$$\begin{aligned}\langle j_1 m_1 j_2 m_2 \mid jm\rangle &= \Big\langle \frac{1}{2}(j_1+j_2+m), \frac{1}{2}(j_1-j_2+m_1-m_2), \frac{1}{2}(j_1+j_2-m),\\ &\quad \frac{1}{2}(j_1-j_2-m_1+m_2) \mid j, j_1-j_2\Big\rangle \quad \text{(Regge symmetry)}\end{aligned} \tag{69}$$

6°. By using (63) we can connect the Clebsch-Gordan coefficients with the Hahn polynomials $h_n^{(\alpha\beta)}(x)$, by using the Rodrigues formula (12.22) for these polynomials (see Table 3):

$$\left.\begin{aligned} &h_n^{(\alpha,\beta)}(x) = \frac{B_n}{\rho(x)}\,\nabla^n \rho_n(x),\\ &\text{where}\\ &B_n = \frac{(-1)^n}{n!}, \quad \rho(x) = \frac{(N+\alpha-1-x)!(\beta+x)!}{x!(N-1-x)!},\\ &\rho_n(x) = \frac{(N+\alpha-1-x)!(\beta+n+x)!}{x!(N-n-1-x)!}.\end{aligned}\right\} \tag{70}$$

To establish the connection we put $x = j_1 - m_1$ in (63). Since

$$\Delta_{m_1} f(m_1) = f(m_1+1) - f(m_1) = f(j_1-x+1) - f(j_1-x) = -\nabla_x f(j_1-x),$$

we have

$$\begin{aligned}\langle j_1 m_1 j_2 m_2 \mid jm\rangle &= (-1)^x \mid D \mid \left[\frac{(j+m)!}{(j-m)!}\right]^{1/2}\left[\frac{x!(j_1+j_2-m-x)!}{(2j_1-x)!(j_2-j_1+m+x)!}\right]\\ &\quad\times \nabla^{j-m}\left[\frac{(2j_1-x)!(j-j_1+j_2+x)!}{x!(j_1+j_2-j-x)!}\right].\end{aligned} \tag{71}$$

It is clear from comparing (70) and (71) that the Clebsch-Gordan coefficients can be expressed in terms of the Hahn polynomials $h_n^{(\alpha,\beta)}(x)$ for

$$n = j-m, \quad N = j_1+j_2-m+1, \quad \alpha = j_1-j_2+m, \quad \beta = j_2-j_1+m:$$

$$\sqrt{\rho(x)}h_n^{(\alpha,\beta)}(x) = \frac{(-1)^{j_1-m_1+j-m}}{|D|\,[(j-m)!(j+m)!]^{1/2}}\langle j_1m_1j_2m_2 \mid jm\rangle.$$

It is easy to see that

$$|D|\,[(j-m)!(j+m)!]^{1/2} = \frac{1}{d_n},$$

where d_n^2 are the squared norms of the Hahn polynomials.

Consequently we have a simple connection between the Clebsch-Gordan coefficients and the Hahn polynomials:

$$(-1)^{j_1-m_1+j-m}\langle j_1m_1j_2m_2 \mid jm\rangle = \frac{1}{d_n}\sqrt{\rho(x)}h_n^{(\alpha,\beta)}(x) \tag{72}$$

with

$$x = j_1 - m_1, \quad n = j - m, \quad N = j_1 + j_2 - m + 1,$$

$$\alpha = m + m', \quad \beta = m - m' \quad (m' = j_1 - j_2),$$

$$\rho(x) = \frac{(j_1+m_1)!(j_2+m_2)!}{(j_1-m_1)!(j_2-m_2)!}.$$

We have established (72) under the restrictions $\alpha > -1$, $\beta > -1$. These inequalities will be satisfied if

$$j_1 \geq j_2, \quad m \geq j_1 - j_2. \tag{73}$$

This is possible for any Clebsch-Gordan coefficient if we use the symmetry relations (67)–(69).

The Clebsch-Gordan coefficients $\langle j_1m_1j_2m_2 \mid jm\rangle$ can be different from zero only under the selection rules given above,

$$|j_1 - j_2| \leq j \leq j_1 + j_2, \quad m = m_1 + m_2. \tag{74}$$

Conditins (74) are necessary. If they are not satisfied, the Clebsch-Gordan coefficients are zero. However, there are cases when the coefficients are zero for special values of the angular momenta and components, even when (74) is satisfied. The existence of such zeros leads to supplementary selection rules which forbid quantum transitions whose amplitudes are proportional to the vanishing Clebsch-Gordan coefficients.

The connection between the Clebsch-Gordan coefficients and the Hahn polynomials provides an interpretation of the zeros of the coefficients. As we noticed above, every Clebsch-Gordan coefficient can be carried by a symmetry relation to a coefficient expressible in terms of a polynomial $h_n^{(\alpha,\beta)}(x)$ by (72). All the zeros of the Clebsch-Gordan coefficients are zeros of the Hahn polynomial (72), and so are at points $x = x_i = 0, 1, \ldots, N-1$.

Example. Let us consider the zeros of $\langle j_1 m_1 j_2 m_2 \mid j, j-1\rangle$, which corresponds to the first-degree Hahn polynomial

$$\begin{aligned} h_1^{(\alpha,\beta)}(x) &= -\tau(x) = (\alpha+\beta+2)x - (\beta+1)(N-1) \\ &= 2jx - (j_2 - j_1 + j)(j_1 + j_2 - j + 1). \end{aligned}$$

Under the selection rule (74) the coefficient $\langle j_1 m_1 j_2 m_2 \mid j, j-1\rangle$ vanishes when $h_1(x)$ has a zero at $x = j_1 - m_1$. This leads to the condition

$$j(m_1 - m_2) = (j_1 - j_2)(j_1 + j_2 + 1).$$

For example, there will be such zeros for $\langle 1,0,1,0|1,0\rangle$ and $\langle 3,2,2,0|3,2\rangle$.

8°. By using the connection between the Clebsch-Gordan coefficients and the Hahn polynomials, given in (72), one can obtain many properties of the coefficients which follow from corresponding properties of the polynomials. As an example, we shall obtain an asymptotic representation of the Clebsch-Gordan coefficient $\langle j_1 m_1 j_2 m_2 \mid jm\rangle$ as $j_1 \to \infty$, $j_2 \to \infty$ for given $m' = j_1 - j_2$, j, and m, assuming that $m' \geq 0$ and $m \geq m'$. This corresponds for the Hahn polynomials $h_n^{(\alpha,\beta)}(x)$ to having $N \to \infty$ with given α, β, and n. Since $(z+a)! \approx z^a z!$ as $z \to \infty$, we can obtain the following asymptotic formula for the weight function $\rho(x)$ of the Hahn polynomials as $N \to \infty$:

$$\rho(x) \approx (N/2)^{\alpha+\beta}(1-s)^\alpha(1+s)^\beta, \quad \text{where } x = \frac{1}{2}N(1+s).$$

Moreover, as was shown in §12, part 6, as $N \to \infty$ we have

$$h_n^{(\alpha,\beta)}(x) \approx N^n P_n^{(\alpha,\beta)}(s).$$

Therefore as $j_1 \to \infty$ with fixed j, m, and $j_1 - j_2$, we have

$$\begin{aligned} &(-1)^{j_1 - m_1 + j - m}(j_1 + j_2 - m + 1)^{1/2}\langle j_1 m_1 j_2 m_2 \mid jm\rangle \\ &\approx \frac{1}{2^m}\left[\frac{(2j+1)(j-m)!(j+m)!}{(j-m')!(j+m')!}\right]^{1/2} \\ &\qquad \times (1-s)^{(m+m')/2}(1+s)^{(m-m')/2} P_{j-m}^{(m+m',m-m')}(s) \end{aligned}$$

$$\left(s = \frac{(j_1 - m_1) - (j_2 - m_2) - 1}{(j_1 - m_1) + (j_2 - m_2) + 1},\ m \geq m' \geq 0,\ m' = j_1 - j_2\right).$$

If also $n = j - m$ is sufficiently large, we may use the asymptotic formula (18) of §19 for the Jacobi polynomials $P_n^{(\alpha,\beta)}(x)$ as $n \to \infty$, and obtain

$$(-1)^{j_1-m_1+j-m}(j_1+j_2-m+1)^{1/2}\langle j_1m_1j_2m_2 \mid jm\rangle$$
$$\approx \left[\frac{2(2j+1)(j-m-1)!(j+m)!}{\pi(j-m')!(j+m')!}\right]^{1/2} \frac{\cos\left[\left(j+\frac{1}{2}\right)\theta - \left(m+m'+\frac{1}{2}\right)\pi/2\right]}{\sqrt{\sin\theta}},$$

where

$$\cos\theta = \frac{(j_1-m_1)-(j_2-m_2)-1}{(j_1-m_1)+(j_2-m_2)+1}, \qquad m \ge m' \ge 0,\ m' = j_1 - j_2,$$
$$0 < \delta \le \theta \le \pi - \delta.$$

The properties of the Clebsch-Gordan coefficients $\langle j_1m_1j_2m_2 \mid jm\rangle$ when j is varied can be obtained by using the connection between the Hahn polynomials and the dual Hahn polynomials. As a result we obtain the following relation between the Clebsch-Gordan coefficients and the dual Hahn polynomials:

$$\langle j_1m_1j_2m_2 \mid jm\rangle = \left[\frac{(2j+1)\rho(j)}{d_n^2}\right]^{1/2} w_n^{(c)}(x,a,b),$$

where $x = j(j+1)$, $a = m$, $b = j_1 + j_2 + 1$, $c = j_2 - j_1$, $n = j_1 - m_1$, and $m \ge j_1 - j_2 \ge 0$. Here $\rho(j)$ and d_n^2 are the weight function and the squared norm of the dual Hahn polynomial $w_n^{(c)}(x,a,b)$.

By using the symmetry properties

$$\langle j_1m_1j_2m_2 \,|\, jm\rangle = \langle j_2,-m_2;j_1,-m_1 \,|\, j,-m\rangle$$
$$= \left\langle \frac{j_1+j_2-m}{2}, \frac{j_2-j_1-m_2+m_1}{2}; \frac{j_1+j_2+m}{2}, \frac{j_2-j_1+m_2-m_1}{2} \,\Big|\, j, j_2-j_1\right\rangle$$

we can obtain the analogous formula

$$\langle j_1m_1j_2m_2 \mid jm\rangle = \left[\frac{(2j+1)\rho(j)}{d_n^2}\right]^{1/2} w_n^{(c)}(x,a,b).$$

Here $x = j(j+1)$, $a = j_2 - j_1$, $b = j_1 + j_2 + 1$, $c = m$, $n = j_1 - m_1$, $j_2 - j_1 \ge m \ge 0$.

6. The Wigner 6j-symbols and the Racah polynomials. In the composition of three angular momenta $\mathbf{J}_1$, $\mathbf{J}_2$ and $\mathbf{J}_3$ in quantum mechanics, there are various possible connection schemes. For example,

$$\mathbf{J}_1 + \mathbf{J}_2 = \mathbf{J}_{12}, \quad \mathbf{J}_{12} + \mathbf{J}_3 = \mathbf{J} \tag{75}$$

or

$$\mathbf{J}_2 + \mathbf{J}_3 = \mathbf{J}_{23}, \quad \mathbf{J}_{23} + \mathbf{J}_1 = \mathbf{J}. \tag{76}$$

In the usual notation of quantum mechanics, we denote the eigenfunctions for the total momentum J corresponding to (75) or (76) by $| j_{12} j m\rangle$ or $| j_{23} j m\rangle$. The transformation from $| j_{12} j m\rangle$ to $| j_{23} j m\rangle$ has the form

$$|j_{23} j m\rangle = \sum_{j_{12}} \begin{Vmatrix} j_1 & j_2 & j_3 \\ j_{12} & j & j_{23} \end{Vmatrix} |j_{12} j m\rangle. \tag{77}$$

The transformation matrix

$$\begin{Vmatrix} j_1 & j_2 & j_3 \\ j_{12} & j & j_{23} \end{Vmatrix}$$

is related to the Wigner $6j$-symbols

$$\begin{Bmatrix} j_1 & j_2 & j_{12} \\ j_3 & j & j_{23} \end{Bmatrix}$$

by

$$\begin{Vmatrix} j_1 & j_2 & j_3 \\ j_{12} & j & j_{23} \end{Vmatrix} = (-1)^{j_1+j_2+j_3+j} \left[(2j_{12}+1)(2j_{23}+1)\right]^{1/2} \begin{Bmatrix} j_1 & j_2 & j_{12} \\ j_3 & j & j_{23} \end{Bmatrix}. \tag{78}$$

Because the transformation matrix between two orthonormal systems of functions $| j_{12} j m\rangle$ and $| j_{23} j m\rangle$ must be unitary, we obtain the following orthogonality relation for the $6j$-symbols:

$$\sum_{j_{23}} (2j_{12}+1)(2j_{23}+1) \begin{Bmatrix} j_1 & j_2 & j_{12} \\ j_3 & j & j_{23} \end{Bmatrix} \begin{Bmatrix} j_1 & j_2 & j'_{12} \\ j_3 & j & j_{23} \end{Bmatrix} = \delta_{j_{12}, j'_{12}}. \tag{79}$$

Let us show that the $6j$-symbols can be expressed in terms of the Racah polynomials. It follows directly from (77) that the $6j$-symbols can be expressed in terms of the Clebsch-Gordan coefficients:

$$\begin{aligned} \langle j_1 m_1 j_{23} m_{23} \mid j m\rangle \begin{Vmatrix} j_1 & j_2 & j_3 \\ j_{12} & j & j_{23} \end{Vmatrix} &= \sum_{m_2} \langle j_{12} m_{12} j_3, m - m_{12} \mid j m\rangle \langle j_1 m_1 j_2 m_2 \mid j_{12} m_{12}\rangle \\ &\quad \times \langle j_2 m_2 j_3, m_{23} - m_2 \mid j_{23} m_{23}\rangle. \end{aligned} \tag{80}$$

Let us elucidate the dependence of the right-hand side of this equation on the variable j_{23}. For this purpose we use the relation between the Clebsch-Gordan coefficients and the dual Hahn polynomials $w_n^{(c)}(x, a, b)$, $x = x(s) = s(s+1)$, which we established in part 4:

$$\langle j_2 m_2 j_3 m_3 \mid j_{23} m_{23} \rangle$$
$$= \left[\frac{(2j_{23}+1)\overline{\rho}(j_{23})}{\overline{d}^2_{j_2-m_2}} \right]^{1/2} w^{(m_{23})}_{j_2-m_2} [j_{23}(j_{23}+1), j_3 - j_2, j_2 + j_3 + 1] \quad (81)$$
$$(j_3 - j_2 \geq m_{23} \geq 0),$$

where $\overline{\rho}(s)$ and $\overline{d}^2_n$ are the weight function and the squared norm of the polynomials $w_n^{(c)}(x, a, b,)$ (see Table 12):

$$\overline{\rho}(s) = \frac{\Gamma(a+s+1)\Gamma(c+s+1)}{\Gamma(s-a+1)\Gamma(b-s)\Gamma(b+s+1)\Gamma(s-c+1)},$$
$$\overline{d}^2_n = \frac{\Gamma(a+c+n+1)}{n!(b-a-n-1)!\Gamma(b-c-n)}.$$

It is easy to see by substituting (81) into (80) that the right-hand side of (80) is, up to the factor $[(2j_{23}+1)\overline{\rho}(j_{23})]^{1/2}$, a polynomial in $x = j_{23}(j_{23}+1)$. The degree of this polynomial is $\max(j_2 - m_2) = j_{12} + m_1 + j_2$. We choose this as small as possible, i.e. we set $m_1 = -j_1$. Then if we substitute into (80) the special values of the Clebsch-Gordan coefficients, we can calculate the weight function and the leading coefficent of the polynomial, if we set $j_{12} = j_1 - j_2$. Then, by using the orthogonality property (79) of the $6j$-symbols, we obtain

$$\begin{Bmatrix} j_1 & j_2 & j_{12} \\ j_3 & j & j_{23} \end{Bmatrix} = (-1)^{j_1+j+j_{23}} \left[\frac{\rho(j_{23})}{(2j_{12}+1) d^2_{j_{12}-j_1+j_2}} \right]^{1/2}$$
$$\times\, u^{m-m', m+m'}_{j_{12}-j_1+j_2} \left[j_{23}\,(j_{23}+1)\,; j_3 - j_2, j_3 + j_2 + 1 \right], \quad (82)$$

where $m = j_1 - j_2$, $m' = j_3 - j$; $j_1 \geq j$, $j_3 \geq j_2$, $j_1 + j \geq j_2 + j_3$, and $j_3 + j \geq j_1 + j_2$.

In (82), $\rho(s)$ and d^2_n are the weight function and squared norm of the Racah polynomials $u_n^{(\alpha,\beta)}(x, a, b)$, $x = s(s+1)$ (see Table 11).

§ 27 Application of special functions to some problems of numerical analysis

1. Quadrature formulas of Gaussian type. In the approximate calculation of definite integrals and of sums of a large number of terms, numerical analysis makes extensive use of quadrature formulas of Gaussian type, which depend on properties of orthogonal polynomials.

Quadrature formulas of Gaussian type for definite integrals are formulas of the form

$$\int_a^b f(x)\rho(x)dx \approx \sum_{j=1}^{n} \lambda_j f(x_j), \tag{1}$$

where $\rho(x) > 0$, and the coefficients λ_j and abscissas $x_j (j = 1, 2, \ldots, n)$ are chosen so that (1) is exact for all polynomials of degree $2n - 1$.

If the moments of the weight function,

$$C_k = \int_a^b x^k \rho(x)dx,$$

are known, the numbers λ_j and x_j can be found from the system of equations

$$\sum_{j=1}^{n} \lambda_j x_j^k = C_k \qquad (k = 0, 1, \ldots,\ 2n - 1).$$

However, we generally use a different method. It turns out that the $x_j (j = 1, 2, \ldots, n)$ are the zeros of a polynomial $p_n(x)$ that is orthogonal on (a, b) with weight function $\rho(x)$. For the proof, we consider the function

$$f(x) = x^k \tilde{p}_n(x),$$

where

$$\tilde{p}_n(x) = (x - x_1)(x - x_2) \ldots (x - x_n) = \prod_{j=1}^{n} (x - x_j)$$

is a polynomial of degree n whose zeros are the abscissas of the integration formula. When $k = 0, 1, \ldots, n - 1$, $f(x)$ is a polynomial of degree at most $2n - 1$. Therefore if $f(x)$ is substituted into (1), the integration formula ought to produce the exact value of the integral for every $k < n$. Hence, for

$k = 0, 1, \ldots, n-1$, we have

$$\int_a^b f(x)\rho(x)dx = \int_a^b x^k \tilde{p}_n(x)\rho(x)dx$$

$$= \sum_{j=1}^n \lambda_j x^k (x - x_1)(x - x_2)\ldots(x - x_n)|_{x=x_j} = 0.$$

Therefore the polynomial $\tilde{p}_n(x)$ is orthogonal to every power of order less than n, and consequently must be the same, up to a constant factor, as the polynomial $p_n(x)$ of degree n that is orthogonal with weight $\rho(x)$ on (a, b). Hence we can conclude that to determine the abscissas x_j of the integration formula it is enough to construct $p_n(x)$ and find its zeros.

To find the coefficients λ_j, which are known as the *Christoffel numbers*, it is convenient to take $f(x)$ in (1) to be a polynomial of degree less than $2n$ that is zero at all the abscissas except $x = x_j$. We obtain

$$\lambda_j = \frac{1}{f(x_j)} \int_a^b f(x)\rho(x)dx.$$

If we take, for example, one of the functions $f(x) = [(x - x_j)^{-1} p_n(x)]^2$ or $f(x) = (x - x_j)^{-1} p_n(x) p_{n-1}(x)$, we obtain the following expressions for the coefficients λ_j:

$$\lambda_j = \int_a^b \left[\frac{p_n(x)}{p_n'(x_j)(x - x_j)}\right]^2 \rho(x)dx, \tag{2}$$

$$\lambda_j = \frac{1}{p_n'(x_j)p_{n-1}(x_j)} \int_a^b \frac{p_n(x)}{x - x_j} p_{n-1}(x)\rho(x)dx. \tag{3}$$

It is clear from (2) that $\lambda_j > 0$. The integral on the right of (3) is easily evaluated. Since

$$\frac{p_n(x)}{x - x_j} = \frac{a_n}{a_{n-1}} p_{n-1}(x) + q_{n-2}(x),$$

where a_n is the leading coefficient of $p_n(x)$, and $q_{n-2}(x)$ is a polynomial of degree $n - 2$, we obtain, by the properties of the orthogonal polynomials $p_n(x)$,

$$\lambda_j = \frac{a_n d_{n-1}^2}{a_{n-1} p_n'(x_j) p_{n-1}(x_j)}, \tag{4}$$

where $d_n^2 = \int_a^b p_n^2(x)\rho(x)dx$ is the square of the norm.

Notice that our whole discussion of Gaussian integration formulas is still valid if the integral $\int_a^b f(x)\rho(x)dx$ is replaced by a sum $\sum_i f(x_i)\rho(x_i)$ (see §12, part 3). In that case we obtain the abscissas and the Christoffel numbers by using the properties of orthogonal polynomials of a discrete variable. This idea is often applied in evaluating sums that contain a function $f(x)$ that is not easily calculated, in order to use sums that have significantly fewer terms.

Let us look at some typical examples of the use of Gaussian integration formulas.

Example 1. Since many special functions can be represented by definite integrals, the use of Gaussian integration formulas for evaluating the integrals can lead to convenient and reasonably accurate approximate formulas for the special functions.

The most frequently encountered Bessel functions are $J_0(z)$ and $J_1(z)$. In many cases it is convenient to use simple approximate formulas (for example, in machine calculations, where formulas are preferable to tables). We shall use Poisson's integral to derive some approximate formulas for $J_0(z)$ and $J_1(z)$. We have

$$J_m(z) = \frac{1}{\sqrt{\pi}\Gamma\left(m+\frac{1}{2}\right)}\left(\frac{z}{2}\right)^m \int_{-1}^{1}(1-s^2)^{m-1/2}\cos zs\, ds.$$

The integral on the right can be evaluated for $m = 0$ by a Gaussian integration formula:

$$\int_{-1}^{1} f(s)\frac{1}{\sqrt{1-s^2}}ds \approx \sum_{j=1}^{n} \lambda_j f(s_j).$$

Here the numbers s_j are the zeros of the Chebyshev polynomials of the first kind, which are orthogonal on (-1,1) with weight $(1-s^2)^{-1/2}$, and λ_j are the Christoffel numbers:

$$s_j = \cos\frac{2j-1}{2n}\pi, \quad \lambda_j = \frac{\pi}{n}.$$

Taking $f(s) = \cos zs$, we obtain the following approximate formula for $J_0(z)$:

$$J_0(z) = \frac{1}{\pi}\int_{-1}^{1}\frac{\cos zs}{\sqrt{1-s^2}}ds \approx \frac{1}{n}\sum_{j=1}^{n}\cos\left(z\cos\frac{2j-1}{2n}\pi\right).$$

Table 14. Application of Gaussian integration formulas to the calculation of $J_0(z)$ and $J_1(z)$

z	$J_0(z)$	$\tilde{J}_0(z)$	$J_1(z)$	$\tilde{J}_1(z)$
0.4	0.9604	0.9604	0.1960	0.1960
1.2	0.6711	0.6711	0.4983	0.4983
2.0	0.2239	0.2239	0.5767	0.5767
2.8	−0.1850	−0.1850	0.4097	0.4097
3.6	−0.3918	−0.3918	0.09547	0.09548
4.4	−0.3423	−0.3423	−0.2028	−0.2027
5.2	−0.1103	−0.1105	−0.3432	−0.3427
6.0	0.1506	0.1496	−0.2767	−0.2748

By using the formula $J_1(z) = -J_0'(z)$, we also obtain an approximate formula for $J_1(z)$:

$$J_1(z) \approx \frac{1}{n}\sum_{j=1}^{n} \cos\frac{2j-1}{2n}\pi \sin\left(z\cos\frac{2j-1}{2n}\pi\right).$$

For even values of n, the set of abscissas s_j in the integration formula is symmetric with respect to $s = 0$. Hence for an even $f(s)$ there are only $n/2$ different terms in the integration formula. Taking, for example, $n = 6$, we obtain

$$J_0(z) \approx \frac{1}{3}\sum_{j=1}^{3} \cos zx_j,$$

$$J_1(z) \approx \frac{1}{3}\sum_{j=1}^{3} x_j \sin zx_j.$$

Here

$$x_j = \cos\frac{2j-1}{12}\pi = \begin{cases} \cos(\pi/12) & = 0.965926, \\ \cos(\pi/4) & = 0.707107, \\ \cos(5\pi/12) & = 0.258819. \end{cases}$$

The accuracy of these formulas is indicated in Table 14, in which the exact values of $J_0(z)$ and $J_1(z)$ are compared with the values given by the approximate formulas (denoted by $\tilde{J}_0(z)$ and $\tilde{J}_1(z)$).

Evidently the resulting formulas can also be used for complex z with not too large values of $|z|$.

Example 2. Let us consider the application of Gaussian integration formulas to the calculation of sums of the form

$$S_N = \sum_{k=0}^{N-1} f(k).$$

Table 15. Application of Gaussian summation formulas to the calculation of the sum $\sum_{k=0}^{N-1} \sqrt{l+k}$

$N-1$	10		1000	
$n \backslash l$	1	10	1	10
1	26.944	42.603	22405	22606
3	25.808	42.360	21207	21453
5	25.786	42.360	21148	21405
N	25.785	42.360	21129	21395

An integration formula lets us replace the sum by a sum of fewer terms:

$$S_N \approx \sum_{j=1}^{n} \lambda_j f(x_j).$$

In this case the abscissas x_j are the zeros of polynomials $p_n(x)$ that have the orthogonality properties

$$\sum_{k=0}^{N-1} p_n(k) p_m(k) = 0 \qquad (m \neq n).$$

The corresponding orthogonal polynomials are the Chebyshev polynomials of a discrete variable (see §12). As an illustration we give a comparative table of the results of calculating the sums S_N for $f(k) = (l+k)^{1/2}$ (with an integer l) for different numbers of integration points (Table 15). Note that when $n = N$ the integration formula gives the exact value of the original sum.

Example 3. Calculating coefficients of absorption of light in spectral lines requires the evaluation of integrals of the form (see [S4])

$$K(x,y) = \frac{1}{\pi} \int_{-\infty}^{\infty} \frac{y e^{-s^2}}{(x-s)^2 + y^2} ds \qquad (y > 0). \tag{5}$$

Because of the factor e^{-s^2}, only values with $|s| < 1$ are significant in the integration. For these values the function $1/[(x-s)^2 + y^2]$ with a given x is quite smooth if y is comparatively large. Consequently, when $y > 1$ we may calculate $K(x,y)$ by using a Gaussian formula based on the Hermite polynomials:

$$K(x,y) \approx K_1(x,y) = \frac{1}{\pi}\sum_{j=1}^{n} \lambda_j \frac{y}{(x-s_j)^2+y^2}. \tag{6}$$

However, for small y the function $y/[(x-s)^2+y^2]$ will have a sharp maximum at $x = s$ and, for x of moderate size, the integration formula will not give satisfactory results. To avoid this difficulty, we can first transform $K(x,y)$ into a form that is more suitable for the application of a Gaussian formula. We have

$$K(x,y) = \frac{1}{\pi}\mathrm{Im}\int_{-\infty}^{\infty} \frac{e^{-s^2}}{(x-s)-iy}ds.$$

If we use Cauchy's theorem to replace integration along the real axis by integration along the line $s = ai + t$ ($a > 0, -\infty < t < \infty$) parallel to the real axis, we obtain the following expression for $K(x,y)$:

$$K(x,y) = \frac{1}{\pi}\mathrm{Im}\int_{-\infty}^{\infty} \frac{e^{-(ai+t)^2}}{(x-t)-i(a+y)}dt$$

$$= \frac{1}{\pi}e^{a^2}\int_{-\infty}^{\infty} \frac{e^{-t^2}[(a+y)\cos 2at - (x-t)\sin 2at]}{(x-t)^2+(a+y)^2}dt. \tag{7}$$

Thanks to the transformation, we have been able to replace the function $1/[(x-s)^2+y^2]$ with its sharp maximum by an integrand involving the much smoother function $1/[(x-t)^2+(a+y)^2]$, which can be replaced in the essential part of the interval of integration by an approximating polynomial of low degree. To be sure, the integrand also involves an additional oscillating factor. However, if $a \approx 1$, the function $[(a+y)\cos 2at - (x-t)\sin 2at]$ is also rather smooth in the essential part of the interval of integration. If we now apply a Gaussian integration formula based on Hermite polynomials, and use (7), we obtain

$$K(x,y) \approx K_2(x,y) = \frac{e^{a^2}}{\pi}\sum_{j=1}^{n} \lambda_j \frac{(a+y)\cos 2as_j - (x-s_j)\sin 2as_j}{(x-s_j)^2+(a+y)^2}. \tag{8}$$

This analysis shows that the application of Gaussian integration to the integral (7) can give good results for all x and y if $a \approx 1$. As an illustration, we give a comparative table of the results of calculating (5) and (7) by formulas (6) and (8) for various numbers of abscissas when $a = 1$ and for various values of x and y (Table 16, p.360).

In conclusion we give tables (Tables 17, 18 and 19) of the Christoffel numbers λ_j and abscissas x_j for computing various types of integrals by Gaussian formulas.* Here the x_j are the zeros of the Legendre, Laguerre or Hermite polynomials. For the Legendre and Hermite polynomials, we tabulate only the nonnegative x_j. It must be remembered that, for each positive x_j, there is also an abscissa $-x_j$ with the same λ_j.

The formulas have the following forms:

1) $$\int_{-1}^{1} f(x)dx = \sum_{j=1}^{n} \lambda_j f(x_j)$$

(x_j the zeros of the Legendre polynomials $P_n(x)$; see Table 17, p.360).

2) $$\int_{0}^{\infty} e^{-x} f(x)dx = \sum_{j=1}^{n} \lambda_j f(x_j)$$

(x_j the zeros of the Laguerre polynomials $L_n^0(x)$; see Table 18, p.361).

3) $$\int_{-\infty}^{\infty} e^{-x^2} f(x)dx = \sum_{j=1}^{n} \lambda_j f(x_j)$$

(x_j the zeros of the Hermite polynomials $H_n(x)$; see Table 19, p.362).

* When $\lambda_j \ll 1$ we use the convention that, for example, 0.(4)233699 means 0.0000233699.

Table 16. Calculation of the Voigt integral

n	$x=0,\ y=0.01$		$x=1,\ y=0.01$	
	$K(x,y)=0.989$		$K(x,y)=0.369$	
	$K_1(x,y)$	$K_2(x,y)$	$K_1(x,y)$	$K_2(x,y)$
3	37.6	1.013	0.0225	0.387
5	30.1	0.991	0.693	0.370
7	25.8	0.9895	0.0434	0.369
n	$x=0,\ y=1$		$x=1,\ y=1$	
	$K(x,y)=0.428$		$K(x,y)=0.305$	
	$K_1(x,y)$	$K_2(x,y)$	$K_1(x,y)$	$K_2(x,y)$
3	0.451	0.441	0.293	0.317
5	0.434	0.428	0.305	0.305
7	0.430	0.428	0.306	0.305

Table 17. Zeros x_j of the Legendre polynomials $P(x)$ and the Christoffel numbers λ_j

n	x_j	λ_j
2	0.5773502692	1
3	0	0.8888888888
	0.7745966692	0.5555555555
4	0.3399810436	0.6521451549
	0.8611363116	0.3478548451
	0	0.5688888888
5	0.5384693101	0.4786286705
	0.9061798459	0.2362688506
	0.2386191861	0.4679139346
6	0.6612093865	0.3607615731
	0.9324695142	0.1713244924
	0	0.4179591837
7	0.4058451514	0.3818300505
	0.7415311856	0.2797053915
	0.9491079123	0.1294849662

n	x_j	λ_j
	0.1834346422	0.3626837834
8	0.5255324099	0.3137066459
	0.7966664774	0.222381035
	0.9602898565	0.1012285363
	0	0.3302393550
9	0.3242534234	0.3123470770
	0.6133714327	0.2606106964
	0.8360311073	0.1806481607
	0.9681602395	0.08127438836
	0.1488743390	0.2955242247
10	0.4333953941	0.2692667193
	0.6794095683	0.2190863625
	0.8650633667	0.1494513491
	0.9739065285	0.06667134430

Table 18. Zeros x_j of the Laguerre polynomials $L_n^0(x)$ and the Christoffel numbers λ_j

n	x_j	λ_j
1	1	1
2	0.5857864376	0.8535533906
	3.4142135624	0.1464466094
3	0.4157745567	0.7110930099
	2.2942803603	0.2785177336
	6.2899450829	0.0103892565
4	0.3225476896	0.6031541043
	1.7457611011	0.3574186924
	4.5366202969	0.03888790851
	9.3950709123	0.(3)53929447056
5	0.2635603197	0.5217556106
	1.4134030591	0.3986668111
	3.5964257710	0.07594244968
	7.0858100059	0.(2)3611758679
	12.6408008443	0.(4)2336997239
6	0.2228466042	0.4589646740
	1.1889321017	0.4170008308
	2.9927363261	0.1133733821
	5.7721435691	0.01039919745
	9.8374674184	0.(3)2610172028
	15.9828739806	0.(6)8985479064
7	0.1930436766	0.4093189517
	1.0266648953	0.4218312779
	2.5678767450	0.1471263487
	4.90003530845	0.02063351447
	8.1821534446	0.(2)1074010143
	12.7341802918	0.(4)1586546435
	19.3957278623	0.(7)3170315479

n	x_j	λ_j
8	0.1702796323	0.3691885893
	0.9037017768	0.4187867808
	2.2210866299	0.1757949866
	4.2667001703	0.03334349226
	7.0429054024	0.(2)2794536235
	10.7585160102	0.(4)9076508773
	15.7406786413	0.(6)8485746716
	22.8331317369	0.(8)1048001175
9	0.1523222277	0.3361264218
	0.8072200227	0.4112139804
	2.0051351556	0.1992875254
	3.7834739733	0.04746056277
	6.2049567778	0.(2)5599626611
	9.3729852517	0.(3)3052497671
	13.4662369110	0.(5)6592123026
	18.8335977889	0.(7)4110769330
	26.3740718909	0.(10)3290874030
10	0.1377934705	0.3084411158
	0.7294545495	0.4011199292
	1.8083429017	0.1180682876
	3.4014336979	0.06208745610
	5.2224961400	0.(2)9501516975
	8.3301527468	0.(3)7530083886
	11.8437858379	0.(4)2825923350
	16.2792578314	0.(6)4249313985
	21.9965858120	0.(8)1839564824
	29.9206970122	0.(12)9911827220

Table 19. Zeros of the Hermite polynomials $H_n(x)$ and the Christoffel numbers λ_j

n	x_j	λ_j
1	0	1.772453851
2	0.7071067812	8.8862269255
3	0	1.181635901
	1.2247448714	0.2954089752
4	0.5246476233	0.8049140900
	1.6506801239	0.08131283545
5	0	0.9453087205
	0.9585724646	0.3936193232
	2.0201828705	0.01995324206
6	0.4360774119	0.7246295952
	1.3358490740	0.1570673203
	2.3506049737	0.(2)4530009906
7	0	0.8102646176
	0.8162878829	0.4256072526
	1.6735516288	0.05451558282
	2.6519613568	0.(3)9717812451
8	0.3811869902	0.6611470126
	1.1571937124	0.2078023258
	1.9816567567	0.01707798301
	2.9306374203	0.(3)1996040722
9	0	0.7202352156
	0.7235510188	0.4326515590
	1.4685532892	0.08847452739
	2.2665805845	0.(2)4943624276
	3.1909932018	0.(4)3960697726
10	0.3429013272	0.6108626337
	1.0366108298	0.2401386111
	1.7566836493	0.03387439446
	2.5327316742	0.(2)1343645747
	3.4361591188	0.(5)7640432855

2. Compression of information by means of classical orthogonal polynomials of a discrete variable. The problem of storing information is one of the fundamental problems of science and technology. In this connection, an important problem is the compression of information for signal processing, in particular, in processing electrocardiograms, data obtained by satellites, etc.

At present, much use is being made of spectral methods of information processing. Instead of recording a table of the values of the function $f(t)$ that describes a signal, we record the initial Fourier coefficients c_n of the expansion of $f(t)$ in a series of functions $y_n(t)$ $(n = 0, 1, \dots)$ that form a complete orthogonal system:

$$f(t) = \sum_n c_n y_n(t), \qquad c_n = d_n^{-2}(f(t), y_n(t)),$$

where (f, g) is the scalar product,

$$(y_n, y_m) = d_n^2 \delta_{mn}, \quad \delta_{mn} = \begin{cases} 0 \ (m \neq n), \\ 1 \ (m = n). \end{cases} \tag{9}$$

In the usual case, the scalar product of $f(t)$ and $g(t)$ is given by an integral:

$$(f, g) = \int_\alpha^\beta f(t)g(t)\rho(t)dt,$$

where $\rho(t) \geq 0$ is the weight function with respect to which $y_n(t)$ are orthogonal (see, for example, §8). However, if $f(t)$ is given by a table of values $f(t_i)$ $(i = 0, 1, \dots, N-1)$, it is more convenient to calculate the coefficients c_n by using a system of orthogonal functions $y_n(t)$ for which the scalar product is a sum:

$$(f, g) = \sum_{i=0}^{N-1} f(t_i)g(t_i)\rho_i. \tag{10}$$

Since the orthogonality condition (9) with the scalar product (10) is what we have for the classical orthogonal polynomials of a discrete variable, these polynomials can be used for compressing information.

As an example, we present the results of applying the Chebyshev polynomials $t_n(x)$ to the processing of electrocardiograms. For reconstructing the signal, the curve $f(t)$ was broken into pieces and only the first three coefficients of the Chebyshev polynomial expansion of $f(t)$ were used for each piece. The sizes of the pieces were chosen so that the mean square error was at most 1%. Even such a simple algorithm proved to be very effective: the coefficient of compression was between 6 and 12. Figure 14 shows part of an original electrocardiogram and the reconstruction.

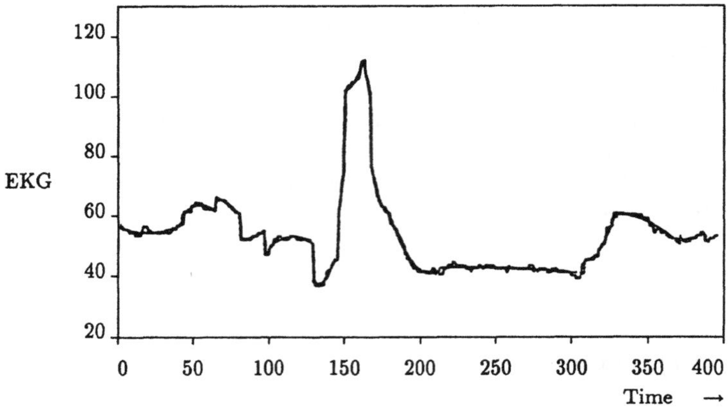

Figure 14. Compressed EKG, Compression coefficient 8.6

3. Application of modified Bessel functions to problems of laser sounding. In recent decades mathematics has penetrated extensively into very varied branches of science. This penetration has arisen because of the construction of mathematical models that provide quantitative descriptions in mathematical language of the phenomena that are studied. The role of mathematical models is especially important in the first stages of formulating a problem, as the first step in investigating a phenomenon that interests us. It is precisely at this stage that there often arises, in various branches of science and technology, the necessity of applying special functions. With their aid one may succeed in obtaining the solution of a problem in an analytic form that is suitable for study, and without much expenditure of time and effort (although possibly also not without appeal to the computer), display the fundamental regularity of the phenomenon, and estimate the influence of various factors. All this applies particularly to Bessel functions, which appear to be perhaps the most widely used of the special functions.

As an example, we discuss the application of the modified Bessel functions $I_n(z)$ to a problem in the laser sounding of the atmosphere, where we can use information about the absorption of a laser pulse in a particular spectral line of a substance that is of interest.

The study of the absorption of light in spectral lines can often provide significant information about the physical properties of a substance. For example, from the Doppler shift we can infer the velocity of an object; and from the width of spectral lines, its temperature and density.

At the present time, extensive use is being made of laser methods for determining the chemical and aerosol content of the atmosphere, in particular for discovering small unlocalized concentrations of gases. The most promising method is apparently that of comparative absorption, based on the use of laser lidar.* Laser pulses are sent into the atmosphere at two nearly equal frequencies ν_1 and ν_2, one of which (ν_1) almost coincides with the center of an absorption line ν_a of the substance under investigation, whereas the other lies outside this line, and the interval (ν_1, ν_2) does not overlap the other absorption lines. The radiation reaches a receiver after being reflected from some sort of reflector.

It can be shown that the ratio of the signal strengths received at frequencies ν_1 and ν_2 can be determined from the absorption of the laser signal at frequency ν_1 by the substance under investigation, provided that we suppose that the effects of all other interactions of the signals with the substance are effectively the same at frequencies ν_1 and ν_2. In fact, let

$k_\Lambda(\nu)$ be the profile of the emitted line, i.e. the spectral intensity of the laser pulse;

$k_a(\nu)$, the profile of the absorption line of the substance under investigation, i.e. the mass absorption coefficient for light at the frequency $\nu = \nu_a$ of the line;

$k(\nu)$, the absorption coefficient for all other interactions with the substance.

Then the power of the laser signal registered at the receiver, assuming a homogeneous atmosphere, will be

$$\int_0^\infty k_\Lambda(\nu) e^{-[\mu_a k_a(\nu) + k(\nu)]m}\, d\nu,$$

where μ_a is the percentage (by mass) of the absorbing substance in the atmosphere, and m is the mass of the absorbing substance in the path of the laser pulse: $m = LS\rho_0$ (L = length of the path traversed by the pulse from laser to receiver, S = area of the receiving antenna, and ρ_0 = density of the atmosphere).

For long-range sounding of the atmosphere one uses a narrow-band signal, for which $k_\Lambda(\nu)$ is appreciably different from zero only for a narrow range of frequencies $\nu \approx \nu_\Lambda$. For $\nu_\Lambda = \nu_1$ or $\nu_\Lambda = \nu_2$, the variation of the frequency of the distorted function $k_\Lambda(\nu)$ in the corresponding interval can be neglected; that is, in either case we may suppose $k(\nu) = \text{const}$. Moreover, when $\nu_\Lambda = \nu_2$ we may suppose, because of the way ν_2 was chosen, that $k_a(\nu)$ is effectively

* "Lidar" (Light Detection And Ranging) is to light as radar is to radio (Translator).

zero over the range of integration. Hence the ratio of the signal intensities at frequencies ν_1 and ν_2 is given by

$$T = \frac{P_1}{P_2} = \frac{\int_0^\infty k_\Lambda^{(1)}(\nu) e^{-\mu_a k_a(\nu) m} d\nu}{\int_0^\infty k_\Lambda^{(2)}(\nu) d\nu}. \tag{11}$$

In many cases that are important in practice, the actual line profile for absorption is close to Lorentzian, namely

$$k_a(\nu) = \frac{I_0}{\pi} \frac{\gamma_a}{\gamma_a^2 + (\nu - \nu_a)^2}$$

(I_0 is the intensity of the line, γ_a the half-width).

The intensity $k_\Lambda(\nu)$ of the radiated line at frequency ν is, as a rule, described by a similar profile,

$$k_\Lambda^{(i)}(\nu) = \frac{P_0}{\pi} \frac{\gamma}{\gamma^2 + (\nu - \nu_i)^2}$$

(P_0 is the power of the emitted pulse; $i = 1, 2$). In this case

$$T = \frac{\dfrac{1}{\pi} \displaystyle\int_0^\infty \frac{\gamma}{\gamma^2 + (\nu - \nu_1)^2} \exp\left[-\frac{I_0 m \mu_a}{\pi} \frac{\gamma_a}{\gamma_a^2 + (\nu - \nu_a)^2} \right] d\nu}{\dfrac{1}{\pi} \displaystyle\int_0^\infty \frac{\gamma}{\gamma^2 + (\nu - \nu_2)^2} d\nu}.$$

Usually $\gamma \ll \nu_i$ and then the integral over $(0, \infty)$ can be extended to $(-\infty, \infty)$ without changing T appreciably. Putting $t = 2 \arctan((\nu - \nu_a)/\gamma_a)$ we have

$$\begin{aligned} T &= T(z, a, \delta) \\ &= \frac{a e^{-z}}{\pi} \int_{-\pi}^{\pi} \frac{e^{-z \cos t} dt}{1 + a^2(1 + \delta^2) + [1 - a^2(1 - \delta^2)] \cos t + 2a^2 \delta \sin t}, \end{aligned} \tag{12}$$

where

$$z = \frac{I_0 m \mu_a}{2\pi \gamma_a}, \quad a = \frac{\gamma_a}{\gamma}, \quad \delta = \frac{\nu_a - \nu_1}{\gamma_a}.$$

In this expression for T, everything except μ_a is easily determined. Hence if T is measured experimentally, μ_a can be found, for example from a graph of $T = T(\mu_a)$ constructed by using (12).

To evaluate the integral in (12) we first expand $e^{-z\cos t}$ in a Fourier series. We find the coefficients by using (16.12) with z replaced by iz and ϕ by $\frac{1}{2}\pi - t$:

$$e^{-z\cos t} = \sum_{n=-\infty}^{\infty} (-1)^n I_n(z) e^{-int}.$$

Since $I_{-n}(z) = I_n(z)$, we have

$$e^{-z\cos t} = I_0(z) + 2\sum_{n=1}^{\infty}(-1)^n I_n(z)\cos nt.$$

Hence

$$T(z,a,\delta) = e^{-z}\left[I_0(z)\, S_0(a,\delta) + 2\sum_{n=1}^{\infty}(-1)^n I_n(z)\, S_n(a,\delta)\right].$$

where

$$S_n(a,\delta) = \frac{a}{\pi}\int_{-\pi}^{\pi} \frac{\cos nt\, dt}{1 + a^2(1+\delta^2) + [1 - a^2(1-\delta^2)]\cos t + 2a^2\delta\sin t}.$$

The substitution $\xi = e^{it}$ transforms $S_n(a,\delta)$ into a contour integral around the unit circumference; this can be evaluated by residues:

$$S_n(a,\delta) = (-1)^n \rho^n \cos n\alpha,$$

where

$$\rho = \left\{\frac{(a-1)^2 + a^2\delta^2}{(a+1)^2 + a^2\delta^2}\right\}^{1/2} \qquad (0 \le \rho < 1),$$

$$\cos\alpha = \frac{1 - a^2(1-\delta^2)}{r}, \quad \sin\alpha = \frac{2a^2\delta}{r},$$

$$r = \sqrt{[(a-1)^2 + a^2\delta^2][(a+1)^2 + a^2\delta^2]}.$$

We obtain

$$T(z,a,\delta) = e^{-z}\left[I_0(z) + 2\sum_{n=1}^{\infty} \rho^n \cos n\alpha I_n(z)\right]. \tag{13}$$

Since $\rho < 1$ and, as $n \to \infty$,

$$I_n(z) \approx (z/2)^n/n!,$$

the series (13) converges very rapidly, so that it is convenient for calculation; there are extensive tables of $I_n(z)$.

Our formulas make it possible to calculate T for arbitrary values of γ_a, γ, μ_a, ν_1, ν_2 and ν_a, and to investigate various limiting cases.

For example, let $\nu_1 = \nu_a$, i.e. let the signal frequency be the center of the absorption line. Then

$$\delta = 0, \quad \rho = \left|\frac{a-1}{a+1}\right|,$$

$$\alpha = \begin{cases} 0 & \text{for } \gamma_a < \gamma \quad (a < 1), \\ \pi & \text{for } \gamma_a > \gamma \quad (a > 1). \end{cases}$$

Then (13) reduces to

$$T(z, a, 0) = e^{-z}\left[I_0(z) + 2\sum_{n=1}^{\infty}\left(\frac{1-a}{1+a}\right)^n I_n(z)\right].$$

In particular, if $\gamma_a = \gamma$ $(a = 1)$, then

$$T(z, 1, 0) = e^{-z} I_0(z).$$

Appendices

A The Gamma Function

The gamma function is related to many quite simple, but at the same time important, special functions. Familiarity with its properties is essential for further study of special functions. Moreover, many integrals that are met with in analysis can be evaluated in terms of gamma functions. In particular, this is the case for the beta function integral.

1. Definition of $\Gamma(z)$ and $\mathrm{B}(u,v)$. The gamma function $\Gamma(z)$ and the beta function $\mathrm{B}(u,v)$ are defined as follows:

$$\Gamma(z) = \int_0^\infty e^{-t} t^{z-1} dt, \quad \mathrm{Re}\, z > 0; \tag{1}$$

$$\mathrm{B}(u,v) = \int_0^1 t^{u-1}(1-t)^{v-1} dt, \quad \mathrm{Re}\, u > 0, \quad \mathrm{Re}\, v > 0. \tag{2}$$

By Theorem 2 (see §3), $\Gamma(z)$ is analytic where it is defined. In fact, (1) converges uniformly in z in the domain $0 < \delta \le \mathrm{Re}\, z \le A$ for arbitrary A and δ, since

$$|e^{-t} t^{z-1}| \le \begin{cases} t^{\delta-1} & \text{for } 0 < t \le 1, \\ e^{-t} t^{A-1} & \text{for } t > 1 \end{cases}$$

and the integrals $\int_0^1 t^{\delta-1} dt$ and $\int_1^\infty e^{-t} t^{A-1} dt$ converge.

It can be shown similarly that $\mathrm{B}(u,v)$ is analytic in both arguments when $\mathrm{Re}\, u > 0$ and $\mathrm{Re}\, v > 0$.

The beta function can be expressed in terms of gamma functions. To show this it is enough to calculate the integral

$$I(u,v) = \iint e^{-(\xi^2+\eta^2)} \xi^{2u-1} \eta^{2v-1} \, d\xi d\eta$$

in two ways; the integration is over $\xi > 0, \eta > 0$. On one hand

$$I(u,v) = I(u)I(v),$$

where

$$I(u) = \int_0^\infty e^{-\xi^2} \xi^{2u-1} d\xi = \frac{1}{2} \int_0^\infty e^{-t} t^{u-1} dt = \frac{1}{2}\Gamma(u).$$

On the other hand, if we transform $I(u,v)$ to polar coordinates $\xi = r\cos\phi$, $\eta = r\sin\phi$, we obtain

$$I(u,v) = \int_0^\infty e^{-r^2} r^{2u+2v-1} dr \int_0^{\pi/2} \cos^{2u-1}\phi \, \sin^{2v-1}\phi \, d\phi$$

$$= \frac{1}{2}\Gamma(u+v) \int_0^{\pi/2} \cos^{2u-1}\phi \, \sin^{2v-1}\phi \, d\phi.$$

With the substitution $\cos^2\phi = t$, the ϕ-integral is expressible in terms of $\mathrm{B}(u,v)$:

$$\int_0^{\pi/2} \cos^{2u-1}\phi \, \sin^{2v-1}\phi \, d\phi = \frac{1}{2}\mathrm{B}(u,v).$$

Substituting into $I(u,v)$, we obtain

$$\mathrm{B}(u,v) = \frac{\Gamma(u)\Gamma(v)}{\Gamma(u+v)}. \tag{3}$$

2. Functional equations. The function $\Gamma(z)$ satisfies the following functional equations:

$$\Gamma(z+1) = z\Gamma(z), \tag{4}$$

$$\Gamma(z)\Gamma(1-z) = \frac{\pi}{\sin \pi z}, \tag{5}$$

$$2^{2z-1}\Gamma(z)\Gamma\left(z+\frac{1}{2}\right) = \Gamma\left(\frac{1}{2}\right)\Gamma(2z). \tag{6}$$

These play an important role in various transformations involving gamma functions. Equation (5) is the ***addition formula***, and (6) is the ***duplication formula***.

To prove (4)–(6), it is convenient to write them, using (3), as functional equations for the beta function:

$$\mathrm{B}(z,1) = 1/z, \tag{7}$$

$$\mathrm{B}(z,1-z) = \pi/\sin \pi z, \tag{8}$$

$$2^{2z-1}\mathrm{B}(z,z) = \mathrm{B}\left(z,\frac{1}{2}\right). \tag{9}$$

Equations (7)–(9) can be derived by direct calculation of the beta integral (2). We have

$$\mathrm{B}(z,1) = \int_0^1 t^{z-1}\,dt = \frac{1}{z},$$

which is (7).

To prove (8), we put $u = z, v = 1 - z$ in (2):

$$\mathrm{B}(z,1-z) = \int_0^1 \left(\frac{t}{1-t}\right)^{z-1} \frac{dt}{1-t}, \quad 0 < \mathrm{Re}\, z < 1.$$

Now put $s = t/(1-t)$. This yields

$$\mathrm{B}(z,1-z) = \int_0^\infty \frac{s^{z-1}}{1+s} ds.$$

The resulting integral can be evaluated by residue calculus. For this purpose we replace integration along the real axis by integration around the closed contour C shown in Figure 15. The function

$$f(s) = \frac{s^{z-1}}{1+s} \qquad (0 < \arg s < 2\pi)$$

has a pole at $s = e^{i\pi}$ and no other singular points in the domain bounded by C. Hence when $R > 1$ (see Fig. 15, p.372)

$$\int_C f(s)ds = 2\pi i \operatorname*{Res}_{s=e^{i\pi}} f(s) = -2\pi i e^{i\pi z}.$$

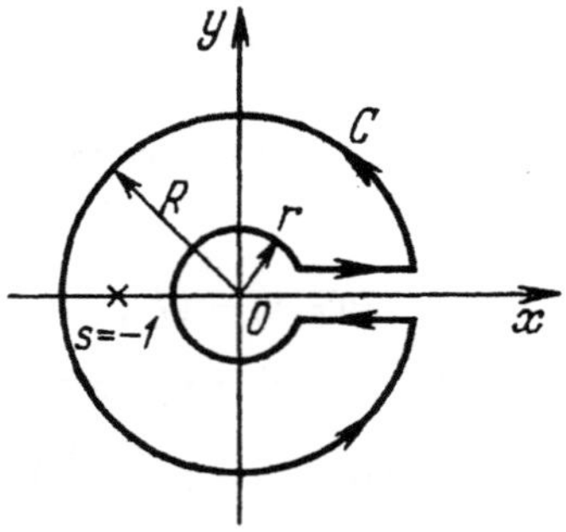

Figure 15.

On the other hand, since $0 < \operatorname{Re} z < 1$, the integrals over the circumferences of radii r and R tend to 0 as $r \to 0$ and $R \to \infty$, and the integral over the lower side of the cut differs from the integral over the upper side by the factor $-e^{2\pi i z}$. Hence when $r \to 0$ and $R \to \infty$ we obtain

$$\mathrm{B}(z,\, 1-z)(1-e^{2\pi i z}) = -2\pi i e^{i\pi z},$$

which is equivalent to (8) for $0 < \operatorname{Re} z < 1$.

To prove (9), we put $u = v = z$ in (2):

$$\mathrm{B}(z,\, z) = \int_0^1 [t(1-t)]^{z-1} dt, \quad \operatorname{Re} z > 0.$$

Since the parabola $y = t(1-t)$ is symmetric with respect to the line $t = 1/2$, we have

$$\mathrm{B}(z,\, z) = 2 \int_0^{1/2} [t(1-t)]^{z-1} dt,$$

whence after substituting $s = 4t(1-t)$ we obtain

$$\mathrm{B}(z,\, z) = \frac{1}{2^{2z-1}} \int_0^1 s^{z-1}(1-s)^{-1/2} ds = \frac{\mathrm{B}(z,\, 1/2)}{2^{2z-1}},$$

and this is equivalent to (9) when $\operatorname{Re} z > 0$.

Consequently we have established the functional equations (4)–(6) for the gamma function.

As an illustration, we calculate the values of $\Gamma(z)$ for integral and half-integral arguments. It follows from (4) that

$$\Gamma(n+1) = n!,$$

since $\Gamma(1) = 1$. We see that the gamma function generalizes the factorial. Also, if we put $z = 1/2$ in (5), we obtain

$$\Gamma(1/2) = \sqrt{\pi}.$$

Consequently we can rewrite (6) in the form

$$2^{2z-1}\Gamma(z)\Gamma(z+1/2) = \sqrt{\pi}\Gamma(2z). \tag{6a}$$

Taking $z = n + 1/2$, we obtain

$$\Gamma(n+1/2) = \frac{\sqrt{\pi}\Gamma(2n+1)}{2^{2n}\Gamma(n+1)} = \frac{\sqrt{\pi}}{2^{2n}}\frac{(2n)!}{n!}. \tag{10}$$

We can continue $\Gamma(z)$ analytically into the domain $\operatorname{Re} z > -(n+1)$ by using the formula

$$\Gamma(z) = \frac{\Gamma(z+n+1)}{z(z+1)\ldots(z+n-1)(z+n)}, \tag{11}$$

which follows from (4). Since n can be chosen arbitrarily, we obtain an analytic continuation of $\Gamma(z)$ for all z. We see from (11) that $\Gamma(z)$ *is analytic except at* $z = -n$ $(n = 0, 1, 2, \ldots)$, *where* $\Gamma(z)$ *has simple poles with residues*

$$\operatorname*{Res}_{z=-n} \Gamma(z) = \frac{(-1)^n}{n!}.$$

By the principle of analytic continuation, (4)–(6) are valid for all values of z for which they make sense. The analytic continuation of the beta function can be obtained from (3).

It follows from (5) that $\Gamma(z)$ *has no zeros in the complex plane.* In fact, suppose $\Gamma(z_0) = 0$. Evidently $z_0 \neq n+1$ $(n = 0, 1, 2, \ldots)$ since $\Gamma(n+1) = n! \neq 0$. Therefore $\Gamma(1-z)$ would be analytic at z_0. On the other hand

$$\lim_{z\to z_0} \Gamma(1-z) = \frac{\pi}{\sin \pi z_0} \lim_{z\to z_0} \frac{1}{\Gamma(z)} = \infty,$$

which contradicts the analyticity of $\Gamma(1-z)$ for $z \neq n+1$.

Figure 16 (p. 374) shows the graph of $y = \Gamma(x)$.

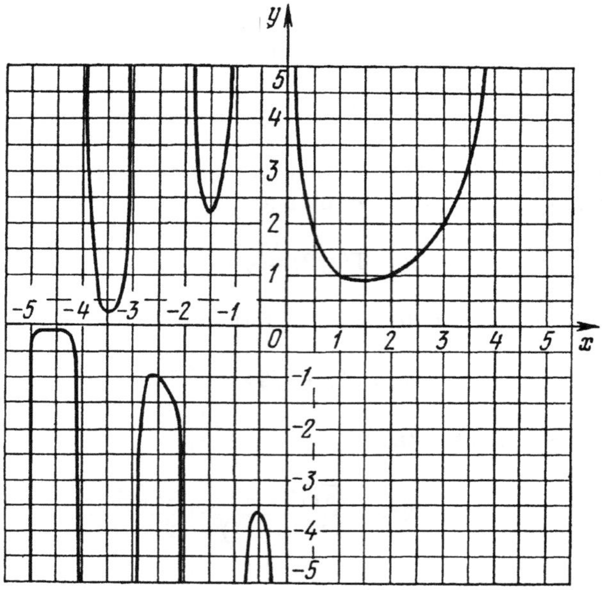

Figure 16.

3. The logarithmic derivative of the gamma function. The function

$$\psi(z) = \frac{d}{dz} \ln \Gamma(z) = \frac{\Gamma'(z)}{\Gamma(z)}$$

also has many applications. This function is analytic in the complex plane except for the points $z = -n$ ($n = 0, 1, \ldots$), at which it has simple poles.

The following *functional equations* for $\psi(z)$ follow from the functional equations (4)–(6) for the gamma function:

$$\psi(z+1) = 1/z + \psi(z), \tag{12}$$

$$\psi(z) = \psi(1-z) - \pi \cot \pi z, \tag{13}$$

$$2 \ln 2 + \psi(z) + \psi\left(z + \frac{1}{2}\right) = 2\psi(2z). \tag{14}$$

We also note the formula

$$\psi(z+n) = \psi(z) + \sum_{k=1}^{n} \frac{1}{z+k-1}. \tag{15}$$

which is easily obtained from (12).

Formulas (12)–(15) let us evaluate $\psi(z)$ for integral or half-integral values of the argument. Let us write

$$\psi(1) = \Gamma'(1) = -\gamma.$$

The number γ is *Euler's constant* ($\gamma = 0.5772156649\ldots$). If we put $z = 1/2$ in (14), we find

$$\psi\left(\frac{1}{2}\right) = -\gamma - 2\ln 2.$$

For $z = 1$ and $z = 1/2$, (15) yields

$$\psi(n+1) = -\gamma + \sum_{k=1}^{n} \frac{1}{k}, \tag{16}$$

$$\psi\left(n + \frac{1}{2}\right) = -\gamma - 2\ln 2 + 2\sum_{k=1}^{n} \frac{1}{2k-1}. \tag{17}$$

The integral representation for the beta function leads to an integral representation for $\psi(z)$. By definition,

$$\psi(z) = \frac{\Gamma'(z)}{\Gamma(z)} = \lim_{\Delta z \to 0} \frac{\Gamma(z) - \Gamma(z - \Delta z)}{\Gamma(z)\Delta(z)} = \lim_{\Delta z \to 0} \left[\frac{1}{\Delta z} - \frac{\Gamma(z - \Delta z)}{\Gamma(z)\Delta z}\right].$$

For sufficiently small $\Delta z > 0$, the quotient $\Gamma(z - \Delta z)/(\Gamma(z)\Delta z)$ approximates the beta function

$$\mathrm{B}(z - \Delta z, \Delta z) = \frac{\Gamma(z - \Delta z)\Gamma(\Delta z)}{\Gamma(z)},$$

since

$$\lim_{\Delta z \to 0} \Delta z \Gamma(\Delta z) = \lim_{\Delta z \to 0} \Gamma(1 + \Delta z) = 1.$$

The factor $1/\Delta z$ in the limit formula for $\psi(z)$ is easily eliminated by considering the difference $\psi(z) - \psi(1)$. We have

$$\psi(z) - \psi(1) = \psi(z) + \gamma = \lim_{\Delta z \to 0} \left[\frac{\Gamma(1 - \Delta z)}{\Gamma(1)\Delta z} - \frac{\Gamma(z - \Delta z)}{\Gamma(z)\Delta z}\right]$$

$$= \lim_{\Delta z \to 0} \frac{1}{\Delta z \Gamma(\Delta z)} [\mathrm{B}(1 - \Delta z, \Delta z) - \mathrm{B}(z - \Delta z, \Delta z)]$$

$$= \lim_{\Delta z \to 0} \int_0^1 \frac{1 - t^{z-1}}{1 - t} \left(\frac{1 - t}{t}\right)^{\Delta z} dt.$$

After taking the limit under the integral sign, which is permissible since the integral converges uniformly for small Δz, we obtain an *integral representation* for $\psi(z)$:

$$\psi(z) = -\gamma + \int_0^1 \frac{1 - t^{z-1}}{1-t} dt, \quad \text{Re } z > 0. \tag{18}$$

Replacing t by e^{-t} in (18), we obtain another frequently used integral representation,

$$\psi(z) = -\gamma + \int_0^\infty \frac{e^{-t} - e^{-zt}}{1 - e^{-t}} dt, \quad \text{Re } z > 0. \tag{19}$$

From (18) we can obtain a simple *series representation* for $\psi(z)$ by expanding $1/(1-t)$ in powers of t and integrating term by term:

$$\psi(z) = -\gamma + \sum_{n=0}^\infty \left(\frac{1}{n+1} - \frac{1}{n+z} \right). \tag{20}$$

4. Asymptotic formulas. We obtain asymptotic formulas for $\Gamma(z)$ and $\psi(z)$ by using the asymptotic properties of the Laplace integral (see Appendix B)

$$F(z) = \int_0^\infty e^{-zt} f(t) dt.$$

We first transform the integral representation (19) of $\psi(z)$ as follows:

$$\psi(z) = -\gamma + \int_0^\infty \frac{e^{-t} - e^{-zt}}{t} dt + \int_0^\infty (e^{-t} - e^{-zt}) \left(\frac{1}{1 - e^{-t}} - \frac{1}{t} \right) dt.$$

Hence

$$\psi(z) = \psi_0(z) - F(z),$$

where

$$F(z) = \int_0^\infty e^{-zt} f(t)dt,$$

$$f(t) = \frac{1}{1-e^{-t}} - \frac{1}{t} = 1 - \frac{1}{t} + \frac{1}{e^t - 1},$$

$$\psi_0(z) = \int_0^\infty \frac{e^{-t} - e^{-zt}}{t} dt - \gamma + F(1).$$

We can express $\psi_0(z)$ in terms of elementary functions. In fact,

$$\psi_0'(z) = \int_0^\infty e^{-zt} dt = \frac{1}{z},$$

so that $\psi_0(z) = \ln z + C$. The constant C is evaluated below.

The function $f(t)$ satisfies the hypotheses of Appendix B for $\theta_1 = \theta_2 = \pi/2$. Hence for $|\arg z| \leq \pi - \epsilon$ we have

$$\psi(z) = \ln z + C - \sum_{k=0}^{n-1} \frac{a_k k!}{z^{k+1}} + O\left(\frac{1}{z^{n+1}}\right),$$

where a_k are the coefficients of the expansion of $f(t)$ in powers of t. They can be expressed in terms of the *Bernoulli numbers* B_k, which are multiples of the Maclaurin coefficients of $t/(e^t - 1)$:

$$\frac{t}{e^t - 1} = \sum_{k=0}^{\infty} B_k \frac{t^k}{k!}, \quad |t| < 2\pi.$$

In fact,

$$f(t) = 1 - \frac{1}{t} + \frac{1}{t} \sum_{k=0}^{\infty} B_k \frac{t^k}{k!} = 1 + \sum_{k=1}^{\infty} B_k \frac{t^{k-1}}{k!},$$

whence

$$a_0 = 1 + B_1, \quad a_k = B_{k+1}/(k+1)! \qquad (k \geq 1).$$

Since $f(-t) = 1 - f(t)$, we have $a_0 = 1/2$, $a_k = 0$ for $k = 2m$ $(m = 1, 2, \ldots)$.

Consequently

$$\psi(z) = C + \ln z - \frac{1}{2z} - \sum_{k=1}^{n} \frac{B_{2k}}{2kz^{2k}} + R_n(z),$$

for $|\arg z| \leq \pi - \epsilon$, where $R_n(z) = O(z^{-2n-2})$.

Since

$$\psi(z) = \frac{d}{dz} \ln \Gamma(z),$$

if we integrate the asymptotic formula for $\psi(z)$ we obtain

$$\ln \Gamma(z) = D + (C-1)z + (z-1/2)\ln z + \sum_{k=1}^{n} \frac{B_{2k}}{2k(2k-1)z^{2k-1}} + \bar{R}_n(z).$$

Here D is a constant and

$$\bar{R}_n(z) = -\int_z^\infty R_n(\xi)d\xi,$$

where the integration is over any contour that extends to infinity. If we take this contour to be $\xi = zt$ $(1 \le t < \infty)$, we easily find that $\bar{R}_n(z) = O(1/z^{2n+1})$.

To determine the constants C and D we use (4), (6), and estimates for $\ln \Gamma(z)$ that follow from the asymptotic formulas that we already have for this function:

$$\ln \Gamma(z) = D + (C-1)z + \left(z - \frac{1}{2}\right)\ln z + O(1/z).$$

From

$$\ln \Gamma(z+1) - \ln \Gamma(z) - \ln z = 0$$

it follows that

$$C - 1 + \left(z + \frac{1}{2}\right)\ln(1+1/z) = O(1/z).$$

Since

$$\ln(1+1/z) = 1/z + O(1/z^2),$$

we have $C = 0$. Similarly we obtain $D = (\ln 2\pi)/2$ from (6a).

Using the values of C and D, we now have the following *asymptotic formulas* for $|\arg z| \le \pi - \epsilon$:

$$\psi(z) = \ln z - \frac{1}{2z} - \sum_{k=1}^{n} \frac{B_{2k}}{2kz^{2k}} + O\left(\frac{1}{z^{2n+2}}\right), \tag{21}$$

$$\ln \Gamma(z) = \left(z - \frac{1}{2}\right)\ln z - z + \frac{1}{2}\ln 2\pi + \sum_{k=1}^{n} \frac{B_{2k}}{2k(2k-1)z^{2k-1}} + O\left(\frac{1}{z^{2n+1}}\right). \tag{22}$$

We can obtain a recursion formula for the Bernoulli numbers from the representation

$$t=(e^t-1)\sum_{k=0}^{\infty}B_k\frac{t^k}{k!}=\sum_{m=1}^{\infty}\sum_{k=0}^{\infty}t^{m+k}\frac{B_k}{m!k!}.$$

Put $n=m+k$ and sum the coefficients of t^n:

$$t=\sum_{n=1}^{\infty}t^n\sum_{k=0}^{n-1}\frac{B_k}{(n-k)!k!}.$$

Comparing coefficients of t on the two sides of the equation, we obtain a recursion relation for calculating the B_n:

$$\sum_{k=0}^{n-1}C_n^kB_k=0 \quad \text{for } n>1, \quad B_0=1, \quad C_n^k=\frac{n!}{(n-k)!k!}.$$

For $n=1$, (22) yields

$$\Gamma(z+1)=\sqrt{2\pi z}(z/e)^z\left[1+\frac{1}{12z}+O\left(\frac{1}{z^2}\right)\right].$$

If we take $z=n$, we obtain *Stirling's formula*

$$n!\approx\sqrt{2\pi n}(n/e)^n.$$

We may observe that this formula is quite accurate even for small n. For example, for $n=1$ and $n=2$ it yields 0.92 and 1.92 for 1! and 2!, respectively.

5. Examples. *1) Integrals expressible in terms of gamma functions:*

$$\int_0^{\infty}\exp(-\alpha t^{\beta})t^{\gamma-1}dt=\frac{\Gamma(p)}{\alpha^p\beta}, \quad p=\gamma/\beta \tag{23}$$

$$(\operatorname{Re}\alpha>0, \quad \operatorname{Re}\beta>0, \quad \operatorname{Re}\gamma>0);$$

$$\int_a^b(t-a)^{\alpha-1}(b-t)^{\beta-1}dt=(b-a)^{\alpha+\beta-1}\frac{\Gamma(\alpha)\Gamma(\beta)}{(\alpha+\beta)} \tag{24}$$

$$(\operatorname{Re}\alpha>0, \quad \operatorname{Re}\beta>0).$$

Integral (23) can be evaluated by the substitution $s = \alpha t^{\beta}$ when $\alpha > 0$, $\beta > 0$, $\gamma > 0$, and then generalized to a wider domain of α, β and γ by the principle of analytic continuation. The substitution $t = a + (b - a)s$ converts the integral (24) into a beta function.

2) Some limits:

$$\gamma = \lim_{n\to\infty} \left[\sum_{k=1}^{n} \frac{1}{k} - \ln n \right], \tag{25}$$

$$\lim_{z\to\infty} \frac{\Gamma(z+a)}{\Gamma(z)z^a} = 1, \quad |\arg z| \le \pi - \delta, \tag{26}$$

$$\lim_{z\to -n} \frac{\psi(z)}{\Gamma(z)} = (-1)^{n+1} n!. \tag{27}$$

Here (25) is obtained by taking the limit as $n \to \infty$ in

$$\gamma = \sum_{k=1}^{n} \frac{1}{k} - \psi(n+1)$$

and using (21) for $\psi(n+1)$. To obtain (27) we need to know that the principal term in the Laurent series of $\Gamma(z)$ about $z = -n$ is $(-1)^n/(n!(z+n))$.

B Analytic properties and asymptotic representations of Laplace integrals

A *Laplace integral* is an integral of the form

$$F(z) = \int_a^b e^{zS(t)} f(t)dt.$$

For our purposes it is enough to consider the case when $S(t) = -t$, $a = 0$, $b = +\infty$, i.e.

$$F(z) = \int_0^\infty e^{-zt} f(t)dt. \tag{1}$$

Let us consider the analytic continuation of $F(z)$ and describe the behavior of $F(z)$ as $|z| \to \infty$. To study $F(z)$ as $|z| \to \infty$, we shall need an *asymptotic formula*, i.e. a representation

$$F(z) = \sum_{k=0}^{n-1} C_k \phi_k(z) + O(\phi_n(z)),$$

where $\phi_k(z)$ satisfies

$$\lim_{|z|\to\infty} \frac{\phi_{k+1}(z)}{\phi_k(z)} = 0.$$

Here $\psi(z) = O(\phi_n(z))$ means that $|\psi(z)| \leq C|\phi_n(z)|$, where C is a constant.

In many cases $\phi_k(z)$ is taken to be $1/z^{\mu_k}$, where μ_k is a constant.

1. An asymptotic formula for (1) can be obtained from the following lemma.

Watson's lemma. *Let $f(t)$ satisfy the following hypotheses:*

1) The integral $\int_0^c |f(t)|dt$ exists for $c > 0$, i.e. $f(t)$ is locally absolutely integrable on $(0, \infty)$;

2) as $t \to 0$, $f(t)$ can be represented in the form

$$f(t) = \sum_{k=0}^{n-1} a_k t^{\lambda_k} + O(t^{\lambda_n}),$$

where $-1 < \operatorname{Re} \lambda_0 < \operatorname{Re} \lambda_1 < \ldots < \operatorname{Re} \lambda_n$;

3) $f(t) = O(e^{\nu t})$ as $t \to +\infty$, where $\nu > 0$ is a constant. Then $F(z)$, defined by (1), has the following asymptotic representation for $z \to \infty$, $|\arg z| \leq \pi/2 - \epsilon$:

$$F(z) = \sum_{k=0}^{n-1} a_k \frac{\Gamma(\lambda_k + 1)}{z^{\lambda_k+1}} + O\left(\frac{1}{z^{\lambda_n+1}}\right). \tag{2}$$

Proof. Put

$$f(t) = \sum_{k=0}^{n-1} a_k t^{\lambda_k} + r_n(t).$$

Then

$$F(z) = \sum_{k=0}^{n-1} a_k \int_0^\infty e^{-zt} t^{\lambda_k} dt + R_n(z),$$

where

$$R_n(z) = \int_0^\infty e^{-zt} r_n(t) dt.$$

Since (see Appendix A, part 5)

$$\int_0^\infty e^{-zt} t^{\lambda_k} dt = \frac{\Gamma(\lambda_k + 1)}{z^{\lambda_k+1}}, \quad |\arg z| < \pi/2,$$

the lemma will be established if we verify that $R_n(z) = O(z^{-\lambda_n - 1})$ as $z \to \infty$. We have

$$R_n(z) = \int_0^\delta e^{-zt} r_n(t) dt + \int_\delta^\infty e^{-zt} r_n(t) dt = R_n^{(1)}(z) + R_n^{(2)}(z).$$

It follows from the hypotheses that $r_n(t) = O(t^{\lambda_n})$ as $t \to 0$. Hence there are positive constants M and δ such that $|r_n(t)| \le M t^{\mathrm{Re}\,\lambda_n}$ for $0 \le t \le \delta$. For such δ and for $\mathrm{Re}\, z > 0$, we have

$$|R_n^{(1)}(z)| \le \int_0^\delta |e^{-zt} r_n(t)| dt \le M \int_0^\delta e^{-t\,\mathrm{Re}\,z} t^{\mathrm{Re}\,\lambda_n} dt$$

$$\le M \int_0^\infty e^{-t\,\mathrm{Re}\,z} t^{\mathrm{Re}\,\lambda_n} dt = M \frac{\Gamma(\mathrm{Re}\,\lambda_n + 1)}{(\mathrm{Re}\,z)^{\mathrm{Re}\,\lambda_n + 1}} .$$

If $|\arg z| \le \pi/2 - \epsilon$, we have $\mathrm{Re}\, z \ge |z| \sin \epsilon$ and consequently

$$R_n^{(1)}(z) = O\left(\frac{1}{|z|^{\mathrm{Re}\,\lambda_n + 1}}\right) = O\left(\frac{1}{z^{\lambda_n + 1}}\right).$$

To estimate $R_n^{(2)}(z)$ we put $t = \delta + \tau$:

$$R_n^{(2)}(z) = \int_\delta^\infty e^{-zt} r_n(t) dt = e^{-\delta z} \int_0^\infty e^{-z\tau} r_n(\tau + \delta) d\tau.$$

The integral $\int_0^\infty e^{-z\tau} r_n(\tau + \delta) d\tau$ is uniformly bounded for $\mathrm{Re}\, z \ge \nu + \epsilon$, $|\arg z| \le \pi/2 - \epsilon$. In fact, for such z,

$$\left| \int_0^\infty e^{-z\tau} r_n(\tau + \delta) d\tau \right| \le \int_0^\infty e^{-(\nu+\epsilon)\tau} |r_n(\tau + \delta)| d\tau.$$

The integral on the right converges, since by hypothesis $r_n(t)$ is locally absolutely integrable and $r_n(t) = O(e^{\nu t})$ as $t \to \infty$. Therefore

$$R_n^{(2)}(z) = O(e^{-\delta z}) \quad \text{as} \quad z \to \infty, \qquad |\arg z| \leq \pi/2 - \epsilon.$$

Since $O(e^{-\delta z}) = O(z^{-s})$ as $z \to \infty$ for arbitrary positive s, we obtain the required estimate for the remainder:

$$R_n(z) = R_n^{(1)}(z) + R_n^{(2)}(z) = O\left(\frac{1}{z^{\lambda_n+1}}\right) \quad \text{as} \quad z \to \infty, \quad |\arg z| \leq \frac{\pi}{2} - \epsilon.$$

This completes the proof of the lemma.

Remark. Watson's lemma also holds for an integral of the form

$$F(z) = \int_0^a e^{-zt} f(t)dt, \quad a > 0.$$

Here hypothesis 3) can be omitted.

2. An analytic continuation of the Laplace integral

$$F(z) = \int_0^\infty e^{-zt} f(t)dt,$$

and an asymptotic formula for the analytic continuation, can be obtained, in the cases in which we are interested, by using the following theorem.

Theorem 1. *Let $f(t)$ be analytic in the sector where $|t| > 0$, $-\theta_2 < \arg t < \theta_1 (\theta_1 > 0, \theta_2 > 0)$, and let $f(t)$ be represented in the form*

$$f(t) = \sum_{k=0}^{n-1} a_k t^{\lambda_k} + O(t^{\lambda_n}), \quad \text{where} \ -1 < \operatorname{Re} \lambda_0 < \operatorname{Re} \lambda_1 < \ldots < \operatorname{Re} \lambda_n,$$

as $t \to 0$ in this sector, and in the form $f(t) = O(t^\beta)$ as $t \to \infty$, where β is a constant. Then the function $F(z)$ defined by (1) for $z > 0$ can be continued analytically into the sector $|z| > 0, -\pi/2 - \theta_1 < \arg z < \pi/2 + \theta_2$. For $-\pi/2 - \theta_1 + \epsilon \leq \arg z \leq \pi/2 + \theta_2 - \epsilon (\epsilon > 0)$, the function $F(z)$ has the asymptotic representation

$$F(z) = \sum_{k=0}^{n-1} a_k \frac{\Gamma(\lambda_k + 1)}{z^{\lambda_k+1}} + O\left(\frac{1}{z^{\lambda_n}}\right). \tag{3}$$

Proof. Let us investigate the domain of analyticity of (1). Let $z = re^{i\phi}$. By the theorem on the analyticity of an integral that depends on a parameter (see Theorem 2 of §3), $F(z)$ is analytic in the sector $|z| > 0$, $|\phi| < \pi/2$, since the integral for $F(z)$ converges uniformly for z in the domain $|\phi| \leq \pi/2 - \epsilon$, $|z| \geq \delta > 0$. In fact, in this domain

$$|e^{-zt}| = e^{-tr\cos\phi} \leq e^{-t\delta\sin\epsilon},$$

and the integral $\int_0^\infty e^{-t\delta\sin\epsilon}|f(t)|dt$ converges by our hypotheses.

For the analytic continuation of $F(z)$ into a larger domain, it is convenient to transform (1) from integration with respect to a positive t to integration along the ray $t = \rho e^{i\theta}$ ($\theta = \text{const.}, \rho > 0$) and consider the function

$$F_\theta(z) = \int_0^{\infty e^{i\theta}} e^{-zt} f(t)dt = \int_0^\infty e^{-(ze^{i\theta})\rho} f(\rho e^{i\theta})e^{i\theta}d\rho \tag{4}$$

$$(-\theta_2 < \theta < \theta_1).$$

A study of the domain of analyticity of $F_\theta(z)$, similar to that for $F(z)$, shows that $F_\theta(z)$ is analytic in the domain $|\arg(ze^{i\theta})| = |\phi + \theta| < \pi/2$. Let us prove that $F_\theta(z)$ is an analytic continuation of $F(z)$ for $|\theta| < \pi$. For the proof we have only to appeal to the equality of these functions on some ray $\phi = \phi_0$ where $F(z)$ and $F_\theta(z)$ are analytic, for example when $\phi_0 = -\theta/2$.

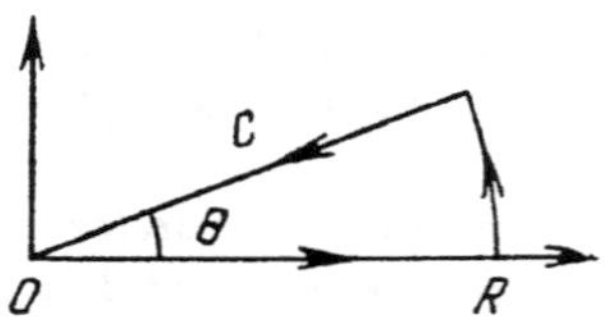

Figure 17.

By Cauchy's theorem, the integral $\int_C e^{-zt} f(t)dt$ is zero over the closed contour of Figure 17. On the arc of radius R we have, with $t = Re^{i\psi}$, $\psi \in [0, \theta]$, and $z = re^{-i\theta/2}$,

$$f(t) = O(R^\beta), \quad |e^{-zt}| = e^{-rR\cos(\psi - \theta/2)}.$$

Since $|\psi - \theta/2| \leq |\theta/2| < \pi/2$, we have $\cos(\psi - \theta/2) \geq \cos(\theta/2) > 0$. We see from these estimates that the integral over the arc of radius R tends to zero as $R \to \infty$, i.e. $F(z) = F_\theta(z)$ for $\phi = -\theta/2$. If the sector $-\theta_2 < \theta < \theta_1$ contains values with $|\theta| \geq \pi$, we can show similarly that $F_{\tilde{\theta}}(z)$ is an analytic continuation of $F_\theta(z)$ if $|\tilde{\theta} - \theta| < \pi$.

Consequently the functions $F_\theta(z)$ for all possible values of θ provide analytic continuations of $F(z)$ to the union of the two sectors $|\phi+\theta| < \pi/2$, $-\theta_2 < \theta < \theta_1$, i.e. in the sector $-\pi/2 - \theta_1 < \phi < \pi/2 + \theta_2$. This establishes the first part of the theorem.

To obtain the asymptotic formula for $F_\theta(z)$ in the sector $|\phi+\theta| \leq \pi/2-\epsilon$ we apply Watson's lemma and formula (4) to $F_\theta(z)$. Since, by hypothesis, $f(\rho e^{i\theta}) = \sum_{k=0}^{n-1} a_k(\rho e^{i\theta})^{\lambda_k} + O(\rho^{\lambda_n})$ as $\rho \to 0$ and $f(\rho e^{i\theta}) = O(\rho^\beta)$ as $\rho \to +\infty$, then by Watson's lemma

$$F_\theta(z) = \sum_{k=0}^{n-1} \frac{a_k(e^{i\theta})^{\lambda_k} e^{i\theta}\Gamma(\lambda_k+1)}{(ze^{i\theta})^{\lambda_k+1}} + O\left(\frac{1}{z^{\lambda_n+1}}\right)$$

$$= \sum_{k=0}^{n-1} a_k \frac{\Gamma(\lambda_k+1)}{z^{\lambda_k+1}} + O\left(\frac{1}{z^{\lambda_n+1}}\right).$$

This establishes the theorem.

Remark 1. If $f(t) = t^\lambda g(t)$ in a neighborhood of $t = 0$, where $g(t)$ is analytic, and Re $\lambda > -1$, then we can take $\lambda_k = \lambda + k$ in the hypotheses of the theorem; the constants a_k are the coefficients in the Maclaurin series of $g(t)$: $g(t) = \sum_{k=0}^{\infty} a_k t^k$.

Remark 2. If the function $f(t)$ in Theorem 1 depends on parameters, is an analytic function of each parameter in a domain D, and is continuous in the parameters collectively in D for $|t| > 0$, $-\theta_2 < \arg t < \theta_1$, then the analytic continuation of $F(z)$ for $z \neq 0$ will also be an analytic function of each parameter in the subset of D where the integral

$$\int_0^\infty e^{-\mu\rho}|f(\rho e^{i\theta})|d\rho$$

converges uniformly in the parameters for each fixed $\mu > 0$.

In fact, in this case $F_\theta(z)$, the analytic continuation of $F(z)$, converges uniformly for $|\phi+\theta| \leq \pi/2 - \epsilon$, $|z| \geq \delta$, since

$$|F_\theta(z)| \leq \int_0^\infty e^{-\mu\rho}|f(\rho e^{i\theta})|d\rho \quad (\mu = \delta \sin\epsilon),$$

in this domain.

Example. Let us find the domain into which

$$F(z,p,q) = \int_0^\infty e^{-zt} t^p (1+at)^q \, dt$$

$$(z > 0, \ |\arg a| < \pi, \ |\arg(1+at)| < \pi, \ \operatorname{Re} p > -1)$$

can be continued analytically in each variable, and find an asymptotic representation for this function as $z \to \infty$.

In this case $f(t) = t^p(1+at)^q$, and $g(t) = (1+at)^q$ has a singularity (a branch point) for $at = -1$. This function is analytic in the sector where $|\arg(at)| < \pi$, i.e. for $-\pi - \arg a < \arg t < \pi - \arg a$. Hence the hypotheses of Theorem 1 will be satisfied if $\theta_1 = \pi - \arg a$, $\theta_2 = \pi + \arg a$, $\lambda_k = p + k$, $\beta = p + q$.

In order to find the domain over which $F(z,p,q)$ can be continued analytically in each variable, we find the domain where the integral

$$\int_0^\infty e^{-\mu\rho} |f(\rho e^{i\theta})| d\rho = \int_0^\infty e^{-\mu\rho} |(\rho e^{i\theta})^p (1 + a e^{i\theta}\rho)^q| d\rho,$$

$$|\arg(a e^{i\theta})| < \pi,$$

converges uniformly in p and q for each given $\mu > 0$ (see Remark 2). This will happen if $\operatorname{Re} p \geq \delta - 1$, $|p| \leq N$ and $|q| \leq N$. In fact, if $t = \rho e^{i\theta}$, $0 \leq \rho \leq 1$, the function $|f(t)/t^{\delta-1}|$ will be bounded, since it is continuous on a closed subset of the variables collectively, i.e. $|f(t)| \leq C_1 \rho^{\delta - 1}$ $(0 < \rho \leq 1)$. For $\rho \geq 1$ we can give a similar argument with ρ replaced by $s = 1/\rho$ $(0 \leq s \leq 1)$. We find that, when $\rho \geq 1$, the function $|t^{-2N} f(t)|$ is bounded, i.e. $|f(t)| \leq C_2 \rho^{2N}$. The constants C_1 and C_2 are evidently independent of p and q. Since the integrals

$$\int_0^1 e^{-\mu\rho} \rho^{\delta-1} d\rho \quad \text{and} \quad \int_1^\infty e^{-\mu\rho} \rho^{2N} d\rho$$

converge, the integral

$$\int_0^\infty e^{-\mu\rho} |f(\rho e^{i\theta})| d\rho$$

converges uniformly in the specified domain. Hence $F(z,p,q)$ is an analytic continuation, in each variable, into the domain

$$\operatorname{Re} p \geq -1 + \delta, \qquad |p| \leq N, \qquad |q| \leq N, \qquad z \neq 0,$$

$$-\frac{3}{2}\pi + \arg a < \arg z < \frac{3}{2}\pi + \arg a.$$

Since δ and N are arbitrary, we can transform this domain into

$$\operatorname{Re} p > -1, \quad z \neq 0, \quad -\frac{3}{2}\pi + \arg a < \arg z < \frac{3}{2}\pi + \arg a.$$

For $|\arg z| < \pi/2$ the analytic continuation of $F(z,p,q)$ can be obtained from the original integral

$$\int_0^\infty e^{-zt} t^p (1+at)^q dt \qquad (\operatorname{Re} p > -1).$$

By Theorem 1, the function $F(z,p,q)$ has the asymptotic representation

$$F(z,p,q) = \frac{\Gamma(q+1)}{z^{p+1}} \left[\sum_{k=1}^{n-1} \frac{\Gamma(p+k+1)}{k!\Gamma(q+1-k)} \left(\frac{a}{z}\right)^k + O\left(\frac{1}{z^n}\right) \right]$$

as $z \to \infty$ in the sector $-(3\pi/2) + \arg a + \epsilon \leq \arg z \leq (3\pi/2) + \arg a - \epsilon$.

Basic formulas

I. Gamma function $\Gamma(z)$

Definition:

$$\Gamma(z) = \int_0^\infty e^{-t} t^{z-1} dt, \quad \operatorname{Re} z > 0.$$

Analytic continuation. The function $\Gamma(z)$ can be continued analytically over the whole complex plane except the points $z = -n$ $(n = 0, 1, 2, \dots)$ at which it has simple poles with residues

$$\operatorname{Res} \Gamma(z) = (-1)^n / n!.$$

Integrals involving the gamma function:

$$B(x,y)=\int_0^1 t^{x-1}(1-t)^{y-1}dt=\frac{\Gamma(x)\Gamma(y)}{\Gamma(x+y)},\quad \mathrm{Re}\, x>0,\quad \mathrm{Re}\, y>0;$$

$$\int_0^\infty \exp(-\alpha t^\beta)t^{\gamma-1}dt=\frac{\Gamma(p)}{\alpha^p\beta},\ p=\frac{\gamma}{\beta},$$

$$\mathrm{Re}\,\alpha>0,\quad \mathrm{Re}\,\beta>0,\quad \mathrm{Re}\,\gamma>0.$$

Functional equations:

$$\Gamma(z+1)=z\Gamma(z),\quad \Gamma(z)\Gamma(1-z)=\pi/\sin\pi z,$$
$$2^{2z-1}\Gamma(z)\Gamma(z+1/2)=\sqrt{\pi}\Gamma(2z).$$

Special values:

$$\Gamma(n+1)=n!,\quad \Gamma(1/2)=\sqrt{\pi},$$
$$\Gamma(n+1/2)=\frac{\sqrt{\pi}}{2^{2n}}\frac{(2n)!}{n!}.$$

Asymptotic formulas and their consequences:

$$\ln\Gamma(z)=(z-1/2)\ln z-z+\frac{1}{2}\ln 2\pi$$
$$+\sum_{k=1}^{n-1}\frac{B_{2k}}{2k(2k-1)z^{2k-1}}+O\left(\frac{1}{z^{2n-1}}\right),\quad |\arg z|\le\pi-\delta,$$

where B_k are the Bernoulli numbers defined by the recursion relations

$$\sum_{k=0}^{n-1}C_n^kB_k=0\quad(n>1),\quad B_0=1,\quad C_n^k=\frac{n!}{k!(n-k)!},$$

$$\Gamma(x+1)=\sqrt{2\pi x}\left(\frac{x}{e}\right)^x\left[1+\frac{1}{12x}+O\left(\frac{1}{x^2}\right)\right]\quad(x>0),$$

$$n!\approx\sqrt{2\pi n}\left(\frac{n}{e}\right)^n\quad\text{(Stirling's formula)},$$

$$\frac{\Gamma(z+a)}{\Gamma(z)}=z^a\left[1+O\left(\frac{1}{z}\right)\right],\quad |\arg z|\le\pi-\delta.$$

Appendix A contains a graph of $y=\Gamma(x)$.

II. $\psi(z)$, the logarithmic derivative of the gamma function.

Definition:

$$\psi(z) = \Gamma'(z)/\Gamma(z).$$

Functional equations:

$$\psi(z+1) = 1/z + \psi(z), \quad \psi(z) = \psi(1-z) - \pi \cot \pi z,$$
$$2\ln 2 + \psi(z) + \psi(z+1/2) = 2\psi(2z).$$

Special values:

$$\psi(1) = \Gamma'(1) = -\gamma, \quad \gamma = 0.57721566\ldots,$$
$$\psi\left(\frac{1}{2}\right) = -\gamma - 2\ln 2, \quad \psi(n+1) = -\gamma + \sum_{k=1}^{n} \frac{1}{k},$$
$$\psi\left(n + \frac{1}{2}\right) = -\gamma - 2\ln 2 + 2\sum_{k=1}^{n} \frac{1}{2k-1}.$$

Integral representations and series expansion:

$$\psi(z) = -\gamma + \int_0^1 \frac{1-t^{z-1}}{1-t} dt = -\gamma + \int_0^\infty \frac{e^{-t} - e^{-zt}}{1-e^{-t}} dt, \quad \operatorname{Re} z > 0,$$

$$\psi(z) = -\gamma + (z-1)\sum_{n=0}^{\infty} \frac{1}{(n+1)(z+n)}, \quad z \neq -n \quad (n = 0, 1, \ldots).$$

Asymptotic formula:

$$\psi(z) = \ln z - \frac{1}{2z} - \sum_{k=1}^{n-1} \frac{B_{2k}}{2kz^{2k}} + O\left(\frac{1}{z^{2n}}\right),$$

$|\arg z| \leq \pi - \delta$, B_k the Bernoulli numbers (part I, above).

III. Generalized equation of hypergeometric type.

Differential equation:

$$u'' + \frac{\tilde{\tau}(z)}{\sigma(z)} u' + \frac{\tilde{\sigma}(z)}{\sigma^2(z)} u = 0,$$

$\sigma(z)$ and $\tilde{\sigma}(z)$, polynomials of degree at most 2; $\tilde{\tau}(z)$ a polynomial of degree at most 1.

Reduction of generalized equation of hypergeometric type to one of hypergeometric type. The substitution $u = \phi(z)y$ carries the original equation to one of the form

$$\sigma(z)y'' + \tau(z)y' + \lambda y = 0,$$

where $\tau(z)$ is a polynomial of degree at most 1, and λ is a constant. Here $\phi(z)$ satisfies

$$\phi'/\phi = \pi(z)/\sigma(z),$$

where $\pi(z)$ is a polynomial of degree at most 1:

$$\pi(z) = \frac{\sigma' - \tilde{\tau}}{2} \pm \sqrt{\left(\frac{\sigma' - \tilde{\tau}}{2}\right)^2 - \tilde{\sigma} + k\sigma},$$

and k is determined by the condition that the quadratic under the square root sign has discriminant zero. Then $\tau(z)$ and λ are determined by

$$\tau(z) = \tilde{\tau}(z) + 2\pi(z), \quad \lambda = k + \pi'(z).$$

Exceptions: 1) If $\sigma(z)$ has a double root, i.e. $\sigma(z) = (z-a)^2$, the original equation can be carried into a generalized equation of hypergeometric type with $\sigma(s) = s$, by the substitution $s = 1/(z-a)$.

2) If $\sigma(z) = 1$ and $(\tilde{\tau}(z)/2)^2 - \tilde{\sigma}(z)$ is a polynomial of degree 1, it will be impossible to reduce the original equation to an equation of hypergeometric type by the method indicated previously. In this case the substitution $\pi(z) = -\tilde{\tau}(z)/2$ reduces the original equation to the form

$$y'' + (az + b)y = 0.$$

The linear transformation $s = az + b$ takes this into a Lommel equation (see §16, part 1).

IV. Equation of hypergeometric type.

Differential equation:

$$\sigma(z)y'' + \tau(z)y' + \lambda y = 0,$$

$\sigma(z)$ and $\tau(z)$ are polynomials of respective degrees at most 2 and 1, and λ is a constant. Solutions of this equation are *functions of hypergeometric type.*

Self-adjoint form:

$$(\sigma\rho y')' + \lambda\rho y = 0,$$

where $\rho(z)$ satisfies

$$[\sigma(z)\rho(z)]' = \tau(z)\rho(z).$$

An equation of hypergeometric type can usually be carried by a linear change of independent variable into one of the following *canonical forms* (see §20, part 1):

hypergeometric equation (if $\sigma(z)$ is a second-degree polynomial)

$$z(1-z)y'' + [\gamma - (\alpha+\beta+1)z]y' - \alpha\beta y = 0,$$

confluent hypergeometric equation (if $\sigma(z)$ is a first-degree polynomial)

$$zy'' + (\gamma - z)y' - \alpha y = 0,$$

Hermite equation (if $\sigma(z)$ is a constant)

$$y'' - 2zy' + 2\nu y = 0.$$

Derivatives of functions $y(z)$ of hypergeometric type. The derivatives $v_n(z) = y^{(n)}(z)$ are functions of hypergeometric type and satisfy the equation

$$\sigma(z)v_n'' + \tau_n(z)v_n' + \mu_n v_n = 0,$$

where $\tau_n(z) = \tau(z) + n\sigma'(z)$, $\mu_n = \lambda + n\tau' + n(n-1)\sigma''/2$.

Self-adjoint form of the equation for $v_n(z)$:

$$(\sigma\rho_n v_n')' + \mu_n\rho_n v_n = 0, \quad \rho_n(z) = \sigma^n(z)\rho(z).$$

Integral representation for particular solutions $y_\nu(z)$:

$$y_\nu(z) = \frac{C_\nu}{\rho(z)}\int_C \frac{\sigma^\nu(s)\rho(s)}{(s-z)^{\nu+1}}ds.$$

Here C_ν is a normalizing constant, $\rho(z)$ satisfies $[\sigma(z)\rho(z)]' = \tau(z)\rho(z)$, the constant ν is a root of $\lambda + \nu\tau' + \nu(\nu-1)\sigma''/2 = 0$, and the contour C is chosen so that

$$\left.\frac{\sigma^{\nu+1}(s)\rho(s)}{(s-z)^{\nu+2}}\right|_{s_1}^{s_2} = 0 \quad (s_1 \text{ and } s_2 \text{ are the endpoints of } C).$$

See §3 for possible choices of C.

Integral representations for $y_\nu^{(k)}(z)$:

$$y_\nu^{(k)}(z) = \frac{C_\nu^{(k)}}{\sigma^k(z)\rho(z)} \int_C \frac{\sigma^\nu(s)\rho(s)}{(s-z)^{\nu+1-k}} ds,$$

$$C_\nu^{(k)} = C_\nu \prod_{s=0}^{k-1} \left(\tau' + \frac{\nu+s-1}{2}\sigma''\right).$$

Recursion relations and differentiation formulas for particular solutions $y_\nu(z)$. Any three functions $y_{\nu_i}^{(k_i)}(z)$ are connected by a linear relation

$$\sum_{i=1}^{3} A_i(z) y_{\nu_i}^{(k_i)}(z) = 0$$

with polynomial coefficients $A_i(z)$, if $\nu_i - \nu_j$ is an integer. For methods of calculating the $A_i(z)$, see §4.

V. Polynomials of hypergeometric type. Polynomials $y_n(z)$ of hypergeometric type are polynomial solutions of the equation

$$\sigma(z)y'' + \tau(z)y' + \lambda y = 0$$

corresponding to

$$\lambda = \lambda_n = -n\tau' - \frac{1}{2}n(n-1)\sigma''.$$

Rodrigues formula for polynomials of hypergeometric type:

$$y_n(z) = \frac{B_n}{\rho(z)} \frac{d^n}{dz^n}[\sigma^n(z)\rho(z)]$$

(B_n is a normalizing constant).

Rodrigues formula for derivatives of polynomials $y_n(z)$ of hypergeometric type:

$$y_n^{(m)}(z) = \frac{A_{mn}B_n}{\sigma^m(z)\rho(z)} \frac{d^{n-m}}{dz^{n-m}}[\sigma^n(z)\rho(z)],$$

$$A_{mn} = \frac{n!}{(n-m)!} \prod_{k=0}^{m-1} \left(\tau' + \frac{n+k-1}{2}\sigma''\right), \quad A_{0n} = 1.$$

By linear changes of variable, we obtain the following *canonical forms:*

1) *Jacobi polynomials*

$$y_n(z) = P_n^{(\alpha,\beta)}(z) = \frac{(-1)^n}{2^n n!}(1-z)^{-\alpha}(1+z)^{-\beta}\frac{d^n}{dz^n}[(1-z)^{n+\alpha}(1+z)^{n+\beta}],$$
$$\sigma(z) = 1 - z^2, \quad \rho(z) = (1-z)^{\alpha}(1+z)^{\beta}.$$

Important special cases of Jacobi polynomials are:

a) *Legendre polynomials* $P_n(z) = P_n^{(0,0)}(z)$;

b) *Chebyshev polynomials of the first and second kinds:*

$$T_n(z) = \frac{n!}{\left(\frac{1}{2}\right)_n}P_n^{(-1/2,-1/2)}(z) = \cos n\phi, \text{ where } \cos\phi = z;$$
$$U_n(z) = \frac{(n+1)!}{\left(\frac{3}{2}\right)_n}P_n^{(1/2,1/2)}(z) = \frac{\sin(n+1)\phi}{\sin\phi}.$$

c) *Gegenbauer polynomials*

$$C_n^{\lambda}(z) = \frac{(2\lambda)_n}{\left(\lambda+\frac{1}{2}\right)_n}P_n^{(\lambda-1/2,\lambda-1/2)}(z).$$

Notation: $(\alpha)_n = \alpha(\alpha+1)\dots(\alpha+n-1)$, $(\alpha)_0 = 1$.

2) *Laguerre polynomials*

$$y_n(z) = L_n^{\alpha}(z) = \frac{1}{n!}e^z z^{-\alpha}\frac{d^n}{dz^n}(z^{n+\alpha}e^{-z}),$$
$$\sigma(z) = z, \quad \rho(z) = z^{\alpha}e^{-z}.$$

3) *Hermite polynomials*

$$y_n(z) = H_n(z) = (-1)^n e^{z^2}\frac{d^n}{dz^n}(e^{-z^2}),$$
$$\sigma(z) = 1, \quad \rho(z) = e^{-z^2}.$$

Differentiation formulas for Jacobi, Laguerre and Hermite polynomials:

$$\frac{d}{dx}P_n^{(\alpha,\beta)}(x) = \frac{1}{2}(n+\alpha+\beta+1)P_{n-1}^{(\alpha+1,\beta+1)}(x),$$
$$\frac{d}{dx}L_n^{\alpha}(x) = -L_{n-1}^{\alpha+1}(x), \quad \frac{d}{dx}H_n(x) = 2nH_{n-1}(x).$$

If $\sigma(z)$ has a double zero, i.e. $\sigma(z) = (z-a)^2$, the corresponding polynomials $y_n(z)$ can be expressed in terms of Laguerre polynomials:

$$y_n(z) = C_n(z-a)^n L_n^\alpha\left(\frac{\tau(a)}{z-a}\right), \quad \alpha = -\tau' - 2n + 1$$

(C_n is a normalizing constant).

VI. General properties of orthogonal polynomials. The polynomials $p_n(x)$ are orthogonal on (a, b) with weight $\rho(x)$ if

$$\int_a^b p_m(x)p_n(x)\rho(x)dx = 0 \quad (m \neq n).$$

Explicit expression for $p_n(x)$:

$$p_n(x) = A_n \begin{vmatrix} C_0 & C_1 & \dots & C_n \\ C_1 & C_2 & \dots & C_{n+1} \\ \dots & \dots & \dots & \dots \\ C_{n-1} & C_n & \dots & C_{2n-1} \\ 1 & x & \dots & x^n \end{vmatrix},$$

$C_n = \int_a^b x^n \rho(x)dx$ are the moments of the weight function, and A_n are normalizing constants.

Recursion relation:

$$xp_n(x) = \frac{a_n}{a_{n+1}}p_{n+1}(x) + \left(\frac{b_n}{a_n} - \frac{b_{n+1}}{a_{n+1}}\right)p_n(x) + \frac{a_{n-1}}{a_n}\frac{d_n^2}{d_{n-1}^2}p_{n-1}(x).$$

Here

$$d_n^2 = \int_a^b p_n^2(x)\rho(x)dx$$

is the *square of the norm*; a_n *and* b_n *are the leading coefficients of* $p_n(x)$:

$$p_n(x) = a_n x^n + b_n x^{n-1} + \dots \qquad (a_n \neq 0).$$

Darboux-Christoffel formula:

$$\sum_{k=0}^n \frac{p_k(x)p_k(y)}{d_k^2} = \frac{1}{d_n^2}\frac{a_n}{a_{n+1}}\frac{p_{n+1}(x)p_n(y) - p_n(x)p_{n+1}(y)}{x-y}.$$

VII. Classical orthogonal polynomials. *The classical orthogonal polynomials* are the polynomials $y_n(x)$ of hypergeometric type for which $\rho(x)$ satisfies

$$\sigma(x)\rho(x)x^k|_{x=a,b} = 0$$

(a and b are real; $k = 0, 1, \ldots$), where $\rho(x) > 0$ on (a, b). These polynomials are orthogonal with weight $\rho(x)$ on (a, b), i.e.

$$\int_a^b y_m(x)y_n(x)\rho(x)dx = 0 \qquad (m \neq n).$$

The classical orthogonal polynomials can be carried by linear transformations into the following *canonical forms*:

1) Jacobi polynomials $P_n^{(\alpha,\beta)}(x)$ if $\alpha > -1$, $\beta > -1$;

2) Laguerre polynomials $L_n^\alpha(x)$ if $\alpha > -1$;

3) Hermite polynomials $H_n(x)$.

Their basic properties are summarized in Tables 1 and 2 (see §§5, 6).

Asymptotic formulas for $n \to \infty$:

$$P_n^{(\alpha,\beta)}(\cos\theta) = \frac{\cos\{[n + (\alpha+\beta+1)/2]\theta - (2\alpha+1)\pi/4\}}{\sqrt{\pi n}(\sin(\theta/2))^{\alpha+1/2}(\cos(\theta/2))^{\beta+1/2}} + O(n^{-3/2})$$
$$(0 < \delta \leq \theta \leq \pi - \delta),$$

$$L_n^\alpha(x) = \frac{1}{\sqrt{\pi}}e^{x/2}x^{-\alpha/2-1/4}n^{\alpha/2-1/4}\cos\left[2\sqrt{nx} - (2\alpha+1)\frac{\pi}{4}\right] + O(n^{\alpha/2-3/4})$$
$$(0 < \delta \leq x \leq N < \infty),$$

$$H_n(x) = \sqrt{2}\left(\frac{2n}{e}\right)^{n/2} e^{x^2/2}\left[\cos\left(\sqrt{2n}x - \frac{\pi n}{2}\right) + O(n^{-1/4})\right]$$
$$(|x| \leq N < \infty).$$

Generating functions:

$$(1-t)^{-\alpha-1}\exp\left(\frac{-xt}{1-t}\right) = \sum_{n=0}^{\infty} L_n^\alpha(x)t^n,$$

$$\exp(2xt - t^2) = \sum_{n=0}^{\infty} H_n(x)\frac{t^n}{n!}.$$

Legendre polynomials. The Legendre polynomials $P_n(x)$ are orthogonal with weight $\rho(x) = 1$ on $(-1,1)$. They are special cases of the Jacobi polynomials $P_n^{(\alpha,\beta)}(x)$ for $\alpha = \beta = 0$ and of the Gegenbauer polynomials $C_n^\nu(x)$ for $\nu = 1/2$.

Differential equation:

$$(1-x^2)y'' - 2xy' + n(n+1)y = 0, \qquad y = P_n(x).$$

Rodrigues formula:

$$P_n(x) = \frac{(-1)^n}{2^n n!}\frac{d^n}{dx^n}(1-x^2)^n.$$

Integral representation:

$$P_n(x) = \frac{1}{2\pi}\int_0^{2\pi}(x + i\sqrt{1-x^2}\sin\phi)^n d\phi.$$

Generating function:

$$\frac{1}{\sqrt{1-2tx+t^2}} = \sum_{n=0}^{\infty} P_n(x)t^n.$$

Special values:

$$P_n(1) = 1, \quad P_n(-1) = (-1)^n,$$

$$P_{2n+1}(0) = 0, \quad P_{2n}(0) = \frac{(-1)^n(2n)!}{2^{2n}(n!)^2}.$$

Square of the norm:

$$d_n^2 = \frac{2}{2n+1}.$$

Recursion relations and differentiation formulas:

$$(1-x^2)P_n'(x) = -(n+1)[P_{n+1}(x) - xP_n(x)],$$

$$P_n(x) = \frac{1}{n+1}[P_{n+1}'(x) - xP_n'(x)] = \frac{1}{2n+1}[P_{n+1}'(x) - P_{n-1}'(x)],$$

$$(n+1)P_{n+1}(x) - (2n+1)xP_n(x) + nP_{n-1}(x) = 0.$$

Asymptotic formula:

$$P_n(\cos\theta) = \left(\frac{2}{\pi n}\right)^{1/2} \frac{\cos[(n+1/2)\theta - \pi/4]}{\sqrt{\sin\theta}} + O(n^{-3/2}).$$

Graphs of the Legendre polynomials $P_n(x)$ are shown for several values of n in Figure 1 (§7).

VIII. Spherical harmonics.

Differential equation:

$$\frac{1}{\sin\theta}\frac{\partial}{\partial\theta}\left(\sin\theta\frac{\partial u}{\partial\theta}\right) + \frac{1}{\sin^2\theta}\frac{\partial^2 u}{\partial\phi^2} + l(l+1)u = 0.$$

Explicit formulas:

$$Y_{lm}(\theta,\phi) = \frac{1}{\sqrt{2\pi}} e^{im\phi}\Theta_{lm}(\cos\theta) \quad (-l \le m \le l),$$

$$\begin{aligned}\Theta_{lm}(x) &= \frac{(-1)^l}{2^l l!}\left[\frac{2l+1}{2}\frac{(l-m)!}{(l+m)!}\right]^{1/2}(1-x^2)^{m/2}\frac{d^{l+m}}{dx^{l+m}}(1-x^2)^l \\ &= \frac{(-1)^{l-m}}{2^l l!}\left[\frac{2l+1}{2}\frac{(l+m)!}{(l-m)!}\right]^{1/2}(1-x^2)^{-m/2}\frac{d^{l-m}}{dx^{l-m}}(1-x^2)^l,\end{aligned}$$

$$\Theta_{l,-m}(x) = (-1)^m\Theta_{lm}(x), \quad Y^*_{lm}(\theta,\phi) = (-1)^m Y_{l,-m}(\theta,\phi).$$

Orthogonality:

$$\int Y_{lm}(\theta,\phi)Y^*_{l'm'}(\theta,\phi)d\Omega = \delta_{ll'}\delta_{mm'}.$$

Recursion relation:

$$\begin{aligned}(\cos\theta)Y_{lm}(\theta,\phi) &= \left[\frac{(l+1)^2 - m^2}{4(l+1)^2 - 1}\right]^{1/2} Y_{l+1,m}(\theta,\phi) \\ &\quad + \left[\frac{l^2 - m^2}{4l^2 - 1}\right]^{1/2} Y_{l-1,m}(\theta,\phi).\end{aligned}$$

Differentiation formulas:

$$\frac{\partial}{\partial\phi}Y_{lm}(\theta,\phi) = imY_{lm}(\theta,\phi),$$

$$e^{\pm i\phi}\left(\mp\frac{\partial Y_{lm}}{\partial\theta} + m\cot\theta\, Y_{lm}\right) = [l(l+1) - m(m\pm 1)]^{1/2}Y_{l,m\pm 1}$$

(when $m = \pm(l+1)$ we must take $Y_{lm}(\theta,\phi) = 0$).

Spherical harmonic expansion of an arbitrary homogeneous polynomial of degree l:

$$u_l(x,y,z) = r^l \sum_{m,n} C_{mn}Y_{l-2n,m}(\theta,\phi).$$

Spherical harmonic expansion of an arbitrary homogeneous harmonic polynomial of degree l:

$$u_l(x,y,z) = r^l \sum_{m=-l}^{l} C_{lm}Y_{lm}(\theta,\phi).$$

Addition theorem:

$$P_l(\cos\omega) = \frac{4\pi}{2l+1}\sum_{m=-l}^{l} Y_{lm}(\theta_1,\phi_1)Y^*_{lm}(\theta_2,\phi_2)$$

(ω is the angle between the vectors $\mathbf{r}_1$ and $\mathbf{r}_2$ in the directions θ_1,ϕ_1 and θ_2,ϕ_2),

$$\begin{aligned}\frac{1}{|\mathbf{r}_1 - \mathbf{r}_2|} &= \sum_{l=0}^{\infty}\frac{r_<^l}{r_>^{l+1}}P_l(\cos\omega)\\ &= 4\pi\sum_{l=0}^{\infty}\left[\frac{1}{2l+1}\frac{r_<^l}{r_>^{l+1}}\sum_{m=-l}^{l} Y_{lm}(\theta_1,\phi_1)Y^*_{lm}(\theta_2,\phi_2)\right],\end{aligned}$$

$$r_< = \min(r_1,r_2), \quad r_> = \max(r_1,r_2).$$

Generalized spherical harmonics. In the coordinate system defined by the Euler angles α,β,γ, the spherical harmonics $Y_{lm}(\theta,\phi)$ transform according to

$$Y_{lm}(\theta,\phi) = \sum_{m'=-l}^{l} D^l_{mm'}(\alpha,\beta,\gamma)Y_{lm'}(\theta',\phi').$$

The coefficients $D^l_{mm'}(\alpha,\beta,\gamma)$ *are the generalized spherical harmonics of order l.*

Explicit formula for $D^l_{mm'}(\alpha,\beta,\gamma)$:

$$D^l_{mm'}(\alpha,\beta,\gamma) = e^{i(m\alpha+m'\gamma)} d^l_{mm'}(\beta),$$

$$d^l_{mm'}(\beta) = \frac{1}{2^m}\left[\frac{(l+m)!(l-m)!}{(l+m')!(l-m')!}\right]^{1/2} \times (1-x)^{(m-m')/2}(1+x)^{(m+m')/2} P^{(m-m',m+m')}_{l-m}(x),$$

where $P^{(\alpha,\beta)}_n(x)$ are Jacobi polynomials and $x = \cos\beta$.

Connections between spherical harmonics and generalized spherical harmonics:

$$D^l_{m0}(\alpha,\beta,\gamma) = \left[\frac{4\pi}{2l+1}\right]^{1/2} Y_{lm}(\beta,\alpha),$$

$$D^l_{0m}(\alpha,\beta,\gamma) = (-1)^m \left[\frac{4\pi}{2l+1}\right]^{1/2} Y_{lm}(\beta,\gamma),$$

$$D^l_{00}(\alpha,\beta,\gamma) = P_l(\cos\beta).$$

IX. Classical orthogonal polynomials of a discrete variable. These are the polynomials $y_n(x)$ that satisfy the difference equation

$$\tilde{\sigma}[x(s)]\frac{\Delta}{\Delta x(s-1/2)}\left[\frac{\nabla y(s)}{\nabla x(s)}\right] + \frac{\tilde{\tau}[x(s)]}{2}\left[\frac{\Delta y(s)}{\Delta x(s)} + \frac{\nabla y(s)}{\nabla x(s)}\right] + \lambda y(s) = 0,$$

where $y(s) = y_n[x(s)]$; $\tilde{\sigma}(x)$ and $\tilde{\tau}(x)$ are polynomials of at most second and first degrees, respectively; $\lambda = \lambda_n$ is a constant; $x(s)$ is a solution of a difference equation of the form $[x(s+1)+x(s)]/2 = \alpha x(s+1/2)+\beta$ (α and β are constants); $\Delta f(s) = f(s+1) - f(s)$, $\nabla f(s) = \Delta f(s-1)$;

$$\frac{\Delta}{\Delta x(s-1/2)} f(s) = \frac{\Delta f(s)}{\Delta x(s-1/2)}.$$

The polynomials $\tilde{y}_n(x)$ satisfy orthogonality relations of the form

$$\sum_{a\le s_i<b} \tilde{y}_m[x(s_i)]\tilde{y}_n[x(s_i)]\rho(s_i)\Delta x(s_i - 1/2) = \delta_{mn} d_n^2$$

$$(s_{i+1} = s_i + 1; \qquad b-a \text{ is an integer})$$

on the interval (a,b), and $\rho(s)$ satisfies the difference equation

$$\frac{\Delta}{\Delta x(s-1/2)}[\sigma(s)\rho(s)] = \tilde{\tau}[x(s)]\rho(s),$$

$$\sigma(s) = \tilde{\sigma}[x(s)] - \frac{1}{2}\tilde{\tau}[x(s)]\Delta x(s-1/2),$$

and the boundary conditions

$$\sigma(s)\rho(s)x^k\left(s-\frac{1}{2}\right)\Big|_{s=a,b} = 0 \qquad (k=0,1,\ldots).$$

The polynomials $\tilde{y}(x)$ have an explicit expression in a form analogous to the Rodrigues formula:

$$\tilde{y}_n[x(s)] = \frac{B_n}{\rho(s)}\frac{\nabla}{\nabla x_1(s)}\cdots\frac{\nabla}{\nabla x_{n-1}(s)}\frac{\nabla}{\nabla x_n(s)}[\rho_n(s)].$$

Here $\rho_n(s) = \rho(s+k)\prod_{i=1}^{n}\sigma(s+i)$, $x_k(s) = x(s+k/2)$.

The classical orthogonal polynomials are considered for the following canonical forms of lattice functions $x(s)$:

I. $x(s) = s$ (linear lattice);

II. $x(s) = s(s+1)$ (quadratic lattice);

III. $x(s) = q^s$ and $x(s) = q^{-s}$, $(q = e^{2\omega})$ (exponential lattices);

IV. $x(s) = \sinh 2\omega s = (q^s - q^{-s})/2$, $(q = e^{2\omega})$;

V. $x(s) = \cosh 2\omega s = (q^s + q^{-s})/2$;

VI. $x(s) = \cos 2\omega s$.

The basic data for the Hahn, Meixner, Kravchuk and Charlier polynomials, $h_n^{(\alpha,\beta)}(x)$, $m_n^{(\gamma,\mu)}(x)$, $k_n^{(p)}(x)$, $c_n^{(\mu)}(x)$, which are orthogonal on a linear lattice, are given in Tables 3–6. The basic data both for the Racah polynomials $u_n^{(\alpha,\beta)}(x)$ and for the dual Hahn polynomials $w_n^{(c)}(x)$ are given in Tables 7, 11, 12 for a quadratic lattice. In addition, analogs of the Hahn, Meixner, Kravchuk and Charlier polynomials are considered in §13 on lattices III and IV, which in the limit $q \to 1 (\omega \to 0)$ take the form of the corresponding polynomials on the uniform lattice $x(s) = s$ (Tables 8, 9). Analogs of the Racah polynomials and the dual Hahn polynomials on lattices V and VI are discussed as well (Table 10).

For the classical orthogonal polynomials of a discrete variable all the properties of the polynomials $p_n(x)$ that are orthogonal on (a,b) with weight $\rho(x)$ remain valid if we replace integration over (a,b) by summation over discrete values of the independent variable.

X. Special functions related to the functions $Q_0(z)$ of the second kind for the classical orthogonal polynomials (see §11).

a) *Incomplete gamma function:*

$$\Gamma(a,x)=\int_x^\infty e^{-t}t^{a-1}dt.$$

After the substitution of t for $x(1+s)$ the integral for $\Gamma(a,x)$ reduces to an integral that represents a confluent hypergeometric function $G(\alpha,\gamma,x)$ of the second kind:

$$\Gamma(a,x)=e^{-x}x^aG(1,1+a,x).$$

Incomplete beta function:

$$\mathrm{B}_x(p,q)=\int_0^x t^{p-1}(1-t)^{q-1}dt.$$

After substituting t for xs the integral for $\mathrm{B}_x(p,q)$ becomes the integral representation for a hypergeometric function $F(\alpha,\beta,\gamma,x)$:

$$\mathrm{B}_x(p,q)=\frac{1}{p}x^pF(p,1-q,1+p,x).$$

b) *Exponential integral:*

$$E_m(z)=\int_1^\infty \frac{e^{-zs}}{s^m}ds,\quad \mathrm{Re}\, z>0\quad (m=1,2,\ldots).$$

Recursion relation and differentiation formula:

$$E_m(z)=\frac{1}{m-1}[e^{-z}-zE_{m-1}(z)],$$

$$E_m'(z)=-E_{m-1}(z).$$

Series expansion for $E_1(z)$:

$$E_1(z)=-\gamma-\ln z+\sum_{k=1}^\infty \frac{(-1)^{k-1}}{k}\frac{z^k}{k!}$$

(γ is Euler's constant).

Asymptotic formula:

$$E_m(z) = \frac{e^{-z}}{z}\left[\sum_{k=0}^{n} \frac{(-1)^k (m)_k}{z^k} + O\left(\frac{1}{z^{n+1}}\right)\right],$$

$$(m)_k = m(m+1)\dots(m+k-1), \quad (m)_0 = 1.$$

Connection between $E_1(z)$ and $\mathrm{Ei}(-z)$:

$$E_1(z) = -\mathrm{Ei}(-z).$$

c) *Integral sine and cosine:*

$$\mathrm{Si}(z) = \int_0^z \frac{\sin s}{s} ds, \qquad \mathrm{Ci}(z) = \int_\infty^z \frac{\cos s}{s} ds.$$

Power series:

$$\mathrm{Si}(z) = \sum_{k=0}^{\infty} \frac{(-1)^k z^{2k+1}}{(2k+1)(2k+1)!},$$

$$\mathrm{Ci}(z) = \gamma + \ln z - \sum_{k=1}^{\infty} \frac{(-1)^{k+1} z^{2k}}{(2k)(2k)!}$$

(γ is Euler's constant).

Asymptotic formulas:

$$\mathrm{Si}(z) = \frac{\pi}{2} - \frac{\cos z}{z} P(z) - \frac{\sin z}{z} Q(z),$$

$$\mathrm{Ci}(z) = \frac{\sin z}{z} P(z) - \frac{\cos z}{z} Q(z),$$

where

$$P(z) = \sum_{k=0}^{n} \frac{(-1)^k (2k)!}{z^{2k}} + O(z^{-2n-2}),$$

$$Q(z) = \sum_{k=0}^{n} \frac{(-1)^k (2k+1)!}{z^{2k+1}} + O(z^{-2n-3}).$$

d) *Error function:*

$$\mathrm{erf}(z) = \frac{2}{\sqrt{\pi}} \int_0^z e^{-s^2} ds.$$

Power series:

$$\operatorname{erf}(z) = \frac{2}{\sqrt{\pi}} \sum_{k=0}^{\infty} \frac{(-1)^k z^{2k+1}}{k!(2k+1)}.$$

Asymptotic formula:

$$\operatorname{erf}(z) = 1 - \frac{1}{\sqrt{\pi}} \frac{e^{-z^2}}{z} \left[1 + \sum_{k=1}^{n} (-1)^k \frac{1 \cdot 3 \cdot \ldots \cdot (2k-1)}{(2z^2)^k} + O(z^{-2n-2}) \right]$$

$$\left(|\arg z| \leq \frac{\pi}{2} - \delta \right).$$

e) *Fresnel integrals*:

$$S(z) = \int_0^z \sin \frac{\pi s^2}{2} ds, \quad C(z) = \int_0^z \cos \frac{\pi s^2}{2} ds.$$

Connection with the error function:

$$C(z) - iS(z) = \int_0^z e^{-i\pi t^2/2} dt = \frac{1}{\sqrt{2i}} \operatorname{erf} \left[z \sqrt{\frac{i\pi}{2}} \right].$$

This lets one obtain asymptotic formulas and power series for the Fresnel integrals.

XI. Bessel functions.

a) Bessel functions in the strict sense.

Bessel's equation:

$$z^2 u'' + z u' + (z^2 - \nu^2) u = 0;$$

$u = Z_\nu(z)$ is a Bessel function of order ν.

Lommel's equation:

$$v'' + \frac{1 - 2\alpha}{z} v' + \left[(\beta \gamma z^{\gamma - 1})^2 + \frac{\alpha^2 - \nu^2 \gamma^2}{z^2} \right] v = 0,$$

$$v(z) = z^\alpha Z_\nu(\beta z^\gamma).$$

Poisson's integral representations for the Bessel function $J_\nu(z)$ of the first kind and the Hankel functions $H_\nu^{(1,2)}(z)$:

$$J_\nu(z) = \frac{(z/2)^\nu}{\sqrt{\pi}\Gamma(\nu+1/2)} \int_{-1}^{1} (1-t^2)^{\nu-1/2} \cos zt\, dt, \qquad \operatorname{Re}\nu > -\frac{1}{2},$$

$$H_\nu^{(1)}(z) = \sqrt{\frac{2}{\pi z}} \frac{e^{i(z-\pi\nu/2-\pi/4)}}{\Gamma(\nu+1/2)} \int_0^\infty e^{-t} t^{\nu-1/2} \left(1+\frac{it}{2z}\right)^{\nu-1/2} dt,$$

$$H_\nu^{(2)}(z) = \sqrt{\frac{2}{\pi z}} \frac{e^{-i(z-\pi\nu/2-\pi/4)}}{\Gamma(\nu+1/2)} \int_0^\infty e^{-t} t^{\nu-1/2} \left(1-\frac{it}{2z}\right)^{\nu-1/2} dt,$$

$$\operatorname{Re}\nu > -\frac{1}{2}.$$

Sommerfeld integrals:

$$J_\nu(z) = \frac{1}{2\pi} \int_{C_1} e^{iz\sin\phi - i\nu\phi} d\phi,$$

$$H_\nu^{(1)}(z) = -\frac{1}{\pi} \int_{C_+} e^{iz\sin\phi - i\nu\phi} d\phi,$$

$$H_\nu^{(2)}(z) = \frac{1}{\pi} \int_{C_-} e^{iz\sin\phi - i\nu\phi} d\phi$$

(the contours C_1, C_+, C_- are shown in Figs. 7 and 8, §16),

$$J_n(z) = \frac{1}{2\pi} \int_{-\pi}^{\pi} e^{iz\sin\phi - in\phi} d\phi,$$

$$e^{iz\sin\phi} = \sum_{n=-\infty}^{\infty} J_n(z) e^{in\phi}.$$

Asymptotic formulas:

$$J_\nu(z) = \left(\frac{2}{\pi z}\right)^{1/2} \left[\cos\left(z - \frac{\pi\nu}{2} - \frac{\pi}{4}\right) + O\left(\frac{1}{z}\right)\sin z + O\left(\frac{1}{z}\right)\cos z\right],$$

$$H_\nu^{(1)}(z) = \left(\frac{2}{\pi z}\right)^{1/2} e^{i(z-\pi\nu/2-\pi/4)} \left[1 + O\left(\frac{1}{z}\right)\right],$$

$$H_\nu^{(2)}(z) = \left(\frac{2}{\pi z}\right)^{1/2} e^{-i(z-\pi\nu/2-\pi/4)} \left[1 + O\left(\frac{1}{z}\right)\right].$$

Connections among Bessel functions:

$$H^{(1)}_{-\nu}(z) = e^{i\pi\nu} H^{(1)}_{\nu}(z), \qquad H^{(2)}_{-\nu}(z) = e^{-i\pi\nu} H^{(2)}_{\nu}(z),$$

$$J_\nu(z) = \frac{1}{2}[H^{(1)}_\nu(z) + H^{(2)}_\nu(z)], \quad Y_\nu(z) = \frac{1}{2i}[H^{(1)}_\nu(z) - H^{(2)}_\nu(z)],$$

$$H^{(1)}_\nu(z) = \frac{J_{-\nu}(z) - e^{-i\pi\nu} J_\nu(z)}{i \sin \pi\nu}, \quad H^{(2)}_\nu(z) = \frac{e^{i\pi\nu} J_\nu(z) - J_{-\nu}(z)}{i \sin \pi\nu},$$

$$Y_\nu(z) = \frac{\cos \pi\nu J_\nu(z) - J_{-\nu}(z)}{\sin \pi\nu}, \quad J_{-n}(z) = (-1)^n J_n(z).$$

Series:

$$J_\nu(z) = \sum_{k=0}^{\infty} \frac{(-1)^k (z/2)^{\nu+2k}}{k!\Gamma(\nu+k+1)},$$

$$H^{(1,2)}_n(z) = J_n(z) \pm \frac{i}{\pi}\left\{2J_n(z)\ln\frac{z}{2} - \sum_{k=0}^{n-1}\frac{(n-k-1)!}{k!}\left(\frac{z}{2}\right)^{2k-n}\right.$$
$$\left. - \sum_{k=0}^{\infty}\frac{(-1)^k (z/2)^{n+2k}}{k!(n+k)!}[\psi(n+k+1)+\psi(k+1)]\right\},$$

$$Y_n(z) = \frac{1}{\pi}\left\{2J_n(z)\ln\frac{z}{2} - \sum_{k=0}^{n-1}\frac{(n-k-1)!}{k!}\left(\frac{z}{2}\right)^{2k-n}\right.$$
$$\left. - \sum_{k=0}^{\infty}\frac{(-1)^k (z/2)^{n+2k}}{k!(n+k)!}[\psi(n+k+1)+\psi(k+1)]\right\}$$

(for $n = 0$ the first sum is to be taken to be zero; $\psi(z)$ is the logarithmic derivative of the gamma function).

Graphs of $J_n(x)$ and $Y_n(x)$ are given for a few values of n in Figs. 9 and 10 (§17).

Recursion relations and differentiation formulas:

$$Z_{\nu-1}(z) + Z_{\nu+1}(z) = \frac{2\nu}{z} Z_\nu(z),$$

$$Z_{\nu-1}(z) - Z_{\nu+1}(z) = 2Z'_\nu(z),$$

$$\left(\frac{1}{z}\frac{d}{dz}\right)^n [z^\nu Z_\nu(z)] = z^{\nu-n} Z_{\nu-n}(z),$$

$$\left(-\frac{1}{z}\frac{d}{dz}\right)^n [z^{-\nu} Z_\nu(z)] = z^{-(\nu+n)} Z_{\nu+n}(z)$$

($Z_\nu(z)$ stands for any of the functions $J_\nu(z)$, $Y_\nu(z)$, $H^{(1,2)}_\nu(z)$).

Bessel functions of half-integral order:

$$J_{1/2}(z) = \left[\frac{2}{\pi z}\right]^{1/2} \sin z, \quad Y_{1/2}(z) = -\left[\frac{2}{\pi z}\right]^{1/2} \cos z,$$

$$H_{1/2}^{(1,2)}(z) = \left[\frac{2}{\pi z}\right]^{1/2} e^{\pm i(z-\pi/2)},$$

$$J_{n-1/2}(z) = \left[\frac{2}{\pi z}\right]^{1/2} z^n \left(-\frac{1}{z}\frac{d}{dz}\right)^n \cos z \quad (n = 0, 1, \ldots),$$

$$H_{n-1/2}^{(1,2)}(z) = \left[\frac{2}{\pi z}\right]^{1/2} z^n \left(-\frac{1}{z}\frac{d}{dz}\right)^n e^{\pm iz} \quad (n = 0, 1, \ldots),$$

$$Y_{n-1/2}(z) = \left[\frac{2}{\pi z}\right]^{1/2} z^n \left(-\frac{1}{z}\frac{d}{dz}\right)^n \sin z \quad (n = 0, 1, \ldots).$$

Fourier-Bessel integrals:

$$f(x) = \int_0^\infty kF(k)J_\nu(kx)dk,$$

$$F(k) = \int_0^\infty xf(x)J_\nu(kx)dx.$$

Graf's addition formula:

$$Z_\nu(kR)e^{i\nu\psi} = \sum_{n=-\infty}^{\infty} J_n(kr)Z_{\nu+n}(k\rho)e^{in\theta} \quad (r < \rho).$$

Gegenbauer's addition formula:

$$\frac{Z_\nu(kR)}{(kR)^\nu} = 2^\nu\Gamma(\nu)\sum_{n=0}^{\infty}(\nu+n)\frac{J_{\nu+n}(kr)}{(kr)^\nu}\frac{Z_{\nu+n}(k\rho)}{(k\rho)^\nu}C_n^\nu(\mu) \quad (r < \rho).$$

Here r, ρ, R are the sides of an arbitrary triangle, ψ is the angle between the sides R and ρ, $\mu = \cos\theta$, θ is the angle between the sides r and ρ, k is an arbitrary number, $C_n^\nu(\mu)$ is a Gegenbauer polynomial, $Z_\nu(z)$ is any of the functions $J_\nu(z)$, $Y_\nu(z)$, $H_\nu^{(1,2)}(z)$, and $r < \rho$.

Legendre polynomial expansion of a spherical wave (see the Gegenbauer addition theorem):

$$\frac{e^{ikR}}{R} = i\pi \sum_{n=0}^{\infty} \left(n + \frac{1}{2}\right) \frac{J_{n+1/2}(kr)}{\sqrt{r}} \frac{H^{(1)}_{n+1/2}(k\rho)}{\sqrt{\rho}} P_n(\mu).$$

Legendre polynomial expansion of a plane wave:

$$e^{i\mathbf{k}\cdot\mathbf{r}} = \sqrt{\frac{2\pi}{kr}} \sum_{n=0}^{\infty} i^n \left(n + \frac{1}{2}\right) J_{n+1/2}(kr) P_n(\mu)$$

($\mathbf{k}$ is the wave vector, $\mu = \cos\theta$, θ is the angle between $\mathbf{k}$ and $\mathbf{r}$).

b) Modified Bessel functions.

Differential equation:

$$z^2 u'' + z u' - (z^2 + \nu^2) u = 0, \qquad u(z) = Z_\nu(iz).$$

When $z > 0$, linearly independent solutions of this differential equation are

$$I_\nu(z) = e^{-i\pi\nu/2} J_\nu(iz), \qquad K_\nu(z) = \frac{\pi}{2} e^{i\pi(\nu+1)/2} H^{(1)}_\nu(iz).$$

Poisson integral representation (Re $\nu > -1/2$):

$$I_\nu(z) = \frac{(z/2)^\nu}{\sqrt{\pi}\Gamma(\nu + 1/2)} \int_{-1}^{1} (1 - s^2)^{\nu - 1/2} \cosh zs\, ds,$$

$$K_\nu(z) = \left[\frac{\pi}{2z}\right]^{1/2} e^{-z} \frac{1}{\Gamma(\nu + 1/2)} \int_0^\infty e^{-s} s^{\nu - 1/2} \left(1 + \frac{s}{2z}\right)^{\nu - 1/2} ds.$$

Sommerfeld integral representations for $K_\nu(z)$:

$$K_\nu(z) = \frac{1}{2} \int_{-\infty}^{\infty} e^{-z\cosh\psi + \nu\psi} d\psi = \int_0^\infty e^{-z\cosh\psi} \cosh\nu\psi\, d\psi, \quad \text{Re}\, z > 0,$$

$$K_\nu(z) = \frac{1}{2} \left(\frac{z}{2}\right)^\nu \int_0^\infty e^{-t - z^2/4t} t^{-\nu-1} dt, \quad \text{Re}\, z > 0.$$

Asymptotic behavior as $z \to +\infty$:

$$I_\nu(z) = \frac{e^z}{(2\pi z)^{1/2}}\left[1 + O\left(\frac{1}{z}\right)\right], \quad K_\nu(z) = \left[\frac{\pi}{2z}\right]^{1/2} e^{-z}\left[1 + O\left(\frac{1}{z}\right)\right].$$

Connection between $I_\nu(z)$ and $K_\nu(z)$ for different values of ν:

$$I_{-n}(z) = I_n(z), \quad K_{-\nu}(z) = K_\nu(z),$$

$$K_\nu(z) = \frac{\pi}{2}\frac{I_{-\nu}(z) - I_\nu(z)}{\sin \pi\nu}.$$

Series:

$$I_\nu(z) = \sum_{k=0}^{\infty} \frac{(z/2)^{\nu+2k}}{k!\Gamma(k+\nu+1)},$$

$$K_n(z) = (-1)^{n+1} I_n(z) \ln\frac{z}{2} + \frac{1}{2}\sum_{k=0}^{n-1} \frac{(-1)^k(n-k-1)!}{k!}\left(\frac{z}{2}\right)^{2k-n}$$

$$+ \frac{1}{2}(-1)^n \sum_{k=0}^{\infty} \frac{(z/2)^{2k+n}}{k!(k+n)!}[\psi(n+k+1) + \psi(k+1)]$$

(when $n = 0$ the first sum is to be taken to be zero).

Recursion and differentiation formulas:

$$I_{\nu-1}(z) - I_{\nu+1}(z) = \frac{2\nu}{z} I_\nu(z),$$

$$I_{\nu-1}(z) + I_{\nu+1}(z) = 2I'_\nu(z), \qquad I'_0(z) = I_1(z),$$

$$K_{\nu-1}(z) - K_{\nu+1}(z) = -\frac{2\nu}{z} K_\nu(z),$$

$$K_{\nu-1}(z) + K_{\nu+1}(z) = -2K'_\nu(z), \quad K'_0(z) = -K_1(z).$$

$I_\nu(z)$ and $K_\nu(z)$ of half-integral order:

$$I_{n-1/2}(z) = \left(\frac{2}{\pi z}\right)^{1/2} z^n \left(\frac{1}{z}\frac{d}{dz}\right)^n \cosh z \qquad (n = 0, 1, \ldots),$$

$$K_{n-1/2}(z) = \left(\frac{\pi}{2z}\right)^{1/2} z^n \left(-\frac{1}{z}\frac{d}{dz}\right)^n e^{-z} \qquad (n = 0, 1, \ldots).$$

Graphs of $I_n(x)$ and $K_n(x)$ for a few values of n are shown in Figs. 11 and 12 (see §17).

XII. Hypergeometric functions $F(\alpha,\beta,\gamma,z)$.

Differential equation:

$$z(1-z)y'' + [\gamma - (\alpha+\beta+1)z]y' - \alpha\beta y = 0.$$

Particular solutions $(\gamma \neq 0, \pm 1, \pm 2, \ldots)$:

a) $y_1 = F(\alpha,\beta,\gamma,z)$, $y_2 = z^{1-\gamma}F(\alpha-\gamma+1, \beta-\gamma+1, 2-\gamma, z)$;

b) $y_1 = F(\alpha,\beta,\alpha+\beta-\gamma+1, 1-z)$,
$y_2 = (1-z)^{\gamma-\alpha-\beta}F(\gamma-\alpha, \gamma-\beta, \gamma-\alpha-\beta+1, 1-z)$;

c) $y_1 = z^{-\alpha}F(\alpha, \alpha-\gamma+1, \alpha-\beta+1, 1/z)$,
$y_2 = z^{-\beta}F(\beta, \beta-\gamma+1, \beta-\alpha+1, 1/z)$.

Integral representation:

$$F(\alpha,\beta,\gamma,z) = \frac{\Gamma(\gamma)}{\Gamma(\alpha)\Gamma(\gamma-a)} \int_0^1 t^{\alpha-1}(1-t)^{\gamma-\alpha-1}(1-zt)^{-\beta}\,dt,$$

$$\mathrm{Re}\,\gamma > \mathrm{Re}\,\alpha > 0.$$

Series:

$$F(\alpha,\beta,\gamma,z) = \sum_{n=0}^{\infty} \frac{(\alpha)_n(\beta)_n}{(\gamma)_n}\frac{z^n}{n!}, \quad |z| < 1,$$

$$(a)_n = \frac{\Gamma(a+n)}{\Gamma(a)} = a(a+1)\ldots(a+n-1).$$

Differentiation formula:

$$\frac{dF(\alpha,\beta,\gamma,z)}{dz} = \frac{\alpha\beta}{\gamma}F(\alpha+1, \beta+1, \gamma+1, z).$$

Recursion relations. Any three hypergeometric functions $F(\alpha_1,\beta_1,\gamma_1,z)$, $F(\alpha_2,\beta_2,\gamma_2,z)$ and $F(\alpha_3,\beta_3,\gamma_3,z)$ having integral differences $\alpha_i-\alpha_k$, $\beta_i-\beta_k$, $\gamma_i-\gamma_k$ are connected by a linear relation whose coefficients are polynomials in z (for the method of finding recursion relations, see §21).

Functional equations:

$$F(\alpha,\beta,\gamma,z)=F(\beta,\alpha,\gamma,z),$$

$$F(\alpha,\beta,\gamma,z)=(1-z)^{\gamma-\alpha-\beta}F(\gamma-\alpha,\gamma-\beta,\gamma,z),$$

$$F(\alpha,\beta,\gamma,z)=\frac{\Gamma(\gamma)\Gamma(\gamma-\alpha-\beta)}{\Gamma(\gamma-\alpha)\Gamma(\gamma-\beta)}F(\alpha,\beta,\alpha+\beta-\gamma+1,1-z)$$
$$+\frac{\Gamma(\gamma)\Gamma(\alpha+\beta-\gamma)}{\Gamma(\alpha)\Gamma(\beta)}(1-z)^{\gamma-\alpha-\beta}F(\gamma-\alpha,\gamma-\beta,\gamma-\alpha-\beta+1,1-z);$$

$$\begin{aligned}F(\alpha,\beta,\gamma,z)&\\ &=\frac{\Gamma(\gamma)\Gamma(\beta-\alpha)}{\Gamma(\beta)\Gamma(\gamma-\alpha)}(-z)^{-\alpha}F\left(\alpha,\alpha-\gamma+1,\alpha-\beta+1,\frac{1}{z}\right)\\ &\quad+\frac{\Gamma(\gamma)\Gamma(\alpha-\beta)}{\Gamma(\alpha)\Gamma(\gamma-\beta)}(-z)^{-\beta}F\left(\beta,\beta-\gamma+1,\beta-\alpha+1,\frac{1}{z}\right),\\ &\qquad|\arg(-z)|<\pi.\end{aligned}$$

These functional equations together with series expansions in terms of $1/z$ lead to *asymptotic formulas* for the functions $F(\alpha,\beta,\gamma,z)$ as $z\to\infty$.

We can obtain many other functional equations by combining the preceding three:

$$\begin{aligned}F(\alpha,\beta,\gamma,z)&\\ &=(1-z)^{-\alpha}\frac{\Gamma(\gamma)\Gamma(\beta-\alpha)}{\Gamma(\gamma-\alpha)\Gamma(\beta)}F\left(\alpha,\gamma-\beta,1+\alpha-\beta,\frac{1}{1-z}\right)\\ &\quad+(1-z)^{-\beta}\frac{\Gamma(\gamma)\Gamma(\alpha-\beta)}{\Gamma(\gamma-\beta)\Gamma(\alpha)}F\left(\gamma-\alpha,\beta,1-\alpha+\beta,\frac{1}{1-z}\right)\\ &\quad(|\arg(-z)|<\pi,\quad|\arg(1-z)|<\pi),\end{aligned}$$

$$\begin{aligned}F(\alpha,\beta,\gamma,z)&\\ &=z^{-\alpha}\frac{\Gamma(\gamma)\Gamma(\gamma-\alpha-\beta)}{\Gamma(\gamma-\alpha)\Gamma(\gamma-\beta)}F\left(\alpha,1+\alpha-\gamma,1+\alpha+\beta-\gamma,\frac{z-1}{z}\right)\\ &\quad+z^{\alpha-\gamma}(1-z)^{\gamma-\alpha-\beta}\frac{\Gamma(\gamma)\Gamma(\alpha+\beta-\gamma)}{\Gamma(\alpha)\Gamma(\beta)}\\ &\qquad\times F\left(\gamma-\alpha,1-\alpha,1+\gamma-\alpha-\beta,\frac{z-1}{z}\right)\\ &\quad(|\arg z|<\pi,\quad|\arg(1-z)|<\pi),\end{aligned}$$

$$F(\alpha,\beta,\gamma,z)=(1-z)^{-\alpha}F(\alpha,\gamma-\beta,\gamma,z/(z-1)),\quad|\arg(1-z)|<\pi,$$

$$F(\alpha,\beta,\gamma,z)=(1-z)^{-\beta}F(\gamma-\alpha,\beta,\gamma,z/(z-1)),\quad|\arg(1-z)|<\pi.$$

For special cases see §21 and [A1].

Various functions expressed in terms of hypergeometric functions:

$$F(\alpha, 0, \gamma, z) = 1,$$

$$F(\alpha, \beta, \beta, z) = (1 - z)^{-\alpha},$$

$$P_n^{(\alpha,\beta)}(z) = \frac{\Gamma(n+\alpha+1)}{n!\Gamma(\alpha+1)} F\left(-n, n+\alpha+\beta+1, \alpha+1, \frac{1-z}{2}\right)$$

$$= \frac{(-1)^n\Gamma(n+\beta+1)}{n!\Gamma(\beta+1)} F\left(-n, n+\alpha+\beta+1, \beta+1, \frac{1+z}{2}\right),$$

$$P_n(z) = F\left(-n, n+1, 1, \frac{1-z}{2}\right) = (-1)^n F\left(-n, n+1, 1, \frac{1+z}{2}\right),$$

$$K(z) = \int_0^{\pi/2} (1 - z^2 \sin^2\phi)^{-1/2} d\phi = \frac{\pi}{2} F\left(\frac{1}{2}, \frac{1}{2}, 1, z^2\right),$$

$$E(z) = \int_0^{\pi/2} (1 - z^2 \sin^2\phi)^{1/2} d\phi = \frac{\pi}{2} F\left(\frac{1}{2}, -\frac{1}{2}, 1, z^2\right).$$

XIII. Confluent hypergeometric functions $F(\alpha, \gamma, z)$ and $G(\alpha, \gamma, z)$.

Differential equation:

$$zy'' + (\gamma - z)y' - \alpha y = 0.$$

Particular solutions:

a) $y_1 = F(\alpha, \gamma, z), \quad y_2 z^{1-\gamma} F(\alpha - \gamma + 1, 2 - \gamma, z);$

b) $y_1 = G(\alpha, \gamma, z), \quad y_2 = e^z G(\gamma - \alpha, \gamma, -z).$

Integral representations:

$$F(\alpha, \gamma, z) = \frac{\Gamma(\gamma)}{\Gamma(\alpha)\Gamma(\gamma-\alpha)} \int_0^1 t^{\alpha-1}(1-t)^{\gamma-\alpha-1} e^{zt} dt \quad (\mathrm{Re}\,\gamma > \mathrm{Re}\,\alpha > 0),$$

$$G(\alpha, \gamma, z) = \frac{z^{-\alpha}}{\Gamma(\alpha)} \int_0^\infty e^{-t} t^{\alpha-1} \left(1 + \frac{t}{z}\right)^{\gamma-\alpha-1} dt \quad (\mathrm{Re}\,\alpha > 0),$$

$$G(\alpha, \gamma, z) = \frac{1}{\Gamma(\alpha)} \int_0^\infty e^{-zs} s^{\alpha-1} (1+s)^{\gamma-\alpha-1} ds \quad (\mathrm{Re}\,z > 0,\ \mathrm{Re}\,\alpha > 0).$$

Series:

$$F(\alpha,\gamma,z)=\sum_{n=0}^{\infty}\frac{(\alpha)_n}{(\gamma)_n}\frac{z^n}{n!}.$$

Differentiation formulas:

$$\frac{d}{dz}F(\alpha,\gamma,z)=\frac{\alpha}{\gamma}F(\alpha+1,\gamma+1,z),$$

$$\frac{d}{dz}G(\alpha,\gamma,z)=-\alpha G(\alpha+1,\gamma+1,z),$$

$$\frac{d}{dz}[z^{\alpha}G(\alpha,\gamma,z)]=-\frac{\gamma-\alpha-1}{z^2}[z^{\alpha}G(\alpha,\gamma-1,z)].$$

Recursion relations. Any three of the confluent hypergeometric functions $F(\alpha_1,\gamma_1,z)$, $F(\alpha_2,\gamma_2,z)$ and $F(\alpha_3,\gamma_3,z)$, that have integral differences $\alpha_i-\alpha_k$, $\gamma_i-\gamma_k$, are connected by a linear relation whose coefficients are polynomials in z. The same is true for $G(\alpha,\gamma,z)$ (for the method of finding the recursion relations see §21).

Functional equations:

$$F(\alpha,\gamma,z)=e^{z}F(\gamma-\alpha,\gamma,-z),$$

$$G(\alpha,\gamma,z)=z^{1-\gamma}G(\alpha-\gamma+1,2-\gamma,z),$$

$$G(\alpha,\gamma,z)=\frac{\Gamma(1-\gamma)}{\Gamma(\alpha-\gamma+1)}F(\alpha,\gamma,z)+\frac{\Gamma(\gamma-1)}{\Gamma(\alpha)}z^{1-\gamma}F(\alpha-\gamma+1,2-\gamma,z),$$

$$F(\alpha,\gamma,z)=\frac{\Gamma(\gamma)}{\Gamma(\gamma-\alpha)}e^{\pm i\pi\alpha}G(\alpha,\gamma,z)+\frac{\Gamma(\gamma)}{\Gamma(\alpha)}e^{z\pm i\pi(\alpha-\gamma)}G(\gamma-\alpha,\gamma,-z)$$

(the plus sign corresponds to $\operatorname{Im} z>0$).

For special cases see §21, part 4.

Asymptotic formulas as $z\to\infty$:

$$G(\alpha,\gamma,z)=z^{-\alpha}[1+O(1/z)],$$

$$F(\alpha,\gamma,z)=\frac{\Gamma(\gamma)}{\Gamma(\gamma-\alpha)}(-z)^{-\alpha}\left[1+O\left(\frac{1}{z}\right)\right]+\frac{\Gamma(\gamma)}{\Gamma(\alpha)}e^{z}z^{\alpha-\gamma}\left[1+O\left(\frac{1}{z}\right)\right]$$

$$(|\arg z|\le\pi,\quad |\arg(-z)|\le\pi).$$

Various functions expressed in terms of confluent hypergeometric functions:

$$F(0,\gamma,z) = G(0,\gamma,z) = 1,$$
$$F(\alpha,\alpha,z) = e^z,$$
$$G(\alpha,\alpha+1,z) = z^{-\alpha},$$

$$L_n^{\alpha}(z) = \frac{\Gamma(n+\alpha+1)}{n!\Gamma(\alpha+1)} F(-n, 1+\alpha, z),$$
$$I_\nu(z) = \frac{(z/2)^\nu}{\Gamma(\nu+1)} e^{-z} F\left(\nu+\frac{1}{2}, 2\nu+1, 2z\right),$$
$$K_\nu(z) = \sqrt{\pi}(2z)^\nu e^{-z} G\left(\nu+\frac{1}{2}, 2\nu+1, 2z\right).$$

XIV. Hermite functions $H_\nu(z)$.

Differential equation:

$$y'' - 2zy' + 2\nu y = 0.$$

Particular solutions:

a) $y_1 = H_\nu(z), \qquad y_2 = H_\nu(-z);$

b) $y_1 = e^{z^2} H_{-\nu-1}(iz), \quad y_2 = e^{z^2} H_{-\nu-1}(-iz).$

Connection with confluent hypergeometric functions:

$$H_\nu(z) = 2^\nu G(-\nu/2, 1/2, z^2) \qquad (|\arg z| \le \pi/2),$$
$$H_\nu(z) = \frac{2^\nu \Gamma(1/2)}{\Gamma((1-\nu)/2)} F\left(-\frac{\nu}{2}, \frac{1}{2}, z^2\right) + \frac{2^\nu \Gamma(-1/2)}{\Gamma(-\nu/2)} zF\left(\frac{1-\nu}{2}, \frac{3}{2}, z^2\right).$$

Integral representation:

$$H_\nu(z) = \frac{1}{\Gamma(-\nu)} \int_0^\infty e^{-t^2-2zt} t^{-\nu-1} dt, \quad \operatorname{Re}\nu > 0.$$

Series:

$$H_\nu(z) = \frac{1}{2\Gamma(-\nu)} \sum_{n=0}^{\infty} (-1)^n \Gamma\left(\frac{n-\nu}{2}\right) \frac{z^n}{n!}.$$

Differentiation formula:

$$H'_\nu(z) = 2\nu H_{\nu-1}(z).$$

Recursion relation:

$$H_\nu(z) - 2zH_{\nu-1}(z) + 2(\nu-1)H_{\nu-2}(z) = 0.$$

Functional equations:

$$H_\nu(z) = \frac{2^\nu\Gamma(\nu+1)}{\sqrt{\pi}}e^{z^2}\left[e^{i\pi\nu/2}H_{-\nu-1}(iz) + e^{-i\pi\nu/2}H_{-\nu-1}(-iz)\right],$$

$$H_\nu(z) = e^{i\pi\nu}H_\nu(-z) + \frac{2^{\nu+1}\sqrt{\pi}}{\Gamma(-\nu)}e^{z^2+i\pi(\nu+1)/2}H_{-\nu-1}(-iz),$$

$$H_\nu(z) = e^{-i\pi\nu}H_\nu(-z) + \frac{2^{\nu+1}\sqrt{\pi}}{\Gamma(-\nu)}e^{z^2-i\pi(\nu+1)/2}H_{-\nu-1}(iz).$$

Asymptotic formulas as $z \to \infty$:

$$H_\nu(z) = (2z)^\nu\left[1 + O\left(\frac{1}{z^2}\right)\right] \quad \left(|\arg z| \le \frac{\pi}{2}\right),$$

$$H_\nu(z) = (2z)^\nu\left[1 + O\left(\frac{1}{z^2}\right)\right] + \frac{2^{\nu+1}\sqrt{\pi}}{\Gamma(-\nu)}e^{z^2}(-2z)^{-\nu-1}\left[1 + O\left(\frac{1}{z^2}\right)\right]$$

$$(\pi/2 \le |\arg z| \le \pi, \quad |\arg(-z)| < \pi/2).$$

Parabolic cylinder functions:

$$D_\nu(z) = 2^{-\nu/2}\exp(-z^2/4)H_\nu(z/\sqrt{2}).$$

List of tables

References

[A1] M. Abramowitz and I.A. Stegun, Editors, Handbook of Mathematical Functions, National Bureau of Standards, Applied Mathematics Series, 55, U.S. Government, Washington, D.C., 1964.

[A2] N.I. Ahiezer and V.B. Berestetsky, Quantum Electrodynamics, Interscience, New York, 1965; Moscow, 1969.

[A3] E.L. Ains, Ordinary differential equations, Harkov, 1939.

[A4] G. Arfken, Mathematical Methods for Physicists, 2nd ed., Academic Press, New York and London, 1970.

[A5] R. Askey and J. Wilson, A set of orthogonal polynomials that generalize the Racah coefficients or $6j$-symbols, SIAM J. Math. Anal. 10:5 (1979), 1008–1016.

[A6] R. Askey, The q-gamma and q-beta functions, Applicable Anal. 8 (1978), 125–141.

[B1] P. Beckmann, Orthogonal Polynomials, Golem Press, Boulder, CO, 1973.

[B2] V.B.Berestetsky, E.M. Lifshits, and L.P. Pitayevsky, Relativistic Quantum Theory, vol. 1, Pergamon, Oxford and New York, 1971; Moscow, 1968.

[B3] H. Bethe and E. Salpeter, Quantum Mechanics of One- and Two-Electron Atoms, Springer, Berlin, 1957; Fizmatgiz, Moscow, 1960.

[B4] D.M. Brink and G.R. Satchler, Angular Momentum, 2nd ed., Oxford, 1968.

[B5] R.C. Buck and E.F. Buck, Advanced Calculus, 3rd ed., McGraw-Hill, New York (etc.), 1978.

[C1] G.F. Carrier, M. Krook, and C.E. Pearson, Functions of a Complex Variable, McGraw-Hill, New York (etc.), 1966.

[D1] A.S. Davydov, Quantum Mechanics, Pergamon, Oxford and New York, 1965; Moscow, 1973.

[D2] J.W. Dettman, Mathematical Methods in Physics and Engineering, 2nd ed., McGraw-Hill, New York (etc.), 1969.

[E1] A.R. Edmonds, Angular Momentum in Quantum Mechanics, Princeton, 1953.

[E2] A. Erdélyi, editor, Higher Transcendental Functions, volumes I, II, III, McGraw-Hill, New York (etc.), 1953

[E3] M.A. Evgrafov, Analytic Functions, Nauka, Moscow, 1968.

[F1] U. Fano and G. Racah, Irreducible Tensorial Sets, Academic Press, New York, 1959.

[F2] S. Flügge, Practical Quantum Mechanics, Springer, Berlin and New York, 1971 (two volumes in one).

[G1] I.M. Gelfand, Lectures on Linear Algebra, Moscow, 1971; Interscience, New York–London, 1961.

[G2] I.M. Gelfand, R.A. Minlos, and Z.Ya. Shapiro, Representations of the Rotation Group and the Lorentz Group, Macmillan, New York, 1963; Fizmatgiz, Moscow, 1958.

[G3] I.M. Gelfand, G.E. Shilov, and others, Generalized Functions, 5 vols., various languages and various publishers, 1964 and later.

[G4] V.V. Golubev, Lectures on the Analytic Theory of Ordinary Differential Equations, Moscow, 1941.

[H1] D.R. Hartree, The Calculation of Atomic Structure, Wiley, New York, 1967; Moscow, 1960.

[H2] W. Heitler, The Quantum Theory of Radiation, 2nd ed., Oxford, 1944.

[H3] E.W. Hobson, The Theory of Spherical and Ellipsoidal Harmonics, Cambridge University Press, 1931.

[H4] H. Hochstadt, Special Functions of Mathematical Physics, Holt, Rinehart and Winston, New York, 1961.

[J1] F.H. Jackson, On q-definite integrals, Quart. J. Pure Appl. Math. 10 (1910), 193–203.

[J2] E. Jahnke, F. Emde, and F. Lösch, Tafeln höherer Funktionen / Tables of Higher Functions, Teubner, Stuttgart, 1966; Russian edition, entitled Special Functions, Nauka, Moscow, 1977.

[K1] S. Karlin and J.L. McGregor, The Hahn polynomials, formulas and applications, Scripta Math. 26 (1961), 33–46.

[K2] T.H. Koornwinder, Krawtchouk polynomials, a unification of two different group theoretic interpretations, SIAM J. Math. Anal. 13 (1982), 1011–1023.

[K3] T.H. Koornwinder, Clebsch-Gordan coefficients for SU(2) and Hahn polynomials, Nieuw Arch. Wisk. 29 (1981), 140–155.

[L1] L.D. Landau and E.M. Lifshits, Quantum Mechanics (Nonrelativistic Theory), Pergamon, Oxford and New York, 1977; Moscow, 1963.

[L2] R.E. Langer, On the asymptotic solutions of ordinary differential equations with an application to Bessel functions of large order, Trans. Amer. Math. Soc. 33 (1931), 23–64.

[L3] M.A. Lavryentyev and B.V. Shabat, Methods of the theory of functions of a complex variable, Nauka, Moscow, 1973.

[L4] N.N. Lebedev, Special Functions and their Applications, Fizmatgiz, Moscow, 1963.

[L5] B.M. Levitan and I.S. Sargsyan, Introduction to Spectral Theory: Selfadjoint Ordinary Differential Operators, American Mathematical Society, Providence, 1973; Nauka, Moscow, 1970.

[L6] Y.L. Luke, The Special Functions and their Applications, Academic Press, New York and London, 1969.

[M1] T.M. MacRobert, Spherical Harmonics, 3rd ed., Pergamon, Oxford and New York, 1967.

[M2] V.L. Makarov, Difference schemes with exact and explicit spectra, preprints IM–72–12, IM 74–13, Mathematical Institute of The Academy of Sciences of the Ukrainian SSR, Kiev, 1974.

[M3] W. Miller, Symmetry Groups and Their Applications, Academic Press, New York, 1972.

[M4] P.M. Morse and H. Feshbach, Methods of Theoretical Physics, McGraw-Hill, New York, Toronto, London, 1953.

[N1] A.F. Nikiforov, S.K. Suslov and V.B. Uvarov, Classical orthogonal polynomials of a discrete variable, Nauka, Moscow, 1985.

[N2] A.F. Nikiforov and V.B. Ouvarov, Fonctions spéciales de la physique mathématique, Mir, Moscow, 1983.

[O1] F.W.J. Olver, Asymptotics and Special Functions, Academic Press, New York and London, 1974.

[P1] J.L. Powell and B. Crasemann, Quantum Mechanics, Addison–Wesley, Reading, MA, 1961.

[R1] E.D. Rainville, Special Functions, Macmillan, New York, 1960.

[R2] M.E. Rose, Elementary Theory of Angular Momentum, Wiley, New York, 1957.

[S1] L.I. Schiff, Quantum Mechanics, 3rd ed., McGraw-Hill, New York (etc.), 1968.

[S2] Yu.V. Sidorov, M.V. Fedoryuk, and M.I. Shabunin, Lectures on the Theory of Functions of a Complex Variable, Nauka, Moscow, 1976.

[S3] I.N. Sneddon, Special Functions of Mathematical Physics and Chemistry, Interscience, New York, 1956.

[S4] I.I. Sobelman, Atomic Structure and Radiative Transitions, Springer, Berlin and New York, 1979; Introduction to the Theory of Atomic Spectra, Moscow, 1963.

[S5] V.A. Steklov, On the Asymptotic Behavior of Solutions of Linear Differential Equations, Harkov University Press, 1956.

[S6] G. Strang, Linear Algebra and its Applications, Academic Press, New York (etc.), 1976.

[S7] P.K. Suetin, Classical Orthogonal Polynomials, Nauka, Moscow, 1976.

[S8] A.G. Sveshnikov and A.N. Tihonov, Theory of functions of a complex variable, Nauka, Moscow, 1973.

[S9] G. Szegö, Orthogonal Polynomials, 4th ed., American Mathematical Society, Providence, 1975; Fizmatgiz, Moscow, 1963.

[T1] F.G. Tricomi, Differential Equations, Hafner, New York, 1961; IL, Moscow, 1962.

[U1] V.B. Uvarov, The connection between systems of polynomials that are orthogonal with respect to different distribution functions, Zh. Vychisl. Mat. i Mat. Fiz. 9 (1969), 1253–1262.

[V1] D.A. Varshalovich, A.N. Moskalev, and V.K. Hersonsky, Quantum Theory of Angular Momentum, Nauka, Moscow, 1975.

[V2] N.Ya. Vilenkin, Special Functions and the Theory of Group Representations, American Mathematical Society, Providence, 1968; Nauka, Moscow, 1965.

[V3] V.S. Vladimirov, The Equations of Mathematical Physics, Nauka, Moscow, 1976.

[W1] P.R. Wallace, Mathematical Analysis of Physics Problems, Holt, Rinehart, Winston, New York, 1972.

[W2] G.N. Watson, A Treatise on Bessel Functions, 2nd ed., Cambridge University Press, 1944.

[W3] E.T. Whittaker and G.N. Watson, A Course of Modern Analysis, 4th ed., Cambridge University Press, 1927.

[W4] E.P. Wigner, The Application of Group Theory to the Special Functions of Mathematical Physics, Princeton University Press, 1955.

[W5] J.A. Wilson, Some hypergeometric orthogonal polynomials, SIAM J. Math. Analysis 11:4 (1980), 690–701.

Index of Notations

a_n: leading coefficient of the polynomial $p_n(x)$

$\mathrm{Ci}(x)$: integral cosine

$C_n^\lambda(x)$: Gegenbauer polynomials

$c_n^\mu(x)$: Charlier polynomials

$D_{mm'}^l(\alpha,\beta,\gamma)$: generalized spherical harmonic of order l

d_n^2: squared norm of an orthogonal polynomial

$E_m(z)$: exponential integral

$\mathrm{erf}(z)$: error function

$F(\alpha,\beta,\gamma,z)$: hypergeometric function

$F(\alpha,\gamma,z)$: confluent hypergeometric function of first kind

$G(\alpha,\gamma,z)$: confluent hypergeometric function of second kind

$h_n^{(\alpha,\beta)}(x)$: Hahn polynomials

$H_n(x)$: Hermite polynomials

$H_\nu(x)$: Hermite function

$H_\nu^{(1)}(z)$: Hankel function of first kind, order ν

$H_\nu^{(2)}(z)$: Hankel function of second kind, order ν

$I_\nu(z)$: modified Bessel function of first kind, order ν

$J_\nu(z)$: Bessel function of first kind, order ν

$k_n^{(p)}(x)$: Kravchuk polynomials

$K_\nu(z)$: Macdonald's function

$L_n^\alpha(x)$: Laguerre polynomials

$m_n^{(\gamma,\mu)}(x)$: Meixner polynomials

$P_n(x)$: Legendre polynomials

$P_n^m(x)$: associated Legendre functions

$P_n^{(\alpha,\beta)}(x)$: Jacobi polynomials

$Q_n(z)$: function of second kind for classical orthogonal polynomials

$\mathrm{Si}(z)$: integral sine

$t_n(x)$: Chebyshev polynomials of a discrete variable

$T_n(x)$: Chebyshev polynomials of first kind

$u_n^{(\alpha,\beta)}(x)$: Racah polynomials

$U_n(x)$: Chebyshev polynomials of second kind

$w_n^{(c)}(x)$: dual Hahn polynomials

$Y_{lm}(\theta, \phi)$: spherical harmonic of order l

$Y_\nu(z)$: Bessel function of second kind, order ν

$(\alpha)_n$: $\alpha(\alpha+1)\cdots(\alpha+n-1)$

$\mathrm{B}(x, y)$: beta function

$\mathrm{B}_z(x, y)$: incomplete beta function

$\Gamma(z)$: gamma function

$\Gamma(a, z)$: incomplete gamma function

$\Gamma_q(z)$: q-gamma function

γ: Euler's constant

$\Phi(z)$: erf (z)

$\psi(z)$: logarithmic derivative of gamma function

List of figures

Index

MIX
Papier aus verantwortungsvollen Quellen
Paper from responsible sources
FSC® C105338

If you have any concerns about our products,
you can contact us on
ProductSafety@springernature.com

In case Publisher is established outside the EU,
the EU authorized representative is:
Springer Nature Customer Service Center GmbH
Europaplatz 3, 69115 Heidelberg, Germany

Printed by Libri Plureos GmbH
in Hamburg, Germany